T0319042

INTEGRATED COASTAL MANAGEMENT
IN THE JAPANESE SATOUMI

INTEGRATED COASTAL MANAGEMENT IN THE JAPANESE SATOUMI

Restoring Estuaries and Bays

Edited by

TETSUO YANAGI

ELSEVIER

Elsevier
Radarweg 29, PO Box 211, 1000 AE Amsterdam, Netherlands
The Boulevard, Langford Lane, Kidlington, Oxford OX5 1GB, United Kingdom
50 Hampshire Street, 5th Floor, Cambridge, MA 02139, United States

© 2019 Elsevier Inc. All rights reserved.

No part of this publication may be reproduced or transmitted in any form or by any means, electronic or mechanical, including photocopying, recording, or any information storage and retrieval system, without permission in writing from the publisher. Details on how to seek permission, further information about the Publisher's permissions policies and our arrangements with organizations such as the Copyright Clearance Center and the Copyright Licensing Agency, can be found at our website: www.elsevier.com/permissions.

This book and the individual contributions contained in it are protected under copyright by the Publisher (other than as may be noted herein).

Notices
Knowledge and best practice in this field are constantly changing. As new research and experience broaden our understanding, changes in research methods, professional practices, or medical treatment may become necessary.

Practitioners and researchers must always rely on their own experience and knowledge in evaluating and using any information, methods, compounds, or experiments described herein. In using such information or methods they should be mindful of their own safety and the safety of others, including parties for whom they have a professional responsibility.

To the fullest extent of the law, neither the Publisher nor the authors, contributors, or editors, assume any liability for any injury and/or damage to persons or property as a matter of products liability, negligence or otherwise, or from any use or operation of any methods, products, instructions, or ideas contained in the material herein.

Library of Congress Cataloging-in-Publication Data
A catalog record for this book is available from the Library of Congress

British Library Cataloguing-in-Publication Data
A catalogue record for this book is available from the British Library

ISBN: 978-0-12-813060-5

For information on all Elsevier publications
visit our website at https://www.elsevier.com/books-and-journals

Publisher: Candice Janco
Acquisition Editor: Louisa Munro
Editorial Project Manager: Devlin Person
Production Project Manager: Vignesh Tamil
Cover Designer: Matthew Limbert

Typeset by SPi Global, India

Contents

7. What Can We Learn From Satoumi to Guide International Ocean Policies?
YOSHITAKA OTA AND WILF SWARTZ

8. Conclusions
TETSUO YANAGI AND KENICHI NAKAGAMI

Contributors

T. Asahi Faculty of Agriculture, Kagawa University, Kagawa, Japan

Allam Ayman Faculty of Agriculture, Kagawa University, Kagawa, Japan

Manabu Fujii School of Environment and Society, Tokyo Institute of Technology, Tokyo, Japan

Xinyu Guo Center for Marine Environmental Studies, Ehime University, Matsuyama, Japan

Takeshi Hidaka School of Humanity-oriented Science and Engineering, Kindai University, Iizuka, Japan

Naoki Hirose Ocean Modeling Group, Center for Oceanic and Atmospheric Research, Research Institute for Applied Mechanics, Kyushu University, Kasuga, Japan

K. Ichimi Seto Inland Sea Regional Research Center, Kagawa University, Kagawa, Japan

Hiroaki Ito Kumamoto University, Kumamoto, Japan

Masashi Ito Center for Marine Environmental Studies, Ehime University, Matsuyama, Japan

T. Kasamo Researcher, ECOH Consulting Company Limited, Japan

Haejin Kim Ocean Modeling Group, Center for Oceanic and Atmospheric Research, Research Institute for Applied Mechanics, Kyushu University, Kasuga, Japan

Teruhisa Komatsu Faculty of Commerce, Yokohama College of Commerce, Yokohama, Kanagawa, Japan

Taishi Kubota Center for Marine Environmental Studies, Ehime University, Matsuyama, Japan

T. Mano Graduate School of Science and Engineering, Ehime University, Matsuyama, Japan

Keito Mineo Graduate School of Agriculture, Kyoto University, Kyoto, Japan

Shigeru Montani Graduate School of Environmental Science, Hokkaido University, Sapporo, Japan

Akihiko Morimoto Center for Marine Environmental Studies, Ehime University, Matsuyama, Japan

Hiroki Murata Faculty of Commerce, Yokohama College of Commerce, Yokohama, Kanagawa, Japan

Kenichi Nakagami Professor Emeritus, Policy Science, Ritsumeikan University, Ibaraki, Osaka, Japan

Yuki Nakano Nippon Steel Engineering, Tokyo, Japan

Masafumi Natsuike Hokkaido Hakodate Fisheries Experimental Station, Hakodate, Japan

Wataru Nishijima Environmental Research and Management Center, Hiroshima University, Higashi-Hiroshima, Japan

Osamu Nishimura Graduate School of Engineering, Tohoku University, Sendai, Japan

N. Obata Professor Emeritus, Policy Science, Ritsumeikan University, Ibaraki, Osaka, Japan

Takahiro Ota Graduate School of Fisheries and Environmental Sciences, Nagasaki University, Japan

Yoshitaka Ota School of Marine and Environmental Affairs, University of Washington and Nippon Foundation Nereus Program

Yoichi Sakai Graduate School of Biosphere Science, Hiroshima University, Higashi-Hiroshima, Japan

Takashi Sakamaki Graduate School of Engineering, Tohoku University, Sendai, Japan

Shingo X. Sakamoto Faculty of Commerce, Yokohama College of Commerce, Yokohama, Kanagawa, Japan

Katazakai Saki Graduate School of Science and Engineering, University of Toyama, Toyama, Japan

Ryo Sakurai College of Policy Science, Ritsumeikan University, Osaka, Japan

Shuji Sasa Faculty of Commerce, Yokohama College of Commerce, Yokohama, Kanagawa, Japan

Shuhei Sawayama Faculty of Commerce, Yokohama College of Commerce, Yokohama, Kanagawa, Japan

Satoquo Seino Ecological Engineering Laboratory, Graduate School of Engineering, Kyushu University, Fukuoka, Japan

Hotaka Seko Graduate School of Asian and African Area Studies, Kyoto University, Kyoto, Japan

Ryota Shibano Center for Marine Environmental Studies, Ehime University, Matsuyama, Japan

J. Shibata Graduate School of Biosphere Science, Hiroshima University, Higashi-Hiroshima, Japan

Wilf Swartz   Institute for Oceans and Fisheries, University of British Columbia, Vancouver, BC, Canada

Kuninao Tada   Seto Inland Sea Regional Research Center, Kagawa University, Kagawa, Japan

Katsuki Takao   Specially Appointed Professor, Ritsumeikan University, Osaka, Japan

Katsumi Takayama   Ocean Modeling Group, Center for Oceanic and Atmospheric Research, Research Institute for Applied Mechanics, Kyushu University, Kasuga, Japan

Tetsutaro Takikawa   Graduate School of Fisheries and Environmental Sciences, Nagasaki University, Nagasaki, Japan

Takeshi Tomiyama   Graduate School of Biosphere Science, Hiroshima University, Higashi-Hiroshima, Japan

Takuro Uehara   College of Policy Science, Ritsumeikan University, Osaka, Japan

Yucheng Wang   Center for Marine Environmental Studies, Ehime University, Matsuyama, Japan

H. Yamamoto   International EMECS Center, Kobe, Japan

Tetsuo Yanagi   International EMECS Center, Kobe, Japan

Hajimu Yatabe   Sojitz Corporation, Tokyo, Japan

Naoya Yokoji   Organo Corporation, Tokyo, Japan

Takafumi Yoshida   Regional Activity Center, Northwest Pacific Region Environmental Cooperation Center, Toyam City, Japan

Naoki Yoshie   Center for Marine Environmental Studies, Ehime University, Matsuyama, Japan

Chihiro Yoshimura   School of Environment and Society, Tokyo Institute of Technology, Tokyo, Japan

T. Yoshioka   OIC Senior Researcher, Ritsumeikan University, Osaka, Japan

Jing Zhang   Graduate School of Science and Engineering, University of Toyama, Toyama, Japan

CHAPTER

1

Introduction

Tetsuo Yanagi

International EMECS Center, Kobe, Japan

1.1 BACKGROUND TO THIS BOOK

The Environment Research and Technology Development Fund, administered by the Ministry of the Environment, Japan, provides funding for research and development in virtually all environmental domains. Its purpose is to promote the accumulation of scientific knowledge that is indispensable for furthering the environmental policy aimed at building a sustainable society. Areas include preventing global warming, creating a recycling-based society, achieving coexistence with the natural environment, and ensuring safety by managing environmental risk.

As one part of this effort, a study titled "S-13 Development of Coastal Management Method to Realize the Sustainable Coastal Sea" was conducted in 2014–18. The study involved a comprehensive examination of natural and human activity in coastal seas and the land areas that constitute their hinterlands to determine how these areas should be changed from their current state to an appropriate status in terms of material cycling and ecotones. Specific actions have been proposed for the environmental management of coastal seas near Japan land areas. A management policy using the following as model areas needs to be established. The principal investigator of this project was Prof. Tetsuo Yanagi (principal researcher, International EMECS Center), the project consisted of five themes (Fig. 1.1), and the defrayed fund was about USD 1.5 million per year.

Theme 1: Development of methods for managing nutrients concentrations in the Seto Inland Sea (enclosed coastal sea).
Theme leader: Wataru Nishijima (Professor and Director, Environmental Research and Management Center, Hiroshima University).
Efforts are underway to expand the current uniform method of water quality management in the Seto Inland Sea to bay and open sea management that takes into consideration social and geopolitical characteristics and seasonal fluctuations, to preserve and restore management of nutrients and biological habitat environments, and to develop highly sustainable coastal management methods with the aim of achieving healthy substance circulation and high biological productivity that are not impaired by red tides or the like.
Subtheme 1-1: Developing methods for managing concentrations of nutrients (W. Nishijima).
Subtheme 1-2: Determining the function of tidal flats and seagrass beds in nutrient cycling and biological reproduction (K. Tada, Kagawa University).

Integrated Coastal Management in the Japanese Satoumi
https://doi.org/10.1016/B978-0-12-813060-5.00001-8

1

© 2019 Elsevier Inc. All rights reserved.

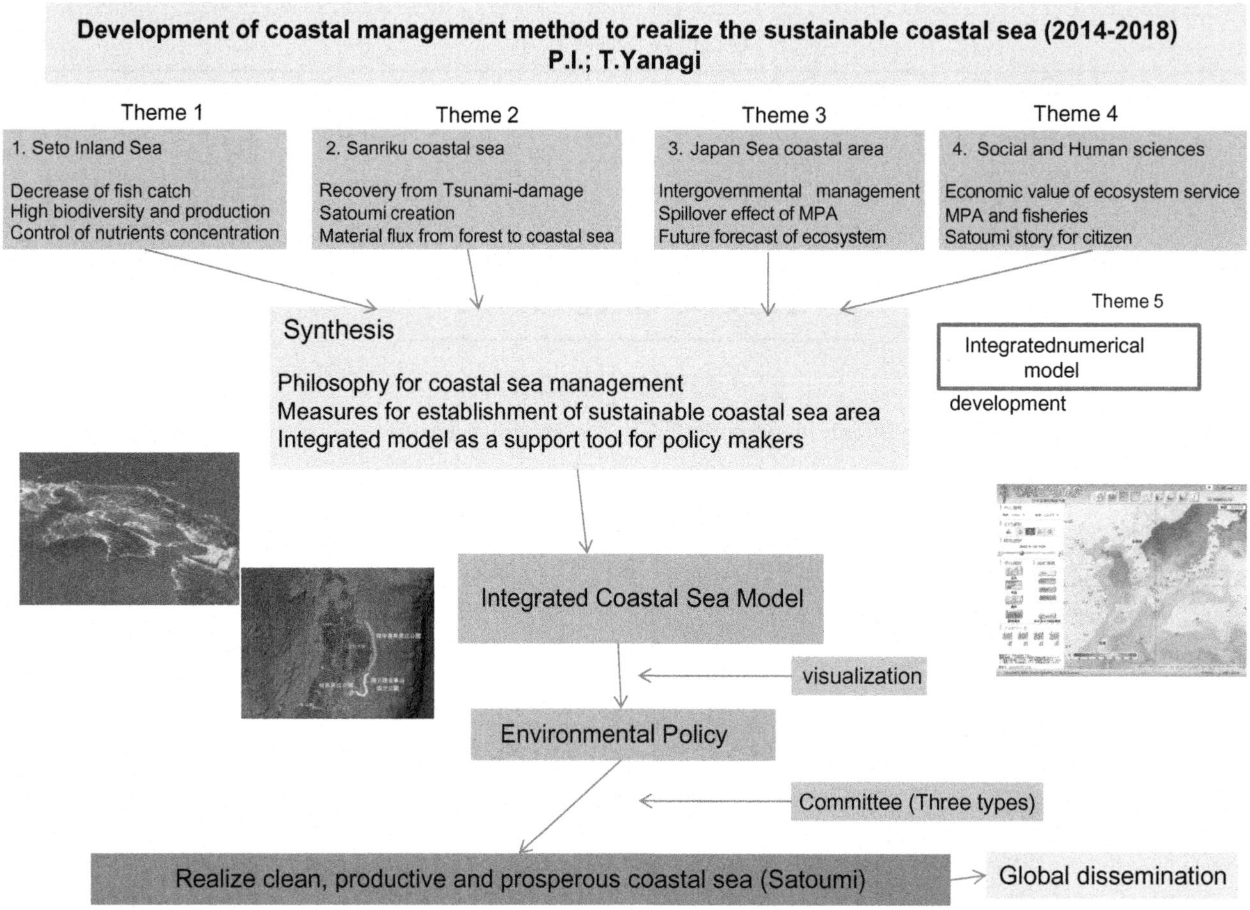

Development of coastal management method to realize the sustainable coastal sea (2014-2018)
P.I.; T.Yanagi

Theme 1	Theme 2	Theme 3	Theme 4
1. Seto Inland Sea	2. Sanriku coastal sea	3. Japan Sea coastal area	4. Social and Human sciences
Decrease of fish catch	Recovery from Tsunami-damage	Intergovernmental management	Economic value of ecosystem service
High biodiversity and production	Satoumi creation	Spillover effect of MPA	MPA and fisheries
Control of nutrients concentration	Material flux from forest to coastal sea	Future forecast of ecosystem	Satoumi story for citizen

Theme 5

Integratednumerical model

development

Synthesis

Philosophy for coastal sea management
Measures for establishment of sustainable coastal sea area
Integrated model as a support tool for policy makers

Integrated Coastal Sea Model

visualization

Environmental Policy

Committee (Three types)

Realize clean, productive and prosperous coastal sea (Satoumi)

Global dissemination

FIG. 1.1   Project S-13 "Development of Coastal Management Method to Realize the Sustainable Coastal Sea" project.

Theme 2: Developing coastal environmental management methods on the Sanriku Coast, which has a succession of open inner bays.
Theme leader: Teruhisa Komatsu (Associate Professor, Atmosphere and Ocean Research Institute, University of Tokyo).
This project will monitor the fluctuations in seaweed bed ecosystems on the Sanriku Coast to determine what human efforts are effective in restoring productive coastal zones. Optimal aquaculture methods for oysters, scallops, and wakame (an edible seaweed) on the Sanriku Coast will be proposed. In addition, quantitative evidence showing that forests are the ocean's best friend will be presented.
   Subtheme 2-1: Monitoring changes in coastal environments by remote sensing for coastal sea management (T. Komatsu).
   Subtheme 2-2: Determining the mechanisms of nutrient transfer among forests, rivers, and coastal seas (S. Montani, Hokkaido University).
   Subtheme 2-3: Determining the role of iron in material transport from forests to the coastal sea (C. Yoshimura, Tokyo Institute of Technology).
   Subtheme 2-4: Determining the role of organic matter in material transport from forests to coastal seas (O. Nishimura, Tohoku University).

Theme 3: Developing methods for ocean management in the Japan Sea, an international enclosed coastal sea that includes continental shelves and islands.
Theme Leader: Takafumi Yoshida (Chief Researcher, Northwest Pacific Region Environmental Cooperation Center).

The environmental changes in the Tsushima Current medial zone that are caused by global environmental changes and fluctuations in the environment of the East China Sea, and the common and unique aspects of the effect on individual bays, will be examined. In addition, the role of Marine Protected Area (MPA) designation in preserving biodiversity will be investigated, and land-sea management methods that integrate land areas with sea areas will be proposed. Moreover, methods for integrated management of the Japan Sea and the Tsushima Current conducted with the cooperation of Japan, China, South Korea, Russia, and other nations will be proposed.

   Subtheme 3-1: Proposed management methods for international enclosed coastal seas (T. Yoshida).
   Subtheme 3-2: Development of a model to predict environmental changes in the lower trophic level in the Japan Sea (A. Morimoto, Ehime University).
   Subtheme 3-3: Development of a model to predict environmental changes in physical condition in the Japan Sea (N. Hirose, Kyushu University).
   Subtheme 3-4: Development of a model to predict a higher trophic level ecosystem for the Japan Sea (X. Guo, Ehime University).

Theme 4: Presentation of a model for economic assessment of ecosystem services in coastal areas and integrated coastal management.
Theme leader: Kenichi Nakagami (Professor, Ritsumeikan University).
The economic value of ecosystem services in coastal zones of Japan will be calculated and a sustainability assessment will be performed. Integrated management methods needed for coastal zones to ensure their sustainable development will be proposed. Satoumi stories that are needed to connect nonfishing residents with coastal areas and that involve them in coastal area management will be discovered or created and passed on. Methods for coordinating fishery activities in the Tsushima and Goto MPAs also will be proposed.

   Subtheme 4-1: Economic assessment of ecosystem services (K. Nakagami).
   Subtheme 4-2: Proposed multi-stage management method for coastal seas (T. Hidaka, Kindai University).
   Subtheme 4-3: Discovery of stories that connect city people with coastal seas based on the perspective of humanity (T. Innami, Aichi University).
   Subtheme 4-4: Coordination of fisheries activities in the Tsushima and Goto marine protected areas (S. Seino, Kyushu University).

Theme 5: General overview and establishment of integrated numerical models for coastal sea management.
Theme leader: Tetsuo Yanagi (Principal Researcher, International EMECS Center).
Research coordination, promotion of collaboration, and control of overall progress of issues will be conducted to achieve the overall objectives of the project. The research achievements produced for each topic will be integrated to provide coastal zone management methods that can achieve sustainable coastal zone use in the 21st century. To this end, the objectives and guidelines for each topic will be presented, and a model that integrates nature, society, and the humanities will be proposed.

   Subtheme 5-1: Design of integrated numerical model (T. Yanagi).
   Subtheme 5-2: Development of an integrated model for Shizukawa Bay (T. Yanagi).
   Subtheme 5-3: Development of an integrated model for Toyama Bay (X. Guo).
   Subtheme 5-4: Development of an integrated model for the Seto Inland Sea (T. Yanagi).

The structure of this project is shown in Fig. 1.2, where Theme 4 relates to Themes 1–3 and all the results will be integrated in Theme 5 as the integrated numerical model.

The experimental sites, Seto Inland Sea (Theme 1), Shizukawa Bay (Theme 2), and Toyama Bay (part of Theme 3), are shown in Fig. 1.3.

This project aims for a quantitative expression of the Satoumi Concept (Yanagi, 2006).

1.2 SATOUMI

Satoumi is defined as "a coastal sea with high biodiversity and productivity under the appropriate human interaction" (Yanagi, 2006). Many kinds of Satoumi construction activities have been carried out under different

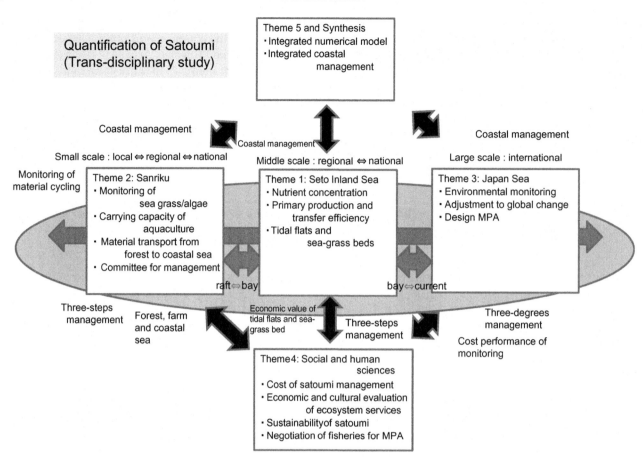

FIG. 1.2   The structure of S-13 project.

definitions in Japan. Hidaka (2016) introduces some of these definitions and provides another definition of Satoumi: "Satoumi is a management structure for conservation of environment and resources in the coastal sea by local people and government or the coastal area that is governed by such structure."

In the special project of S-13, Satoumi is defined as a "clean, productive, prosperous, and sustainable coastal sea."

1.3   EBM, CBM, MSP, MPA, ICM, AND SATOUMI

We hope to disseminate the concept of Satoumi throughout the world in the same way as "tsunami" has been, but it is not easy to achieve. A tsunami can be defined simply as "a long wave generated by an earthquake," but Satoumi includes not only concepts from natural science, such as the "relationship between human activities and biodiversity and productivity," but also concepts from social and human sciences, such as the "relationship between humans and nature." The basic concept of the "relationship between humans and nature" is completely different between European and Asian people.

Generally speaking, Christians believe that "we have to preserve nature separately from human beings because God created nature separately from humans" (Fig. 1.4). The desired attitude of humans to nature is expressed by the word "stewardship." And yet, humans ignore nature in the city, where most human activities are carried out. A buffer zone is set up between areas of nature preservation and the city. As a result, under-use occurs in preservation areas and over-use in the city.

Asian people, on the other hand, generally think that "gods live everywhere in nature, while humans, who live in nature must co-exist with gods and nature, and wise-use is the best attitude" (Fig. 1.4). Moreover, people think that they can be reborn as other animals or even plants in their next lifetime, and they must respect every form of life in nature.

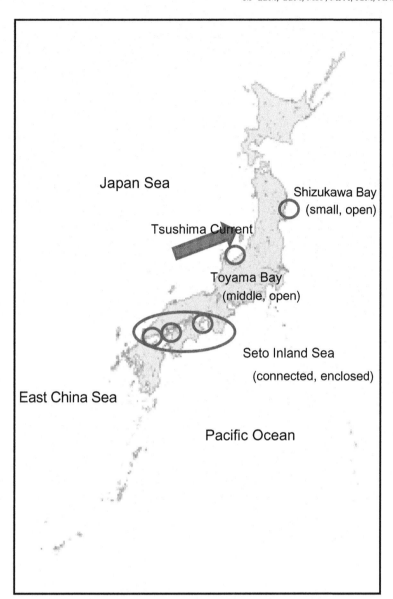

FIG. 1.3 Seto Inland Sea, Shizukawa Bay, and Toyama Bay.

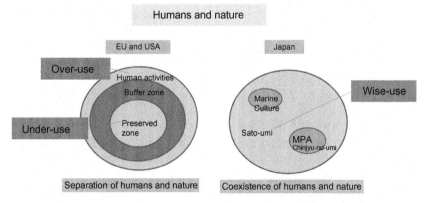

FIG. 1.4 The difference of the relationship between human and nature by Westerners and Asian people.

Westerners believe that ecosystem-based management (EBM) and/or community based-management (CBM) must form the basis of integrated coastal management (ICM) (Fig. 1.5). We have discussed the relationships among EBM, CBM, and Satoumi in the International Workshop on Satoumi, which has been carried out every year since 1998 (EMECS website).

FIG. 1.5 Relationships among Satoumi, EBM, and CBM (A) and Satoumi and ICM (B).

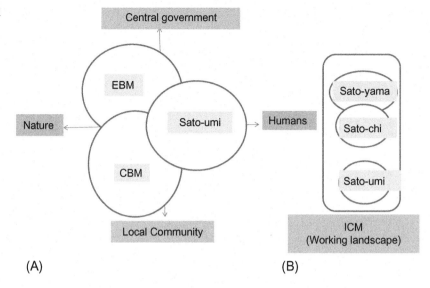

(A)   (B)

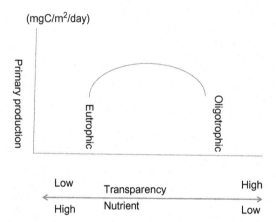

FIG. 1.6 Relationship between transparency (inversely proportional to nutrient concentration) and primary production. *Deformed from Diaz, 2001.*

1.3.1 EBM and Satoumi

Water quality indexes such as chemical oxygen demand (COD), total phosphorus (TP), and total nitrogen (TN) concentrations have been used for a long time in Japan as environmental conservation indexes because they are used to control the load from factories on the land. Moreover, total allowable catch (TAC) has been used for management of fish resources.

Westerners, however, believe that such water quality indexes are not appropriate for conservation of the ecosystem; they proposed the concept of EBM and such a concept applied in the United States for the preservation of owls in 1993 (Mori, 2012). An appropriate index of EBM in the coastal sea does not exist, however, and the submerged vegetable index (SVI) or oyster production index (OPI) have been used for EBM in the Chesapeake Bay region (refer to the Chesapeake Bay Program website).

In Satoumi, the following indexes have been proposed for EBM. We first clarify the most suitable transparency (or nutrient concentration) that achieves the largest primary production (mgC m^{-2} day^{-1}) in the target coastal sea (Fig. 1.6). Higher transparency means a coastal sea with low nutrient concentration and low primary production, while lower transparency has high nutrient concentration and high primary production. Too high a nutrient concentration, however, results in the occurrence of red tides and hypoxia, and low transparency and low primary production per unit water column, which consists of swimming phytoplankton or phytoplankton attached to the bottom or on the leaves of seaweed and/or seagrass, and seaweed and/or seagrass themselves.

The relationship between primary production and fish catch is not simple, as shown in Fig. 1.7. Higher primary production results in a higher fish catch, but the correspondence is not one-to-one. The fish catch in a coastal sea with a tidal current is larger by one order than in a lake without a tidal current. This might be because the organic matter

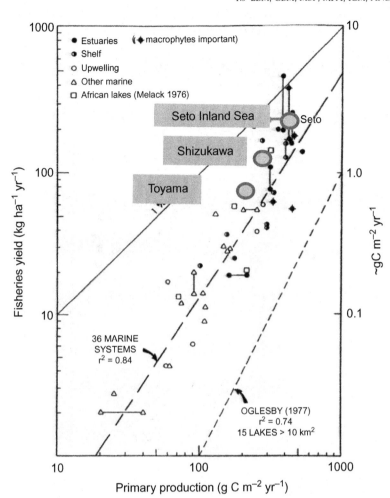

FIG. 1.7 Relationship between primary production and fish catch. *Deformed from Nixon, 1988.*

generated by primary production is effectively higher up the food chain by the advection-diffusion effect of the tidal current. Higher transfer efficiency from lower trophic levels to higher trophic levels plays an important role in a higher fish catch. The differences in primary production and fish catch among the Seto Inland Sea, Shizukawa Bay, and Toyama Bay, which are the experimental sites of the S-13 project, shown in Fig. 1.7, might be because of the differences in nutrient concentrations and their transport mechanism shown in Fig. 1.8.

The desirable transparency and transfer efficiency in the Seto Inland Sea, a concrete example of EBM, will be shown quantitatively in Section 1.4.

1.3.2 CBM and Satoumi

In the past, coastal management was carried out by a top-down approach, that is, the central government showed local people and local government how to manage the coastal area, but this system frequently resulted in failure. Therefore, a bottom-up CBM approach was proposed. CBM itself, however, has certain limitations. For example, local people follow the Sashi system, a good coastal resources management system in Indonesia (Murai, 1998), but people in other districts sometimes violate Sashi (Mosse, 2008). In order to avoid such difficulties, close cooperation among the central government, regional governments, and local governments becomes very important.

We propose a multiphase system of committees for the governance of Satoumi, as shown in Fig. 1.9.

Satoumi includes the concepts of both EBM and CBM (Fig. 1.5A).

1.3.3 MSP and Satoumi

Marine spatial planning (MSP) is the process by which various stakeholders (fishery, navigation, recreation, energy development, and so on), adjust their uses in the marine space (UNESCO website, 2017). The United Kingdom

FIG. 1.8 Schematic transport mechanism of nutrients in Seto Inland Sea, Shizukawa Bay, and Toyama Bay.

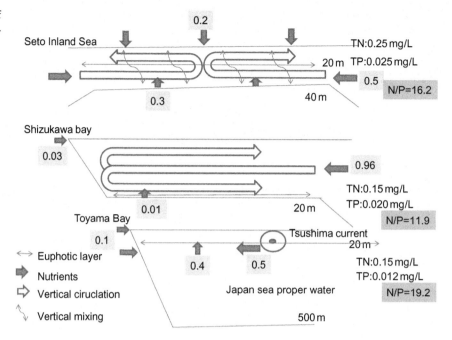

FIG. 1.9 Multiphase committee system for governance of Satoumi (Hidaka, 2016).

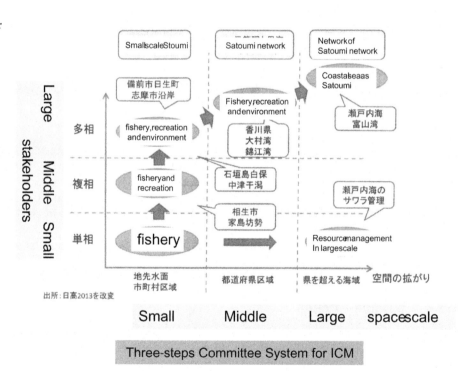

promulgated the Marine and Coastal Access Act in 2009 to adjust such uses in the coastal sea; planning is carried out under close cooperation between central and local governments. In addition, the UK started Marine Management Organization (MMO) for MSP in 2010.

In Satoumi, MSP is carried out by many stakeholders under the committee shown in Fig. 1.9.

The relationship between Marine Protected Area and Satoumi is discussed in detail in Sections 4.3 and 5.8.

1.3.4 ICM and Satoumi

It is clear that management of the coastal sea must be carried out in an integrated manner. In the case of Satoumi, integrated coastal management will be carried out through a variety of different aspects.

FIG. 1.10 Fushino Catchment Area Committee.

(1) Academic integration, namely, integration of natural, social, and human sciences, is needed.
(2) Area integration, namely, forests (Satoyama: forests with high biodiversity and productivity under human interaction), fields (Satochi: fields with high biodiversity and productivity under human interaction), and river, city, and ocean areas because river water quality is affected by forest, field, and city areas and subsequently determines water quality in the coastal sea (Fig. 1.5B).
(3) Stakeholders integration, namely, many different stakeholders such as forestry people, farmers, citizens, fishermen, and so on must be integrated to manage the coastal area.
(4) Governmental integration, namely, central, regional, and local governments must be integrated to achieve successful management.

An example of successful ICM in the Seto Inland Sea is the Fushino Catchment Area Committee in Yamaguchi Prefecture (Fig. 1.10), (Seto Inland Sea Research Institute, 2007). Many local committees from the forest to the coastal sea along the Fushino River cooperate with each other about the conservation of the water environment of the Fushino River and the coastal sea.

1.4 CLEAN AND PRODUCTIVE COASTAL SEA

Year-to-year variations in averaged transparency and yearly fish catch in the Seto Inland Sea (shown in Fig. 1.11) show transparency decreased until the mid-1980s during the period of rapid economic growth in Japan. In contrast, the fish catch increased until the mid-1980s. Since then, transparency has increased because of the inaction of TP and TN loads reduction law, but the fish catch has decreased. It is interesting to note that the variation in transparency does not correspond to that of the fish catch in a one-to-one correction in the Seto Inland Sea, as shown in Fig. 1.12. The fish catch during the eutrophication period is larger than during the oligotrophication period at the same

FIG. 1.11 Year-to-year variations in transparency and fish catch in Seto Inland Sea. *Transparency data before 1971 are from Yanagi (1988); those after 1971 from Setouchi-Net.*

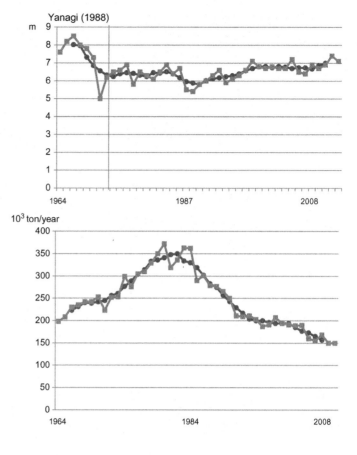

FIG. 1.12 Year-to-year variation in the relationship between transparency and fish catch in Seto Inland Sea. Circle shows the five-year running mean and the target transparency (6 m) and fish catch $(270 \times 10^3$ tons year^{-1}).

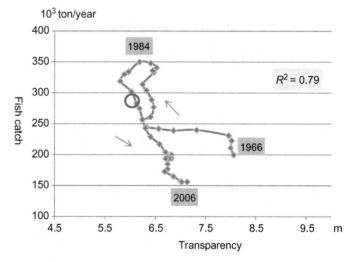

degree of transparency, which means hysteresis exists. Two possible explanations for such hysteresis are that the effect of hypoxia, which was generated during the eutrophication period, remains during the oligotrophication period and the biomass of higher trophic level fish is smaller than during the eutrophication period, and that the increase in jellyfish during the eutrophication period affects the recovery of a healthy fish ecosystem during the oligotrophication period.

Not only the fish catch but also the relationship between transparency and the trophic level (TL) of caught fish differs during both periods in the Seto Inland Sea as shown in Fig. 1.13. Average TL decreased during the eutrophication period because the catch of plankton-feeding fish such as anchovy (TL=2.5) increased. Average TL has increased during the oligotrophication period, however, because the catch of plankton feeders has decreased and that of higher trophic level fish has shown a relative increase.

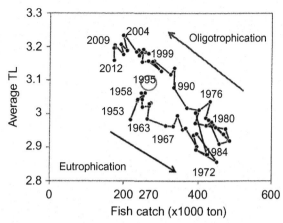

FIG. 1.13 Year-to-year variation in average trophic level of caught fish and fish catch in the Seto Inland Sea (Tanda et al. 2015). Circle shows the target fish catch (270×10^3 tons year^{-1}) and TL (3.1).

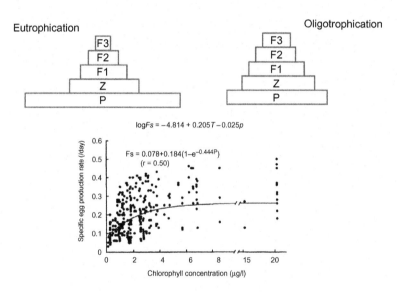

FIG. 1.14 Relationship between the egg production rate of zooplankton and the chl.a concentration (Uye and Shibuno, 1992).

The reason for such a difference in TL in both periods might be explained by the difference in the ecosystem of caught fish, which is mainly determined by the relationship between the growth rate of zooplankton and the biomass of phytoplankton (chlorophyll a concentration) shown in Fig. 1.14. The growth rate of zooplankton is saturated at some phytoplankton biomass concentration (Chl.$a = 4.5\,\mu g\,L^{-1}$) as shown in Fig. 1.14 (Uye and Shibuno, 1992). Therefore, the phytoplankton that remain during the eutrophication period die and sink to the bottom layer, resulting in the generation of hypoxia. Hypoxia kills the eggs of zooplankton and decreases the transfer efficiency to the higher trophic level fish. Most phytoplankton are eaten by zooplankton during the oligotrophication period, and the transfer efficiency to the higher trophic level fish becomes high.

The relationship between transparency and Chl.a concentration in the Seto Inland Sea is shown in Fig. 1.15, where a Chl.a concentration of $4.5\,\mu g\,L^{-1}$ corresponds to a transparency of 6 m. We propose that the target transparency for a clean and productive Seto Inland Sea should be 6 m, based on these results, the target fish catch is 270×10^3 tons year -1 (Fig. 1.12), and the target TL is 3.1 (Fig. 1.13).

The relationships between transparency and TP and TN concentrations in the Seto Inland Sea are shown in Fig. 1.16. $TP = 0.028\,mg\,L^{-1}$ and $TN = 0.28\,mg\,L^{-1}$ correspond to the transparency $= 6$ m.

The relationships between TP concentration and TP load from the land and TN concentration and TN load from the land in the Seto Inland Sea are shown in Fig. 1.17. TP load of 13 tons day^{-1} and TN load of 260 tons day^{-1} correspond to a transparency of 6 m. This suggests that we have to go back to the mid-1980s from the viewpoint of total load reduction policy; in other words, the total reduction policy in the Seto Inland Sea is too advanced now.

In summary, we propose a clean and productive Seto Inland Sea with a transparency of 6 m, a fish catch of 270×10^3 tons year^{-1}, an average trophic level of caught fish of 3.1, a TP concentration of $0.028\,mg\,L^{-1}$, a TN concentration of $0.28\,mg\,L^{-1}$, a TP load from land of 13 tons year^{-1}, and a TN load from land of 260 tons year^{-1}.

FIG. 1.15 Transparency and chl.*a* concentration in the Seto Inland Sea observed in four seasons, 2000. *From http://www.pa.cgr. mlit.go.jp/chiki/suishitu/download/download_ su.htm.*

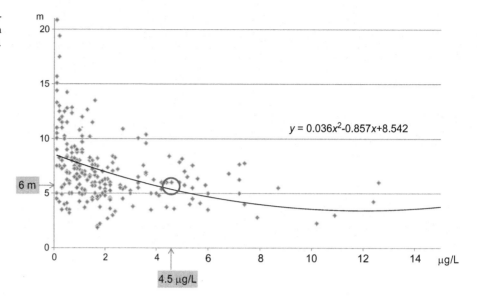

$$y = 0.036x^2 - 0.857x + 8.542$$

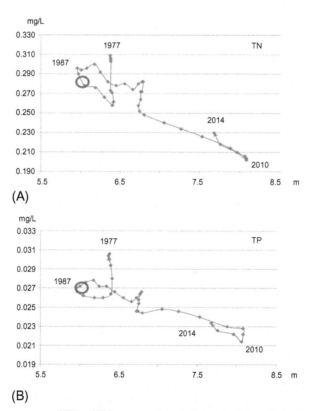

(A)

(B)

FIG. 1.16 Relationship between transparency and TP and TN concentrations in Seto Inland Sea. Circle shows the target transparency (6 m) and TP concentration (0.028 mg L^{-1}) and TN concentration (0.28 mg L^{-1}).

It must be noted that the target transparency and other parameters such as fish catch, TL, TP, and TN concentrations do not correspond one-to-one as shown in Figs. 1.12 and 1.13. Therefore, we have to carry out the necessary action for environmental conservation by adaptive management, that is, a plan-do-check-action-cycle is important on the basis of continuous environmental monitoring.

Moreover, we must rehabilitate tidal flats and seaweed or seagrass beds, which are important for fish resource recruitment and habitat for marine biota, because they have decreased continuously in the Seto Inland Sea as shown in Fig. 1.18 (Yanagi, 2015).

By such combined activities, we will be able to create a clean and productive Seto Inland Sea.

We will discuss the prosperous and sustainable coastal sea in detail in Section 8.1.

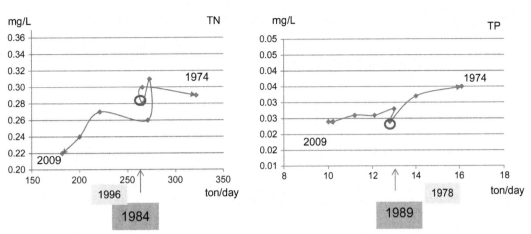

FIG. 1.17 Relationship between TP concentration and TP load, and TN concentration and TN load in Seto Inland Sea. Circle shows the target TP $(0.028\,\mathrm{mg\,L^{-1}})$ and TN $(0.28\,\mathrm{mg\,L^{-1}})$ concentrations and TP $(13\ \mathrm{tons\ year^{-1}})$ and TN $(260\ \mathrm{tons\ year^{-1}})$ loads.

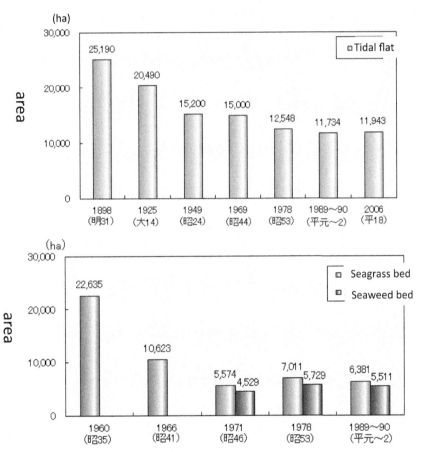

FIG. 1.18 Year-to-year variations in tidal flat area and seagrass area in Seto Inland Sea *From Setouchi Net.*

References

Diaz, R., 2001. Overview of hypoxia around the World. J. Environ. Qual. 30–2, 275–281.

Hidaka, T., 2016. Satoumi and Coastal Management. Norin Tokei Kyokai, Tokyo, 301p. (in Japanese).

Mori, A., 2012. Ed, Ecosystem Management: Integrated Conservation and Management. Kyoritsu Press, Tokyo, 320p. (in Japanese).

Mosse, J.W., 2008. Sasi Laut: History and its role of marine coastal resource management in Maluku archipelago. International workshop on "Sato-Umi" Report. International EMECS Center, Japan, pp. 68–76.

Murai, Y., 1998. Sashi, Asia and Marine World. Commons, Tokyo.

Nixon, S.W., 1988. Physical energy inputs and comparative ecology of lake and marine ecosystem. Limnol. Oceanogr. 33, 1005–1025.

Seto Inland Sea Research Institute, 2007. Satoumi Creation in the Seto Inland Sea. Koseisha-Koseikaku, Tokyo, (109p).

Tanda, M., Yamashita, M., Harada, K., 2015. Trend in trophic level and fish catch in the Seto Inland Sea. Environ. Technol. 44 (3), 122–127 (in Japanese).

UNESCO, 2017. Marine Spatial Planning Initiative. http://www.unesco-ioc-marinesp-be/msp_guide?PHPSESSID=b3dc4277b7018ba9c4d1de91.

Uye, S., Shibuno, T., 1992. Reproductive biology of the planktonic copepod *Paracalanus* sp. in the Inland Sea of Japan. J. Plankton Res. 14 (3), 343–358.

Yanagi, T., 1988. Preserving the Inland Sea. Mar. Pollut. Bull. 19, 51–53.

Yanagi, T., 2006. Sato-Umi; New Concept for Coastal Sea Management. TERRAPUB, Tokyo, 110p.

Yanagi, T., 2015. Oligotrophication in the Seto Inland Sea. In: Yanagi, T. (Ed.), Eutrophication and Oligotrophication in Japanese Estuaries. Springer, pp. 39–68.

Further Reading

Chesapeake Bay Program Website, https://www.chesapeakebay.net/discover/history.

EMECS Website, https://www.emecs.or.jp/.

Seouchi Net, http://www.env.go.jp/water/heisa/heisa_net/setouchiNet/seto/sitemap.html.

2

Toward Realizing the Sustainable use of and Healthy Marine Environments in an Open-Type Enclosed Bay

Teruhisa Komatsu*, Shingo X. Sakamoto*, Shuhei Sawayama*, Hiroki Murata*, Shuji Sasa*, Shigeru Montani§, Hajimu Yatabe¶, Naoya Yokoji‖, Hotaka Seko#, Yuki Nakano**, Manabu Fujii†, Allam Ayman††, Masafumi Natsuike‡, Hiroaki Ito‡‡, Chihiro Yoshimura§§, Takashi Sakamaki¶¶, and Osamu Nishimura¶¶

*Faculty of Commerce, Yokohama College of Commerce, Yokohama, Kanagawa, Japan
†School of Environment and Society, Tokyo Institute of Technology, Tokyo, Japan
‡Hokkaido Hakodate Fisheries Experimental Station, Hakodate, Japan
§Graduate School of Environmental Science, Hokkaido University, Sapporo, Japan
¶Sojitz Corporation, Tokyo, Japan
‖Organo Corporation, Tokyo, Japan
#Graduate School of Asian and African Area Studies, Kyoto University, Kyoto, Japan
**Nippon Steel Engineering, Tokyo, Japan
††Faculty of Agriculture, Kagawa University, Kagawa, Japan
‡‡Kumamoto University, Kumamoto, Japan
§§School of Environment and Society, Tokyo Institute of Technology, Tokyo, Japan
¶¶Graduate School of Engineering, Tohoku University, Sendai, Japan*

© 2019 Elsevier Inc. All rights reserved.

2.1 TOWARD REALIZING THE SUSTAINABLE USE OF AND HEALTHY MARINE ENVIRONMENTS IN AN OPEN-TYPE ENCLOSED BAY

Teruhisa Komatsu

Coastal waters are an important part of the landscape that provide many ecosystem services for human beings. Pressure brought by human activities has accumulated, however, and is beyond the capacity of coastal waters, as noted by the concept of the planetary boundary. No studies have been done about the impact of the open ocean and land-based loads from rivers and aquacultures on the environmental conditions in an open-type enclosed bay. These characteristics are different from those of a typical enclosed bay, which are influenced mainly by human activities. Few studies have been about the management methods of coastal waters, realizing that human beings can live connected with nature. Yanagi (2013) defined Satoumi as coastal waters whose productivity and biodiversity are increased by human interventions that generate fat, long, and smooth material (nutrient) flows. According to Yanagi, material flows can be accomplished through management of the environments in the watersheds from the mountains to the sea and the management of marine bioresources, including human consumers, who are top predators in a marine food chain. Matsuda (2007) noted that sound coastal fisheries are indispensable for maintaining healthy material flows and a sound marine environment. These two objectives form Satoumi activities, but no studies can present how to accomplish these objectives in an open-type enclosed bay. When we obtain an answer, we will be able to put them into practice for realizing a sound, prosperous, and sustainable open-type enclosed bay.

Some fishermen are afforesting broadleaf trees based on the iron hypothesis proposed by Dr. Katsuhiko Matsunaga, professor emeritus of Hokkaido University (Matsunaga, 1993). This hypothesis was introduced in a book titled "Fishermen's afforestation: the forest is a lover of the sea" (Hatakeyama, 2000). The hypothesis is explained as follows. Although dissolved iron is one of the minor elements indispensable for the growth of plants, it is generally not distributed sufficiently in the sea. Iron is rich on land and is transported from the land to the sea in dissolved forms that marine plants can use. Dissolved iron is an iron complex consisting of iron and corrosive substances (dissolved organic matter) such as fluvic or humic acids, which forests of broadleaf trees produce in abundance. Thus, the afforestation of broadleaf trees is good for primary marine production. This is why Hatakeyama proposed that the forest is a lover of the sea. Based on this hypothesis, the Ministry of the Environment of Japan promotes the afforestation of broadleaf trees in Japan as one of the activities of Satoyama and Satoumi. There have been no quantitative studies about the iron budget in an open-type enclosed bay where Hatakeyama started afforestation. The ministry needs scientific evidence of afforestation effects to enhance primary production through an increase in the dissolved iron that travels from the river to the sea. Therefore, scientists must quantitatively study the dynamics of not only the iron, but also of the nutrients and particulate organic matter provided by rivers to the sea from the point of view of land use, including the use of forests.

In this context, the Ministry of the Environment of Japan started the project of the "Development of Coastal Management Method to Realize the Sustainable Coastal Sea" (Research Number S-13), which is headed by Tetsuo Yanagi, professor emeritus of Kyushu University, using the Environment Research and Technology Development Fund from 2014. The project aims to provide scientific evidence to promote coastal management using the Satoumi approach. The project includes "Topic 2: Development of coastal environmental management methods on the Sanriku Coast, which has a succession of open inner bays," the leader of which is Professor Teruhisa Komatsu of Yokohama College of Commerce. The project consists of three subthemes: "Monitoring of changes in coastal environments, and use of the results to develop methods for coastal sea management" by Professor Teruhisa Komatsu of the University of Tokyo (later, Yokohama College of Commerce); "Determination of the mechanism of nutrient transfer among forests, rivers and oceans" by Shigeru Montani, professor emeritus of Hokkaido University (later, Yokohama College of Commerce); and "Determination of the role of organic matter in material transport between forests and oceans" by Professor Chihiro Yoshimura of Tokyo Institute of Technology, who focuses on iron, and Professor Osamu Nishimura of Tohoku University, who focuses on particulate organic matter. Shizugawa Bay was selected as a study site because the bay is a typical open-type enclosed bay along the Sanriku Coast. The bay is within a spatial scale of 5 km, and the watersheds of the rivers flowing into the bay are within only Minamisanriku Town. Moreover, a forest of broadleaf trees that fishermen planted is distributed in the hinterland of Shizugawa Bay.

The Sanriku Coast has been struck by large tsunamis at intervals of several decades. When we consider the ecosystems and marine environments in the Sanriku Coast, we cannot neglect their recovering processes. Therefore, we monitored the conditions of seagrass, seaweed beds, saltmarshes, and tidal flats with remote sensing devises to examine the succession of coastal habitats after the strike of a large tsunami in 2011. We collected basic information about the impacts of high seawalls in the coastal ecosystems recovering from the tsunami. These results are shown in Section 2.1, "Strike of large tsunami on the coastal habitats called blue infrastructure: succession under the impacts of tsunami and the human activities post-tsunami." The nutrient flows from the land to the sea, including aquacultures, are reported in Section 2.2, "Is the forest a lover of the sea? Nutrients." The flows of the iron and particulate organic matter from the land to the sea also are discussed in Section 2.3, "Is the forest a lover of the sea? Iron," and Section 2.4, "Is the forest a lover of the sea? Particulate organic matter."

An ecological model of Shizugawa Bay was established by Professor Tetsuo Yanagi using the data about the biomass in the sea and the landing of aquaculture products, water temperature, salinity, and currents from the first subtheme and the nutrients, iron, particulate organic matter, and the growth and excrements of the oysters obtained by the first, second and third subthemes. The model was run under several scenarios with different numbers and spatial arrangements of aquaculture facilities codesigned by local fishermen and us. The results of the simulations of the model and our in situ observations were presented at "The Council for Future Environments of Shizugawa Bay." It was organized by a local fishermen's cooperative, Minamisanriku Town, Miyagi Prefecture, WWF Japan, and us to examine how we can create a prosperous, rich, and sound Shizugawa Bay through human interventions, such as fishery and land use. The council's discussions are introduced in Section 2.5, "Management of the aquaculture and marine environments in an open-type enclosed bay." We expect that the results shown in this chapter will contribute to the sustainable production of aquacultures and sound marine environments in open-type enclosed bays in Japan and the world even after 50 to 100 years.

2.2 STRIKE OF LARGE TSUNAMI ON THE COASTAL HABITATS CALLED BLUE INFRASTRUCTURE: SUCCESSION UNDER THE IMPACTS OF TSUNAMI AND THE HUMAN ACTIVITIES POST-TSUNAMI

Teruhisa Komatsu, Shingo X. Sakamoto, Shuhei Sawayama, Hiroki Murata and Shuji Sasa

2.2.1 Introduction

A coastal area is located between the land and the sea, forming an ecotone with habitats of macrophyte beds of seagrasses and seaweeds, salt marshes, and tidal flats that provide ecosystem services. Seagrasses and seaweeds produce seascapes similar to meadows and forests on land. The basal and leaf parts provide habitats for the epiphytic and benthic organisms adapted to the seagrass and seaweed beds, resulting in a spawning of substrates for fishes and Cephalopoda, nursery grounds for fish larvae, and feeding grounds for sea urchins and shells. The seagrass and seaweed beds supply particulate organic matter, such as the cultured and absorbed nutrients from seawater, to oysters, scallops, and ascidians. Therefore, they are indispensable for maintaining biodiversity and fishery resources. In tidal flats, the benthic microalgae growing on the surface actively photosynthesize to absorb nutrients and carbon dioxide. Because many shells live in the surface layer of the tidal flats and feed on detritus and microalgae, they serve as a place for migrant birds to to feed on prey and take a rest. In Japan, tidal flats are fishing grounds for the Manila clam (*Ruditapes philippinarum*), which is a filter feeder that takes suspended particulate organic matters (POMs) that are present when the tidal flat is under sea level and converts them to inorganic nutrients. When Manila clams are harvested and eaten by people, the nutrients that flowed from the land to the sea return from the sea to land. In this way, a tidal flat ecosystem plays an important role in biodiversity, fisheries, and material flow. Salt marshes contribute to biodiversity through their unique flora and fauna and a place for migrant birds to rest and feed. The salt marshes also constitute a link in the nutrient flow between the land and the sea through photosynthesis and the accumulation of the produced organic matter on the bottom of the marsh, playing important roles in adjusting the environments through nutrient sinks and the sequestration of carbon dioxide. The coastal ecotones consisting of seagrass and seaweed beds, salt marshes, and tidal flats present many ecosystem services to human society.

According to the European Commission (2013), green infrastructure can be defined broadly as a strategically planned network of high-quality natural and semi-natural areas with other environmental features (meaning habitats) that is designed and managed to deliver a wide range of ecosystem services and protect the biodiversity in both rural and urban settings. Therefore, Komatsu et al. (2017) used the term "blue infrastructures" for coastal habitats, such as seagrass and seaweed beds, salt marshes, and tidal flats, because we can manage the habitats and plan their networks similar to those of green infrastructures. We use the term blue infrastructure with reference to the coastal habitats when we discuss the Satoumi.

Aquacultures, especially no-feeding ones, are developed in open-type rias bays along the Sanriku Coast, which the Miyagi and Iwate prefectures include in their jurisdictions. According to the fisheries statistics of 2014 provided by the Ministry of Agriculture, Forestry and Fisheries of Japan, the fishery productions, in cash, of the Miyagi and Iwate Prefectures were $591 million and $316 million, respectively. The coastal fishery productions of the Miyagi and Iwate prefectures consisted of the cultured marine animals of oysters (*Crassostrea gigas*), scallops (*Mizuhopecten yessoensis*), and ascidians (*Halocynthia roretzi*); edible brown seaweeds of wakame in Japanese (*Undaria pinnatifida* (Harvey) Suringar) and kelp (kombu in Japanese) [*Saccharina japonica* (Areschoug) C. Lane, Mayes, Druehl and G.W. Saunders]; fish caught with set nets and abalone (*Haliotis discus hannai*) and sea urchins (*Mesocentrotus nudus*) taken by dip net from a small boat. The percentages of the coastal fishery productions relative to the total fishery productions in the Miyagi and Iwate prefectures were 50% and 80%, respectively. The processed marine products, in cash, of the Miyagi and Iwate prefectures were $2213 million and $698 million dollars, respectively. Both prefectures supplied marine products indispensable for a quotidian family table. According to the Ministry of Agriculture, Forestry and Fisheries of Japan, Miyagi Prefecture supplied 81.3% of the total seed oysters sold in 2009, followed by Hiroshima Prefecture, which sold 16.6%. Iwate Prefecture produced 75% of the national production of wakame in 2008. Marine products, including abalones and sea urchins from coastal fisheries and aquacultures, are essential for tourism along the Sanriku Coast. The coastal fisheries and aquacultures along the Sanriku Coast form a social infrastructure that establishes a wide range of industrial bases and supplies marine products to Japan.

The large tsunami struck along the Sanriku Coast 120 km to 300 km from the epicenter of the Great East Japan Earthquake on March 11, 2011, (Fig. 2.1) destroyed fishing boats, aquaculture facilities, set nets, ports, fish markets, warehouses, and freezers that were social infrastructures supporting fishery production. Rebuilding the coastal fisheries is key for the recovery of societies along the Sanriku Coast. Although the social infrastructures visible to people have

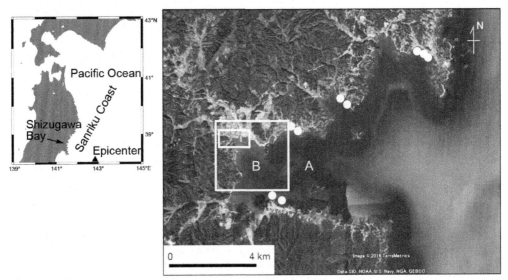

FIG. 2.1 Left map shows Honshu Island and Shizugawa Bay (arrow) with the epicenter of the Great East Japan Earthquake on March 11, 2011 (closed triangle). The right satellite image (Google Earth) shows a closeup of Shizugawa Bay, with the hinterland surrounded by mountains and the areas analyzed with remote sensing represented by a square (A) and a rectangle (B). Shizugawa Bay is a rias-type bay with a deep bay mouth and bay head into which rivers flow.

been rebuilt, the rebuilding of the invisible natural infrastructures (blue infrastructures), such as the seagrass and seaweed beds under the water, is unknown.

Remote sensing on coastal waters has advanced because of an increase in the spatial resolution of satellite images (Komatsu et al., 2012). One of the merits of satellite images is the archive of images that permits us to retrospectively explore the past states of coastal waters. By analyzing the images from the archive before and after the tsunami, it is possible to compare the habitats before and after the destruction.

We observed the succession of the coastal ecotone after the tsunami because these observations will serve to plan a conservation strategy for coastal ecotones struck by future tsunamis. Therefore, we examined the states of the blue infrastructures before the tsunami and examined their damages after the tsunami, as well as the successions affected by their recovery constructions, to avoid or remove the negative impacts on the recovery process. This section introduces our research results about the changes in the blue infrastructures before and after the tsunami discovered by remote sensing and field surveys. We hope that the information will serve as basic knowledge to realize sustainable and sound marine environments in the Sanriku Coast, where large tsunamis strike every several decades.

2.2.2 Influence of the Large Tsunami on Macrophyte Beds in Shizugawa Bay and Their Spatial and Temporal Changes After the Tsunami

Shizugawa Bay, classified as an enclosed bay by the Ministry of the Environment of Japan, is a typical open-type rias bay located in the southern Sanriku Coast (Fig. 2.1). It has a wide bay mouth (6.6 km), a relatively short longitudinal bay length (7.7 km), a deep maximum bay depth (54 m), a deep bottom in the bay mouth (54 m), and a surface area of 46.8 km^2. In the bay, the no-feeding aquacultures of oysters, ascidians, scallops, wakame, and kelp are conducted actively with rope-and-buoy type aquaculture facilities. Wakame culture facilities are deployed from autumn to spring, while those of other no-feeding target species are conducted throughout the year. The feeding aquaculture of coho salmon (*Oncorhynchus kisutch*) is conducted with dozens of cages from winter to summer. The 2011 Tohoku Earthquake Tsunami Joint Survey Group estimated through field observations that the mean run-up height of the tsunami in the bay was 14.4 m, and greatly damaged the bay area.

Field surveys were conducted at 6-month intervals from October 2011 to October 2013 and at intervals of about 3 months from 2014 until 2016. The field surveys consisted of visual observations, drop camera observations, and side-scan sonar observations from a small fishing boat and visual observations from the shore and on land using GPS to locate positions.

We selected the northwest area of the bay where two rivers flow into the bay, shown as Area A of Fig. 2.1, to observe the temporal changes of blue infrastructures, such as seagrass and seaweed beds, tidal flats, and salt marshes. We used satellite images from Digital Globe, provided by Google Earth, from 2010 to 2016. We downloaded the satellite images of the bay head areas in Shizugawa Bay as JPEG images and processed them with Adobe Photoshop CS2 to extract the seagrass and seaweed beds in the sea and ponds in the salt marshes on land. Photoshop CS2 could select neighboring pixels with the chosen pixel values. The range of values also was tuned for the detection of habitats. The resulting habitat maps were checked by fishermen.

We also selected another area west of the bay, shown as Area A of Fig. 2.1, to map the seaweed beds and barren rocky beds by remote sensing. A multiband image of GeoEye-1 taken on February 12, 2015, was analyzed with image analysis software (ENVI 5.2, Exelis VIS provided by Institute for Information Management and Communication, Kyoto University as a shared-use service for universities nationwide) by applying a supervised classification of a maximum likelihood method to classify the pixels into habitats. The resulting habitat map was checked by fishermen.

2.2.2.1 Seagrass Beds

Seagrass species are land plants that returned to the sea. Some mature shoots become longer and reproduce, producing flowers and fruits, while other shoots are immature and remain small. In Shizugawa Bay, broad seagrass beds were distributed near river mouths and in the port on June 25, 2010 (Fig. 2.2A). After the tsunami in 2011, they disappeared until 2014 (Fig. 2.2B and C). In 2015, small patches of seagrass appeared off the river mouth of the Mizushiri River (Fig. 2.2D) for the first time since the tsunami. The seagrass beds recovered slowly because of the lack of light near the river mouths and seashores caused by the turbid water brought by construction, such as seawalls, river embarkments, and grounds elevated >8 m above the sea level for the safety of houses from tsunamis (Komatsu et al., 2014). It is estimated that a decrease in turbidity promoted a recovery of seagrass patches in 2015 (Komatsu et al., 2018a).

Field surveys in October 2011 showed that sites where seagrasses of *Zostera marina* L. and *Z. caulescens* Miki survived the tsunami were in the areas sheltered from the wave (Fig. 2.1) (Komatsu et al., 2017). The contributions of the seeds buried in the sand beds from previous years the seeds produced by remaining seagrasses were important for the recovery of the seagrasses. Seagrass beds can recover faster by protecting the seagrasses distributed in sheltered areas as a source of seed supply (Komatsu et al., 2017).

To secure the growth of germinated seeds to recover seagrass beds, it is necessary to protect the current vegetative and reproductive shoots that produce seeds and remove the debris remaining on the bottom. Because turbid water retards the recovery of seagrass, fences against the diffusion of turbid water are useful for seagrass recovery (Komatsu et al., 2014).

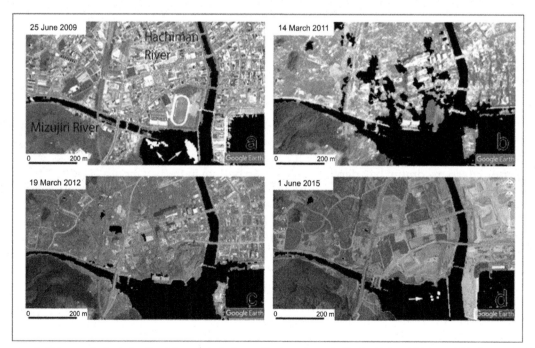

FIG. 2.2   Seagrass (white areas indicated with white arrows in the sea) beds and ponds filled with brackish water (black areas on land) extracted from and overlaid onto satellite images from June 25, 2009 *(upper left panel)*, March 14, 2011 *(upper right panel)*, March 19, 2012 *(lower left panel)*, and June 1, 2015 *(lower right panel)* provided by NASA and Digital Globe by Google Earth.

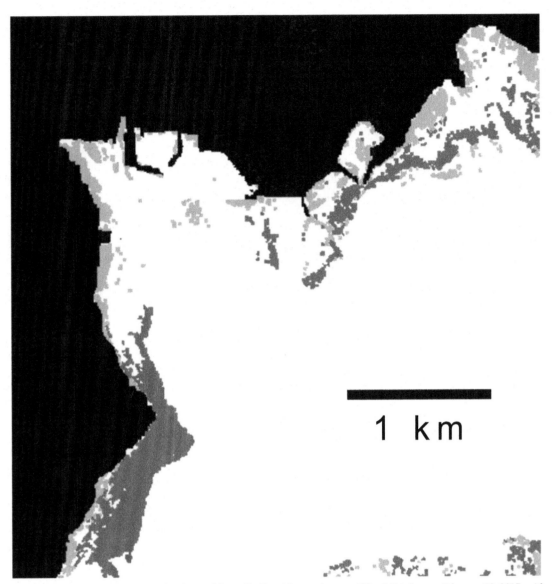

FIG. 2.3 Map showing the classification results obtained from GeoEye-1 image (area a of Fig. 2.1) taken on February 12, 2015, with a supervised classification of the remote sensing analysis. The *black*, *dark gray*, *light gray*, and *white* areas correspond to land, devastated seaweed, seaweed, and the others, respectively.

2.2.2.2 Seaweed Beds

Sargassum species grew on the rocky beds in Shizugawa Bay in October 2011, about 6 months after the tsunami struck. An aerial photo taken that November showed that *Sargassum* beds were distributed along the coast. *Eisenia bicyclis* (Kjellman) Setchell beds also were not significantly damaged by the tsunami. The debris from the seawalls along the coast broken by the tsunami was removed until 2014, showing that the *Sargassum* beds growing on hard debris decreased because of substrates removal. Since 2014, barren rocky coasts have broadly replaced *Eisenia* beds distributed along the coast, especially east coast of the bay head (Fig. 2.3). In Otsuchi Bay of Iwate Prefecture, which neighbors Miyagi Prefecture (to which Shizugawa Bay belongs), the sea urchins and seaweed beds were not damaged by the tsunami (Komatsu et al., 2015). Therefore, it is possible that sea urchins could survive a tsunami in Shizugawa Bay. The sea urchins that survived could reproduce in good nutritional conditions because seaweeds were luxuriant in the summer of 2011 (Komatsu et al., 2015). Sea urchin fisheries stopped harvesting for 3 years after the tsunami. Because the sea urchins born in 2011 were massively recruited in 2013, seaweed beds were replaced by barren rocky beds in the Sanriku Coast because of their feeding pressure.

Because sea urchin fisheries removed the sea urchins that spawned in each summer before 2011, seaweed beds have been protected from mass predations by sea urchins. This relationship is called top-down control. Estes and Palmisano

(1974) found that seaweed beds were not depleted on the coast where sea otters controlled the sea urchin population through predation. Seaweed beds hosting marine organisms contribute to marine biodiversity by acting as a nursery, spawning ground, and feeding ground. Because seaweeds influence the fatness of sea urchins, it is important to control the number of sea urchins to prevent an explosive increase. On the Sanriku Coast, where sea otters are absent, fishermen control the number of sea urchins through fisheries to keep the seaweed beds sound. An absence of sea urchin fisheries for 3 years, however, changed the seaweed beds to barren rocky beds through the lack of top-down control. Matsuda (2007) notes that sound fisheries are important to keep a material flow and a marine environment sound. On the Sanriku Coast, sea urchin fisheries create a fat, long, and smooth material flow through the maintenance of seaweeds.

Since 2014, barren rocky coasts have broadened on the Sanriku Coast because of the predations of seaweeds by sea urchins. Along the coasts, the lack of seaweed makes the sea urchins' ovaries small, which gives them no commercial value. Sea urchins with small ovaries produce small quantities of eggs and eggs of poor quality. Fishermen do not catch sea urchins with small ovaries (Komatsu et al., 2015). Because the lifetime of sea urchins is 14 to 15 years, the rocky coasts remain barren. Therefore, it is necessary to catch the sea urchins without commercial value. Artificial feeding to sea urchins removed from the barren in an aquarium increases their ovary size and then they can be sold. When a cycle of sea urchin use is established, the seaweeds on rocky coasts can recover to bring a sustainable long material flow.

2.2.3 Saltmarshes and Tidal Flats Appeared After the Tsunami

In an inner area northwest of Shizugawa Bay, into which two rivers flow (Fig. 2.1), land directly connected with the sea, tidal flats and saltmarshes appeared after the tsunami (Fig. 2.2B) because the seawalls were destroyed and disappeared, and the ground level dropped by 60 cm because of the earthquake. The population of Manila clams explosively increased on the tidal flats that appeared after the tsunami, and local people caught the clams. The recovery construction and the construction of land-level elevation for houses decreased the number of Manila clams (Komatsu et al., 2017). In March 2012, three new artificial tidal flats were constructed between the Mizushiri River and the Hachiman River, and large seawalls were built on the land behind them. When the tidal flats are separated from the land with such seawalls, the flats cannot play full ecological roles because they cannot link with the land.

Satellite remote sensing showed that the salt marshes that appeared after the tsunami had not connected with the sea by March 19, 2012, 1 year after the tsunami, because of construction (Fig. 2.2C). On June 1, 2015, the ponds disappeared (Fig. 2.2D). Fukushima et al. (2016) measured the concentration of dissolved organic carbon, UV254, and the relative fluorescence intensities of fluvic acid at the salt marsh influx where the tsunami caused a river to merge into a salt marsh, the salt marsh itself, and the salt marsh outlet where the water flows out to the river in Môûné Cove of East Kesen-numa Bay near Shizugawa Bay. When the river transports dissolved iron to the sea, the dissolved iron is oxidized, becomes heavier than the sea water, and falls to the sea bottom. The dissolved organic carbon (DOC), UV254, and relative fluorescence intensity of fluvic acid indicate the concentrations of dissolved organic matter (DOM) that are ligands bound to iron make a complex that is not oxidized easily in the sea. Fukushima et al. revealed that the concentrations at the salt marsh outflux were greater than those at the salt marsh influx. They considered that the DOM produced in the salt marsh increased the ligands bound to iron and supplied more dissolved iron complexes, which increased the primary production to the sea.

On the Sanriku Coast, the ecological succession of a salt marsh created by the tsunami to land occurs through a sedimentation by plants and a slow uplifting of the land level over several decades, synchronized with large tsunami events. Sanriku Fukko (reconstruction) National Park was created on May 24, 2013, to contribute to the reconstruction of the Sanriku region that was devastated by the Great East Japan Earthquake in 2011. The concept of the park is to experience the life and culture nurtured through the bounty and dangers of nature and the coexistence of humans and nature. The park aims to promote the regeneration of a coastal ecotone consisting of tidal flats, salt marshes, and seagrass beds damaged by the tsunami and earthquake by helping the natural resilience with the local people's consent and without preventing social restoration. The current status of the Sanriku Coast, however, places the construction of large seawalls against tsunamis and neglect the preservation of the coastal ecotone. The concept of the coexistence of humans and nature on the Sanriku Coast is forgotten. The active conservation of wetlands and tidal flats where ecological succession can be observed is desired for the coexistence of man and nature and the sustainable development of coastal area based on the prosperous fisheries, rich biodiversity, and sound marine environment.

2.2.4 Blue Infrastructures and Human Pressure

The macrophyte beds, salt marshes, and tidal flats forming the ecotone between the land and the sea are important habitats fostering biodiversity and blue infrastructures supporting the coastal fisheries. The Government of Japan decided to ensure biodiversity, and the Ministry of the Environment of Japan directed the country to restore the natural ecosystems (Ministry of the Environment of Japan, 2014). We want to emphasize that the salt marshes and tidal flats lost by reclamation were recovered by the tsunami (Komatsu et al., 2018a). Their ecosystem values are very expensive, similar to those of the macrophyte beds. Miyagi Prefecture and the Ministry of Land and Transport of Japan decided to construct sizeable seawalls and river embarkments against large tsunamis to secure human life. Seawalls higher than 8 m above the sea level are designed to endure large tsunamis by the construction of a wide base with a gentle slope revetment and the deep driving of steel sheet piles. Such seawalls can influence the underground waterways supporting underwater springs. Because the underwater springs foster seagrasses that grow on the sea bottom and are influenced by fresh water in the shallow bottom, a habitat for seagrasses is diminished by the large seawalls. Moreover, the large seawalls destroy the connections between the saltmarshes or tidal flats and the sea.

Large embarkments have been constructed along the river from the estuary because the last large tsunami ran upstream for a long distance. When the river banks and riverbeds are covered with concrete panels, the quantity of sand supplied from the river to the sea decreases and changes the bottom sediment compositions in the sea, which can decrease the habitats for seagrasses that prefer sand and mud sediments. In the rivers of the Tohoku Region, northeast of Honshu Island, chum salmon (*Oncorhynchus keta*) and masu salmon (*Oncorhynchus masou masou*) migrate upriver from the sea. Because the chum salmon spawn on the riverbed from which groundwater gushes, the spawning grounds for chum salmon have disappeared. The juveniles of salmonids predate aquatic insects. When the river banks and riverbeds are covered with concrete panels, the carrying capacity of the river for the salmonid juveniles is reduced through a decrease in habitats for the aquatic insect populations. Large seawalls and river embarkments destroy natural environments, blue infrastructures, and, consequently, the fisheries that are the most important industries on the Sanriku Coast. The Ministry of Land Construction and Transport of Japan and Miyagi Prefecture neglected the sustainable development of the coastal areas on the Sanriku Coast, not considering the sustainable use of blue infrastructures.

2.2.5 Efforts to Conserve Blue Infrastructure in the Future

In Otsuchi Bay of the northern Sanriku Coast, stormy weather because of the development of low pressure destroyed seagrass beds in October 2006. The seagrass beds recovered after several years. This experience suggests that the seagrass beds in Shizugawa Bay could recover within 10 years. The tsunami, however, left behind debris from aquaculture facilities, such as rope, on the sea bottom. Because water currents and waves move the debris, damaging the seagrass shoots, the debris should be removed to promote seagrass recovery. After a tsunami, when seagrasses are recovering on their own with their buried seeds or their seeds produced by the remaining shoots, human interventions, such as the transplantation of shoots, must be avoided to not change the genetic diversity of the seagrasses (Komatsu et al., 2014). Where sea urchin predation has created barren, rocky beds, it is necessary to remove sea urchins by actively monitoring with remote sensing.

Finally, we consider the conservation of the blue infrastructures that are key for the recovery of coastal marine biodiversity and fisheries. Ongoing restoration practices, such as building large seawalls and river embarkments, do not use blue infrastructures efficiently and, in fact, destroy blue infrastructures. The post-tsunami decision to build seawalls that would damage the sustainable use of blue infrastructures, was justified as a measure to secure human life, although the local people live on grounds >8 m above the sea level and, therefore, are safe from tsunamis. We must now discuss how we construct a sustainable society that enables the coexistence of prosperous lives for people and sound environments in nature. The discussion must be based on points of view consistent with strategies such as the biodiversity strategy to emphasize the profits of Japan and future generations that SDGs emphasize.

2.3 IS THE FOREST A LOVER OF THE SEA? NUTRIENTS

As a model area of Shizugawa Bay, Minamisanriku Town.
Shigeru Montani, Hajimu Yatabe, Naoya Yokoji, Hotaka Seko and Yuki Nakano

2.3.1 Introduction

The interdependency of the relationship between land and coastal areas is becoming generally accepted. In recent years, coastal ecosystems have been figuratively described by the expression "the forest is a lover of the sea." It has

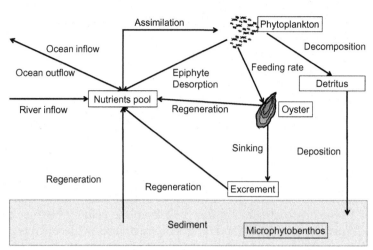

FIG. 2.4   Schematic view of nutrient cycling in the Shizugawa Bay that focuses on cultured oysters.

become apparent that the supply of substances from the land is of prime importance. The idea that nutrient-rich water produced by the forests supporting the productivity of the sea has not yet been quantitatively demonstrated, and many aspects of the idea are unclear.

In this research, we determine standing stocks of nutrients, its origins, magnitudes and seasonal changes in the bay controlling primary production, which is the driving force of biological production in Shizugawa Bay (Fig. 2.1) as a model bay. Possible sources of nutrient inflow to the bay are the land area from the forest to the sea, carried through rivers; the seawater flowing in from the Pacific Ocean; and the regeneration of nutrients from filter feeders such as oysters and sea squirts cultured in the bay. It is necessary to clarify how various primary producers (phytoplankton, attached microalgae, large seaweed, and seagrasses) use nutrients for photosynthesis and how nutrients connect to primary production that drives the bay ecosystem. Therefore, we quantify stocks and fluxes of nutrients shown in Fig. 2.4 to understand nutrient cycling of the bay. In addition, we aim to acquire basic knowledge for establishing an aquaculture management method for an open-type enclosed bay using a multi-layered and comprehensive analysis and a biological production mechanism in the bay.

2.3.2 Observation and Sample Analysis Methods

In order to quantify an inflow of nutrients from the land to Shizugawa Bay, continuous monitoring over a long period of time was carried out. We collected samples once a week from the estuaries of the three major rivers (Hachiman River, Mizushiri River, and Oritate River) from July 9, 2014, to March 2015 (Fig. 2.5) and measured their nutrients and photosynthetic pigment concentrations. Furthermore, at 15 fixed stations set in the Shizugawa Bay (Fig. 2.5), broad field surveys were conducted from July 2014 to April 2015 to cover four seasons. Water samples were taken at 5 m intervals from the sea surface to the bottom at the stations to determine nutrients in seawater and various chemical substances of particulates in seawater. Seawater taken at a station in the central part of the bay was used for estimation of primary production at a land facility by the ^{13}C simulated in situ method. At the same time, to estimate the influence of oyster excrement on the bottom surface environment in the aquaculture area, a sediment trap was deployed at a fixed station at each observation time (four seasons) to measure particle flux. To estimate the origin and the transport process of the primary production caused by a photosynthetic process using nutrients, we measured carbon and nitrogen stable isotope ratios of particulate matters in seawater, the attached microalgae attached to an oyster cultivation facility and the large seaweed and sea grass.

Bivalve cultured in coastal seas mainly feed phytoplanktons and suspended organic matters in seawater. Bivalve cultures are highly sustainable from the viewpoint of managements of coastal environment and nutrient loads. Feeding and excretion of bivalves also play an important role in a material cycle process in an aquaculture area of the bay through high filtering capacity and supplying organic matters in the bottom layer because of the accumulation of excretion and regeneration of nutrients from excretion, etc. Few studies on nutrient cycling, however, have quantitatively showed the influence of bivalve farming on the water quality, sediment environment and biological production process. In Shizugawa Bay, effects of cultured oysters (*Crassostrea gigas*) and sea squirts (*Halocynthia roretzi*) on the nutrient cycling process

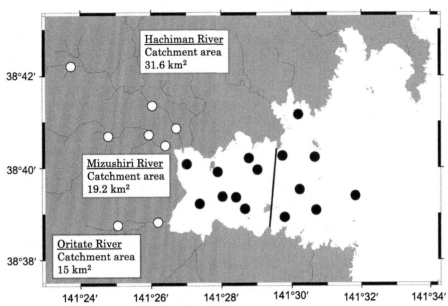

FIG. 2.5 Map showing stations in the rivers and Shizugawa Bay where water samples were taken for nutrient analysis and the inner part of the bay west of the solid line of which material cycle was studied.

through their feeding and excretion were evaluated based on the detailed laboratory experiments described later. Seawater and river water samples were filtered with a Whatman GF/F filter under pressure within several hours after collection. Chlorophyll *a* and pheopigments in seawater samples were extracted with a 90% acetone solution for 24h, and then photosynthetic pigments were measured using a fluorometer. NH_4-N, NO_2-N, $NO_2 + NO_3$-N, PO_4-P and Si $(OH)_4$-Si were measured using an autoanalyzer. Particulate organic carbon (POC) and particulate organic nitrogen (PON) in seawater samples were filtered under pressure with a GF/F filter. The filters were heated at 450°C for 12h to burn organic matter for determining them. Samples obtained by laboratory experiments described later were measured with the same methods.

2.3.3 Composition Ratio of Nutrients and Estimation of Their Standing Stocks

To investigate the nutrient distribution characteristics in Shizugawa Bay, we determined the concentrations of each nutrient in the seawater samples obtained from the sea surface to the bottom at 5 m depth intervals at 15 stations shown in Fig. 2.5. Based on the results, the molar ratios of nutrients were determined. It was found that dissolved inorganic nitrogen (DIN) was generally the restricted nutrient in Shizugawa Bay. The average concentration for each season was calculated using the nutrient concentrations. Based on these values and the water volume of the bay, the standing stocks of nutrients per unit area were calculated (Table 2.1). The DIN level increased 6.7 times, from 184 mg N m^{-2} in autumn (October) to 1287 mg N m^{-2} in winter (January). DIN is a restricted nutrient with a considerable seasonal fluctuation. The dissolved inorganic phosphorus (DIP) level increased by a factor of 3.2, from 73 mg P m^{-2} in autumn to 235 mg P m^{-2} in winter, and its pattern of seasonal change was similar to that of DIN. Silicon (DSi) increased by a factor of only 1.7, from 5876 mg Si m^{-2} in summer (July) to 9792 mg Si m^{-2} in autumn, and its seasonal change was greatly different from DIN and DIP. Since silicon was always abundant in the bay, the silicon is not a nutrient that restricts a biological production in the bay.

TABLE 2.1 Seasonal Changes of Nutrients in Seawater of Shizugawa Bay (mg L^{-1})

	DSi	DIN	DIP
Spring (Apr.)	5898	1160	169
Summer (Jul.)	5876	623	107
Autumn (Oct.)	9792	184	73
Winter (Jan.)	9482	1237	235

2.3.4 Estimate of Nutrient Supply From Rivers and Pacific Ocean

A quantitative evaluation of nutrients supplied from the rivers and the Pacific Ocean to Shizugawa Bay as mentioned in Section 2.3.1 was conducted. There is a high possibility that the supply of nutrients from the Pacific Ocean, in particular, by the Oyashio water system has a great influence on the nutrient cycling process in Shizugawa Bay, which is an open inner bay. First, we estimated the flow rate of the rivers and their supply of nutrients. Because flow data for rivers flowing into Shizugawa Bay have not been measured and published, this study calculated specific flow rates of rivers in Shizugawa Bay using their catchment areas by referring to another river near Shizugawa Bay of which flow rate is available. The class B river has data of water level and flow rate. Okawa River in Iwate Prefecture flowing into Kesennuma Bay is the closest to Shizugawa Bay. The three main rivers in Shizugawa Bay are the Hachiman River, Mizushiri River, and Oritate River, with basin areas of 31.6, 19.2, and 15 km^2, respectively while the catchment area of the Okawa was 168 km^2. The freshwater flow rate Q at the cutting point was calculated from the following empirical formula provided by the Ministry of the Environment using the water level and flow rate data of the Okawa at the cutting point of the Okawa water system from 2002 to 2015.

$$Q = 18.11 \times (H - 0.210)^2$$

where Q and H are freshwater flow rate [$m^3 day^{-1}$] and dayly average water level at the cutting point, respectively. The catchment area at the cutting point was 127 km^2. The flow rates for the three rivers in Shizugawa Bay were calculated from the ratios of the catchment area between that at the cutting point of Okawa River and the catchment areas of the three rivers that poured into Shizugawa Bay. Supply of nutrient i shown with suffix from the river r shown with suffix r to unite area of the inner part of the bay represented as Tri [$mg m^{-1}$] was calculated by the following equation:

$$Tri = Qr \times Mri / A$$

where Qr is the flow rate of the river r flowing into Shizugawa Bay [$L day^{-1}$], Mri is the concentration of each nutrient i in river r pouring into Shizugawa Bay [$\mu g L^{-1}$] and A is an area of the inner part of Shizugawa Bay (13.8 km^2).

2.3.5 Nutrient Supply From Rivers

The maximum and minimum of nutrient concentrations of the rivers in four seasons are as follows: 0.2 to 59.6 $\mu mol L^{-1}$ for NH_4-N, 0.6 to 114.6 $\mu mol L^{-1}$ for $NO_2 + NO_3$-N, 0.1 to 22.6 $\mu mol L^{-1}$ for PO_4-P, and Si $(OH)_4$-Si, 2.3 to 930.4 $\mu mol L^{-1}$. No seasonal trend could be confirmed for each nutrient concentration. The N/P ratio exceeded the Redfield ratio (molar ratio of C: N: Si: P = 106: 16: 15: 1), which promotes efficient absorption of nutrients by phytoplanktons, except for one period of the Oritate River. Rivers supplied nutrients sufficient for the growth of phytoplanktons in the estuary area. It is thought that a high concentration of nitrogen was constantly supplied for the PO_4-P required for the growth of phytoplanktons. Although the N/Si ratio exceeded 1 for some samples from the Mizushiri River and Oritate River, a high concentration of Si $(OH)_4$-Si was always present. The annual mean $NO_2 + NO_3$-N showed the maximum value of 63.2 ± 24.8 $\mu mol L^{-1}$ in the Mizushiri River, and the annual average NH_4-N reached the maximum value of 3.2 ± 2.9 $\mu mol L^{-1}$ in the Hachiman River. Most of the nitrogen supplied as DIN was $NO_2 + NO_3$-N, and both NH_4-N and $NO_2 + NO_3$-N concentrations of the Oritate River were the lowest among the rivers. Conversely, the annual average PO_4-P of Oritate River showed the maximum value of 0.9 ± 0.4 $\mu mol L^{-1}$ among three rivers. The annual mean Si $(OH)_4$-Si of Hachiman River showed the maximum value of 221.1 ± 96.4 $\mu mol L^{-1}$. No correlation was found between the concentrations of nutrients in each river and its specific flow rate.

Throughout the observation period, the flow rate increased from September to October and showed a tendency to decrease in March and April. The Hachiman River, with the largest watershed area, also supplied the most of all nutrient among the three rivers. In addition, a positive correlation was found between the flow rate of each river and the $NO_2 + NO_3$-N, PO_4-P, and Si $(OH)_4$-Si supplies. This correlation indicates that a flow rate of the river controls the nutrient supply. The nutrient supply from all rivers pouring to Shizugawa Bay was calculated using the total river flow rate estimated from the whole catchment area assuming that the concentration of each nutrient in other small and medium rivers had concentrations that were the average of the three major rivers.

The annual average of nutrient supplies per unit area of the inner part of the bay from all rivers were 0.9 ± 0.8 $mg N m^{-2} day^{-1}$ for NH_4-N, 17.6 ± 13.7 $mg N m^{-2}$ for $NO_2 + NO_3$-N, 0.3 ± 0.2 $mg P m^{-2} day^{-1}$ for PO_4-P and 65.0 ± 49.9 $mg Si m^{-2} day^{-1}$ for Si $(OH)_4$-Si. The annual total nutrient supplies from the rivers in 2015 were 407 $mg N m^{-2}$ for NH_4-N, 6661 $mg N m^{-2}$ for $NO_2 + NO_3$-N, 91 $mg P m^{-2}$ for PO_4-P. It was 22,605 $mg Si m^{-2}$, reflecting the characteristics of the volcanic zone where the supply of Si from the river is abundant. Of the annual nutrient supplies from the rivers to the bay, those from the Hachiman River accounts for 43% of NH_4-N, 46% of $NO_2 + NO_3$-N, 39% of PO_4-P and 47% of Si $(OH)_4$-Si.

To evaluate the nutrient supply capacity of the rivers, the nutrient supply per day from the rivers was calculated as a ratio to standing stock of nutrient in the inner part of the Shizugawa Bay. The amount of DIN supplied from the rivers was $26\,mgN\,m^{-2}\,day^{-1}$ in January, which was the greatest in a year, and corresponds to approximately 2% of DIN standing stock of $1442\,mgN\,m^{-2}$ in the inner part of the bay. Nutrient supplies in July 2014 and in July 2015 correspond to approximately 1% and 11%, respectively. This is because the nutrients supplied to the bay fluctuated depending on river discharge that is impacted by unexpected events such as rainfall. In July, the DIN concentration in the bay decreased, and the N/P ratio fell below the Redfield ratio. These facts suggest that the rivers were a major source of DIN. The PO_4-P supplied from the rivers was <1% of the PO_4-P standing stock in the inner part of the bay throughout the year, and the effect of the PO_4-P supplied by the rivers on the bay was extremely small. The supply amount of Si $(OH)_4$-Si, however, corresponded to a maximum of 29% (with respect to its standing stock in the bay) in May.

2.3.6 Calculation of the Amount of Nutrient Supply From the Pacific Ocean

We assumed that the east side of the line shown in Fig. 2.5 was the outside of the inner part of the bay. The cross-sectional areas at 5 m depth layer intervals from the surface to a depth of 20 m along the line were obtained from the waterway map. Based on the results of the cross-sectional current distributions measured by ADCP observations along the line, the average current velocities of ebb and flood tides for each layer on the nutrient observation day were estimated. Inflow and outflow volumes at each cross-sectional layer along the line were calculated by multiplying cross-sectional area of layer by the average tidal current velocity of the layer at the flood tide and at the ebb tide, respectively. Nutrient concentrations $[mg\,m^{-3}]$ of each layer at a station outside the bay were multiplied by the inflow/outflow amount of each layer to obtain the inflow and outflow masses of each nutrient with the inflow mass being positive and the outflow mass being negative. Finally, the total amount of cross-sectional inflow and outflow nutrients along the line were divided by the inner part area of the bay ($13.8\,km^2$) to obtain the mass of inflow and outflow nutrients from the ocean per unit area of the bay.

2.3.7 Primary Production Rate in Shizugawa Bay

Primary production (photosynthesis) rate of microalgae such as phytoplankton in an estuary varies dynamically because of light, nutrient restriction, mixing, salinity gradient and other factors. Likewise, it was considered that nutrients and the light environment changed considerably according to the seasonal changes of seawater exchange, influx of terrestrial-derived substances and activities of aquaculture organisms in Shizugawa Bay. Therefore, we evaluated the role of microalgae supporting the biological production by quantifying the primary production rate of the water column in Shizugawa Bay.

The maximum and minimum primary production rates in a year obtained in this study were $756\,mgC\,m^{-2}\,day^{-1}$ in April 2015 and $230\,mgC\,m^{-2}\,day^{-1}$ in January 2015, respectively (Fig. 2.6). The annual average was $437 \pm 191\,mgC\,m^{-2}\,day^{-1}$. Using the Redfield ratio, the maximum, minimum and annual average values of primary production rate are converted to $114\,mgN\,m^{-2}\,day^{-1}$, $35\,mgN\,m^{-2}\,day^{-1}$ and $66 \pm 29\,mgN\,m^{-2}\,day^{-1}$, respectively. The primary production rate increased and decreased depending on Chl.*a* concentration.

2.3.8 Evaluation Method for Low-Trophic-Level Consumers as Nutrient Cycling Drivers

The annual landing amount of oysters (*C. gigas*) in Miyagi Prefecture accounted for 25% of that for the whole country at 41,653 tons in 2010 before the 2011 earthquake and tsunami. Although it decreased after the earthquake disaster, it recovered to 11,581 tons (7% nationwide) in 2013. *C. gigas* is also a major aquaculture species in Shizugawa Bay and is extremely important as a major member of lower trophic level production in the bay's ecosystem. This study quantitatively evaluated the ecological role of oyster as as an element constituting the fundamental part of the material cycle system in the bay by identifying its food sources and by quantifying food consumption and excretion.

A sediment trap with six acrylic cores with an inner diameter of 7.5 cm and a height of 30 cm was used (Montani et al., 1988) to collect sinking particles by deploying it at the rope inside and outside the aquaculture facilities at a depth of 7 m. The obtained samples were divided using a plankton divider and analyzed for Chl.*a*, POC, and PON, carbon nitrogen stable isotopic ratios ($\delta^{13}C$, $\delta^{15}N$), etc.

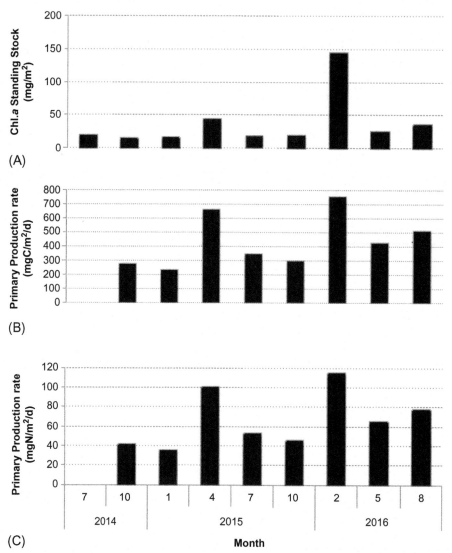

FIG. 2.6   Standing stock of Chl.*a* (A) and primary productions in Shizugawa Bay expressed with carbon (B) and nitrogen (C).

Microalgae adhering to the surface of an oyster were scraped off with a brush. The oyster was cultured in an aquarium filled with seawater filtered through a Whatman GF/F filter (pore size 0.7 μm) for 3 days without feeding to make the oyster egest fecal particles. Thereafter, the molluscous part of oyster was lyophilized and pulverized using an agate mortar. In order to carry out the degreasing treatment, a powdered sample was put into a mixed solution of chloroform and methanol (in a 2:1 ratio) and left for 24h. Then it was lyophilized again to obtain an analytical sample.

2.3.9 Estimation of Oyster's Food Sources From the Stable Isotope Ratio (δ^{13}C and δ^{15}N)

Fig. 2.7 shows the seasonal variations of δ^{13}C and δ^{15}N in algae attached on the shell of *C. gigas* and on particulate organic matter (POM). The δ^{13}C of oyster varied from −20.1‰ to −17.8‰, showing the maximum value in October and the minimum in July. Its δ^{15}N fluctuated in the range of 6.7‰ to 8.6‰, showing the maximum value in November and the minimum value in April. While the length of a vertical rope with oyster clusters which is attached to a horizontal rope was 7m, there was no difference in δ^{13}C and δ^{15}N values of oyster molluscous parts in the surface layer and the deepest layer of 7m deep of the vertical rope. The δ^{13}C and δ^{15}N of attached microalgae fluctuated in the ranges of −21.1‰ to −14.4‰ and 6.1‰ to 9.0‰, respectively, showing the maximum value in October and January and the minimum value in July. These results strongly suggest the possibility that the oysters filtered and fed attached microalgae detached from surfaces of the culture rope and oysters (Fig. 2.7).

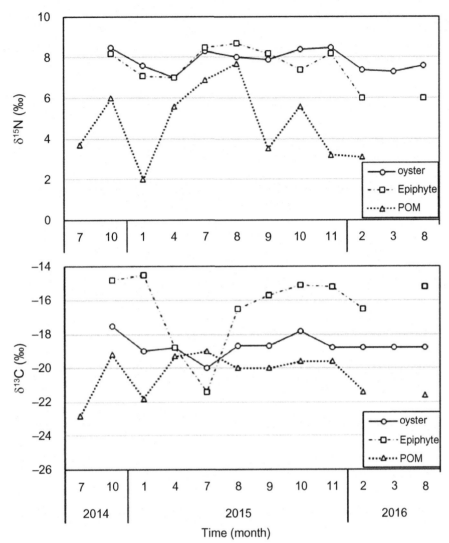

FIG. 2.7 Seasonal changes in stable isotope ratios of δ^{15}N *(upper panel)* and δ^{13}C *(lower panel)* of oysters, attached microalgae (epiphyte), and particulate organic matters (POM: suspended organic matters in water) taken at an area of oyster culture rafts.

Therefore, in order to verify whether attached microalgae is an oyster's food item, we attempted an in situ experiment on growth of attached microalgae and biomass of detached microalgae. The same vertical rope as that used for the oyster culture in the bay was prepared and cut to a length of 14 cm. These rope pieces were attached to a culture raft in Shizugawa Bay, and recovered to collect microalgae attached to surfaces of the rope pices in three periods of two weeks, four weeks and eight weeks after the deployment between May and November 2016. The retrieved two ropes were cut in half to obtain four rope pieces. One of the rope pieces was rubbed carefully with a brush in a container filled with 1 L of filtered seawater to remove the attached microalgae. The removed attached microalgae were collected by a Whatman GF/F filter. The Chl.*a*, POC, PON, and carbon nitrogen stable isotope ratio were analyzed. The other three rope pieces were separately suspended in a 500 mL bottle filled with filtered seawater that was sampled at the in situ experimental site. These culture experiments were conducted for 4, 8, and 12 days under light condition of 12 h day and 12 h night and water temperature condition of 15°C to 20°C. After the culture experiments, these ropes were treated as the same as the first rope piece above-mentioned to obtain adherent microalgae, which were also analyzed as the same as the first one. The seawater in the bottle was filtered, collected, frozen, and stored at −30°C.

Assuming that the main food sources of *C. gigas* were only water column phytoplankton and attached microalgae, we compared the stable isotopic ratios of the oysters's molluscous parts in the bay with their dietary sources (phytoplankton in water column and attached microalgae) in August 2016. The δ^{13}C value of a *C. gigas* molluscous part was between those of phytoplankton and attached microalgae, strongly suggesting that oysters consumed microalgae attached to the oysters as an important food source. A significant correlation (y=0.0342x−0.004) was obtained between the attached amount (x: μg Chl.a cm^{-2}) and the detached amount (y: μgChl.a cm^{-2} day^{-1}) of microalgae obtained from the experiments (R^2=0.3665, n=19, α<0.01). The amount of attached microalgae was saturated at 3.5 μg Chl.a cm^{-2} after about two months from the start of culture experiment. It is thought that the same amount of algae always attached to the cultivation rope and oyster shell surfaces in summer. From the saturation amount (3.5 μg Chl.a cm^{-2}) and the regression equation, it was suggested that 0.051 μg Chl.a cm^{-2} was detached per day from the cultivation rope. The microalgae of 0.038 g Chl.a/oyster culture raft/day were detached from the whole rope surface area of one oyster cultivation raft (753,600 cm^2). In August, the microalgae of 0.35 μg Chl.a cm^{-2} were detached from the average oyster shell surface area of 112 cm^2. Assuming that 245,000 oyster individuals are cultured on an oyster raft, it is estimated that the microalgae of 0.12 g Chl.a per day is detached from the surface of the oyster shells.

In this way, it is clear that the *C. gigas* cultivates attached microalgae which cultured oysters feed on by themselves as if oysters do gardening by themselves. Thus, this relation between oyster culture and attached microalgae makes a considerable contribution to increasing the biological production efficiency in the bay. Finally, we can say that attached microalgae have a symbiotic relationship with the oyster cultures.

2.3.10 Quantification of Nutrient Cycling System Driven by Filter Feeders Through Nutrient Regeneration

The functional role of an oyster as a primary consumer is that the oyster not only converts primary production (plants) to flesh but also supplies nutrients to primary producers through nutrient regeneration by excretion. Therefore, we quantify the feeding and excretion abilities of *C. gigas*.

Experiments about the feeding and nutrient regeneration ability of *C. gigas* were designed. Then, two aquariums for each of feeding, nonfeeding and control groups were prepared with use of diatoms (*Chaetoceros gracilis*) as feed. The experiment was conducted for 72 h under 12-h light (day), 12-h dark (night) condition and water temperature conditions of 10°C and 20°C. During the experiment, seawater and egested feces in the aquarium were sampled at appropriate intervals for later analysis. Changes in liquid excretion nutrient amounts per elapsed time were fitted with a straight line using the least-squares method. A nutrient regeneration rate was calculated from its slope (Fig. 2.8). By multiplying a nutrient regeneration rate of fecal pellet by an excretion rate of feces calculated during the pretreatment of the experiment, we obtained the nutrient eluted rate per gram of dry weight of *C. gigas* molluscous part per hour (Magni et al., 2000).

Applying the measured dry weight of the *C. gigas* molluscous body and the culture water temperature to a well-known formula to estimate water volume filtered by an oyster (Akashige et al., 2005), filtered water volumes per unit dry weight of molluscous part per day at a water temperature of 10°C and 20°C were 20 and 200 L gDW^{-1}/day, respectively. The amounts of feed per day per unit oyster dry weight in the aquaculture area at water temperatures of 10°C and 20°C converted to Chl.a were 30 and 200 μg Chl.a gDW^{-1} per day, respectively, by multiplying filtered water volume per day per gDW of molluscous body with average Chl.a concentration (μgL^{-1}) in the bay.

The nutrient regeneration rates of DIN and DIP were 0.12–2.66 μmol h^{-1} gDW^{-1} and 0.008–0.19 μmol h^{-1} gDW^{-1}, respectively. We compare the nutrient regeneration rates from a series of experiments with the nutrient standing stocks in Shizugawa Bay to obtain the proportion of regeneration nutrients through cultured oysters per day. It is estimated

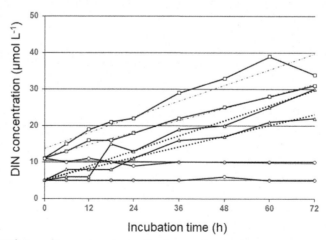

FIG. 2.8   Scatter plot of DIN (μmol L^{-1}) depending on time obtained by the experiments about nutrient regeneration speeds of oyster conducted in October 2014. Triangles, squares and circles correspond to oysters under nonfeeding, feeding and control (no oysters in an aquarium) conditions, respectively. Straight lines are linear regressions. Two sets of experiments were conducted for three conditions.

that DIN and DIP calculated on an annual basis are regenerated at 2.68 kmol day^{-1} and 0.19 kmol day^{-1} from liquid excretion, respectively and at 0.36 kmol day^{-1} and 0.013 kmol day^{-1} from solid excretion, respectively produced by all oysters cultured at oyster rafts in the inner bay in a year. In addition, the seasonal fixed-point observation results show that the annual average standing stocks of DIN and DIP in the inner bay were estimated to be 38.5 and 2.81 kmol, respectively. A percentage of nutrient regeneration per day of cultured oysters is equivalent to approximately 7% to 8% of the nutrient pool in the inner bay. It has become clear that *C. gigas* culture has a considerable impact on the nutrient circulation process in Shizugawa Bay.

In order to calculate feeding rate and nutrient excretion by *C. gigas*, we estimated the monthly standing stock of *C. gigas* in the bay. The weight of *C. gigas* and attached organisms per oyster cultivation rope in Shizugawa Bay was 84.4 kg in wet weight measured in October 2014, of which the oysters and other organisms were 50.1 kg and 34.3 kg mainly consisting of mussels, respectively. The estimated dry weight of *C. gigas* molluscous parts per cultivation rope was 1.1 kg. The dry weight of molluscous part of oyster in Shizugawa Bay was 15% of its wet weight. There were 583 rafts in the bay. One raft consists of two horizontally oriented ropes to which vertically hanged 200 cultivation ropes were attached. Thus, the standing stock of *C. gigas* in the inner part of Shizugawa Bay was estimated to be 129.0 tons in dry weight. Because the carbon and nitrogen contents were 47% and 16% of the dry weight of oyster molluscous part in October 2014, the carbon and nitrogen standing stocks in dry weight were 60.6 tons C and 20.6 tons N, respectively.

2.3.11 Estimation of the C. *gigas* Dietary Demand and Its Impact

The amounts of nitrogen and carbon in suspended matters captured by the *C. gigas* ranged from 0.4 to 2.3 mg N day^{-1}ind.$^{-1}$ and from 3.2 to 18.9 mg C day^{-1}ind.$^{-1}$ in a year, respectively. They reached the maximum values in July. The number of oysters per vertical rope suspended from the horizontal rope was determined by dividing standing stock of oysters in dry weight per vertical rope (0.4 kg DW in January; 1.1 kg DW in October) by mean dry weight of oyster (1.1 g DW). Assuming that one aquaculture rope is hanged per 1 m square of sea surface in the inner bay, dietary demand of *C. gigas* in a unit area in the inner bay was given by multiplying the number of oysters per vertical rope, the amount of suspended matters captured by an oyster and assimilation rate (0.746) of captured suspended matters. The results were 1.7 g C m^{-2} day^{-1} and 0.2 g N m^{-2}day^{-1} in January, and 9.3 g C m^{-2}day^{-1} and 1.1 g N m^{-2} day^{-1} in October.

The oysters of Shizugawa Bay release the sperms and eggs from July to August and are harvested in January. Therefore, the standing stock of oysters in January and October were the minimum and the maximum values, respectively. Assuming that one vertical rope of oyster culture is hung per square meter of the sea surface and oysters evacuate suspended organic matters, which are captured but not assimilated by oysters, as excrement, the suspended organic matter was estimated to be 564 mg C m^{-2} day^{-1} and 73 mg N m^{-2} day^{-1}, in January, and 3194 mg C m^{-2} day^{-1} and 394 mg N m^{-2} day^{-1} in October. These values are similar to those of the sediment flux just below the oyster culture raft (275 mg C m^{-2} day^{-1} and 29 mg N m^{-2} day^{-1} in January; 4831 mg C m^{-2} day^{-1} and 427 mg N m^{-2} day^{-1} in October). Therefore, most of the settling particles collected immediately under the oyster rafts are thought to be brought from oysters' feces.

2.3.12 Mass Transport Capability of Another Important Aquaculture Species: Ascidians

In addition to oysters, an ascidian (*Halocynthia roretzi*) is one of the main cultured species in the bay. Therefore, the mass transport capacity of *H. roretzi* known as ascidians was estimated with the feeding rate calculated from the filtering rate and food consumption rate including food sources of ascidians. We quantitatively evaluated the influence of cultured ascidians on nutrient cycling in the aquaculture area of the inner part of the bay.

Laboratory experiments on the feeding and excretion of ascidians were carried out using a system similar to that of the previously described on the experiments on oysters. Experiments to estimate the filtering and feeding rates of *H. roretzi* over a short time were conducted using water tanks of which feed concentrations were set by adding a cultured diatom, Ch. gracilis, into the tanks as feed. The feeding rate of ascidians varied from 0.28 to 1.53 µg Ch^{-1} g DW^{-1}. Regarding the excreted nutrients, NH$_4$-N and NO$_3$-N concentrations increased significantly in each tank, and NO$_2$-N concentrations changed very little. DIN increased in all aquariums with a high excretion rate of 1.9 µmol h^{-1} ind.$^{-1}$ and 6.7 µg N h^{-1} g DW^{-1}. The N/P ratio was approximately 20, which was higher than the Redfield ratio (N/P = 16).

Using these data, we estimated the nutrient cycling driven by ascidians in the inner part of Shizugawa Bay. The average annual Chl.*a* concentration in the waters where ascidians were cultured was 2.2 µg Chl.*a* L^{-1}. The feeding rate of ascidians calculated from laboratory experiments was 91.9 µg Chl.*a*$^{-1}$ g DW^{-1}. Considering the contribution rate of POM, attached microalgae and zooplankton to this feeding rate, the carbon and nitrogen circulation driven by

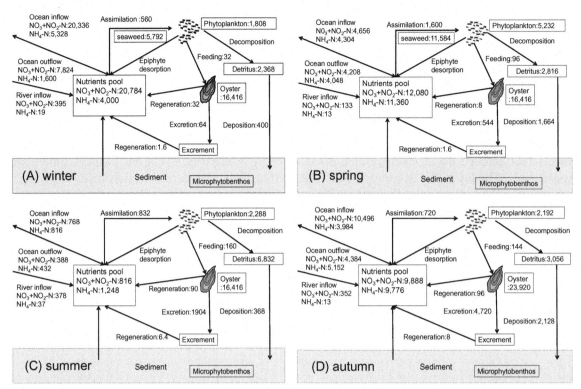

FIG. 2.9   Nitrogen flows in Shizugawa Bay in winter (A), spring (B), summer (C) and autumn (D) indicated with standing stocks (kg N) and fluxes (kg day^{-1}).

ascidians' mean annual standing stock of 904 tons in the inner part of Shizugawa Bay (13.8 km^2) in 2016 were calculated. The nitrogen circulation in Shizugawa Bay was greatly influenced by the seawater exchange with the open ocean. At least, 16% of the net DIN inflow from the ocean was used for primary production. On the other hand, ascidians and oysters consumed some of primary production and directly reduced it to nitrogen equivalent to 6.4% of the net DIN inflow as liquid or 17.6% of the river inflow to the inner part of the bay. The regeneration of nitrogen by ascidians and oysters during two weeks can support the primary production in the inner part of the bay. Although both the standing stock and feeding rate of *H. roretzi* were about 30% of those of *C. gigas*, the amount of excreted feces of *H. roretze* reached as much as three times that of *C. gigas*. While the amount of nitrogen of feces occupies only 6.8% of all nitrogen in excreta for *C. gigas*, it was 64% for *H. roretzi*. The results show that *H. roretzi* greatly affects nitrogen transport to the seabed.

2.3.13 Summary: Nitrogen Cycle in Four Seasons in Shizugawa Bay

Based on the data obtained in this study, seasonal nitrogen circulation maps of the inner part of Shizugawa Bay are shown in Fig. 2.9. These figures represent nitrogen flux [mg m^{-2} day^{-1}] and nitrogen standing stock [mg m^{-2}] in four seasons in Shizugawa Bay in 2015. The nutrient regeneration rate of *C. gigas* per unit area was calculated for an area of the inner part of Shizugawa Bay (13.8 km^2) using the data of liquid and solid excretion. Standing stock of phytoplankton as POC was estimated by multiplying the sum of standing stock of Chl.*a* from the surface to a depth of 20 m with a POC/Chl.*a* weight ratio which was assumed 50. That as PON was obtained by using phytoplankton POC above-mentioned and Redfield ratio (C/N = 6.6) shown as the second term of right hand side of the following equation. Standing stock of nitrogen of detritus was calculated from the difference between all PON of water samples from the surface to a depth of 20 m and the PON of phytoplankton shown as the following equation.

$$\text{Nitrogen standing stock of detritus} = \Sigma\text{PON} - 50/6.6\Sigma\text{Chl.}a$$

Standing stock of nutrients were also integrated from the surface to a depth of 20 m. We used standing stock of oysters in the inner part of the bay in October 2014 as that in October 2015. That in other month after October 2015 was defined as the difference between that in October 2015 and cumulative landing from October 2015 to one month before a current month. Because *C. gigas* growth in October is the maximum in a year, it is possible to determine standing stock of oysters in a month after October by subtracting cumulative landing oyster weights from October from standing stock of oysters in October without considering the growth rate. The units of standing stock and flux of nitrogen are $mgN m^{-2}$ and $mgN m^{-2} day^{-1}$ with a positive value as inflow from the open ocean to the inner part of the bay and a negative value as outflow from the inner part of the bay to the open sea, respectively.

As a result of this calculation, the nitrogen pool reached the maximum in January 2015, indicating that organisms do not actively uptake nutrients. Because the concentration of $NO_2 + NO_3$-N in the open ocean was high during this period, the flux from the open ocean was high and the primary production rate (per day) was approximately 2% of the total nitrogen standing stock of nutrient pool in the inner part of the bay. In April 2015, the standing stock of phytoplankton contributed to a nitrogen stock and to assimilation of nitrogen with a high speed. The primary production rate at this time was consuming nearly 20% of the net inflow of nitrogen from the open ocean. Also, because of low supply of nutrients from the rivers due to low precipitation, it seems that inflow from the open ocean was the main source of nitrogen in April. In July 2015, there were considerable detritus in the inner part of the bay, and the nutrient pool was nearly exhausted. In summer, water mass originated from the Kuroshio current consisting of oligotrophic water approaches the Sanriku Coast. The open ocean off Shizugawa Bay is predominantly occupied with this water. Therefore, it is natural that nutrients are depleted in Shizugawa Bay as a semi-enclosed bay greatly influenced by the open ocean during this period. Thus, in summer, the nutrient supply from the rivers to the bay has a relatively important contribution to the nutrient pool of the bay over that from the open ocean, and the primary production in Shizugawa Bay might be partly supported with the supply of nutrients from the rivers. The high nitrogen standing stock of the detritus estimated from the POC/Chl.a ratio and the C/N ratio is thought to be from the increase in bacteria decomposing suspended particles and growing at high water temperatures. In addition, the amount of sinking particles from the oyster farming rafts to the bottom was less than in autumn. This is because the particles were trapped in the surface layer above the thermocline layer due to stratification caused by heating from the sea surface. In October 2015, when the stratification was more relaxed than in July, the deposition from oyster farming rafts increased due to increase in excrements by the active feeding activity of oysters.

The characteristics of nitrogen circulation in Shizugawa Bay that is an open-type enclosed bay are summarized as follows: The main nitrogen source is the inflow from the open ocean; throughout the year, penetrations of water mass from the open ocean change greatly hydrographic and nutrient environments in the bay. Although the supply of nutrients from the forest to the bay through the rivers had a little influence during the summer rainfall, most of the nutrient influx from the rivers was used by phytoplankton near the estuaries. In other words, the primary production system within Shizugawa Bay depends on the open ocean in terms of nutrients. Thus, we conclude that "the forest is a close friend of the sea" and not "the forest is a lover of the sea" (Hatakeyama, 2000). As a nitrogen source, the total standing stock of massively farmed organisms such as oysters and sea ascidians is greater than that of the nutrient pool in summer and autumn. The results show that the aquaculture of filter-feeding organisms including bivalves play an important nutrient sink in the nutrient cycling in an open-type enclosed bay. Aquaculture of bivalves (oysters) provides substrates such as surfaces of ropes and shells for adherent diatoms, which grow and detach from the substrates. Detached diatoms are fed by oysters, which provide nutrients through excretion after digestion for adherent diatoms. The relation between oysters and diatoms is mutualistic symbiosis. This self-gardening effect by cultured oysters increases the speed of nitrogen circulation. In the future, a detailed evaluation about this self-gardening will be necessary.

2.4 IS THE FOREST A LOVER OF THE SEA? IRON

Manabu Fujii, Allam Ayman, Masafumi Natsuike, Hiroaki Ito and Chihiro Yoshimura

2.4.1 Introduction

Iron (Fe) is an essential micronutrient for the growth of marine and freshwater organisms. Iron is crucially important to phytoplankton for a range of metabolic activities, including photosynthetic and respiratory electron transport, as well as biosynthesis of chlorophyll. The bioavailability of iron limits the growth of phytoplankton in some area of the open ocean, namely high nutrients low chlorophyll [HNLC] region (Martin et al., 1994). In contrast, the estuaries and coastal regions, where freshwater and seawater are primarily mixed, accommodate high biological productivity and form rich aquatic ecosystems. Nutrients (i.e., essential trace metals such as iron and copper, major nutrients such as nitrogen and phosphorus) are supplied from river basins and play an important role in maintaining such high biological productivity in the coastal areas. A fisherman of the Kesennuma Bay introduced the famous phrase, "The forest is a lover of the sea," based on the hypothesis that the iron supply from river basins support the productivity of coastal area (Hatakeyama, 2000). Fishermen and residents are engaged in reforestation activities planting broadleaf trees, which they believe maintains rich coastal production.

In water, iron generally exists in the two oxidation states, Fe(II) and Fe(III). Dissolved iron (DFe) in natural water generally is composed of organic Fe(II) and Fe(III), soluble Fe(II) and Fe(III) complexed with inorganic ligands, iron hydroxide colloid (Fe(III)OOH), and others. Because Fe(II) is oxidized rapidly to Fe(III) by dissolved oxygen at neutral pH, Fe(III) is usually dominant in coastal and oceanic euphotic zones. In addition, because Fe(III) has high affinity to hydroxide ions and the solubility of Fe(III) at neutral pH is very low, 99% or more of the DFe is complexed with the organic ligand in the open ocean. Humic substances (HS), which are important organic ligands derived from terrestrial systems, contain acidic functional groups, such as carboxyl group, that are generated by oxidative decomposition processes. Consequently, such functional groups keep iron in dissolved form by complexing at neutral pH. Although a certain portion of the complexation of HS and iron is precipitated under a high-salinity condition by coagulation and flocculation, the remaining DFe in the water column and dissociated iron from sedimentary particulate iron might play an important role in maintaining high primary production in estuaries and coastal areas (Bundy et al., 2015).

Although several studies shows that HS and iron supplied from the land area plays an important role in the primary production in the coastal areas, few reports quantitatively investigate the transport of DFe from river basins and the bioavailability of iron in coastal areas (e.g., characteristics of organic ligands and reactivity of organic iron). This section will describe the dynamics of DFe derived from the river basins and its contribution to primary production in coastal zone, based on our research targeted in Shizugawa Bay and adjacent river basins in northeast Japan.

2.4.2 Dissolved Iron Transport From River Basins

Water quality in rivers is affected by land use in river basins. The close connection between river basins and coastal areas has been hypothesized (Matsunaga, 1993), and various forest conservation activities take place to maintain a rich coastal ecosystem (Hatakeyama, 2000). For example, the Amur-Okhotsk project investigated the importance of forests and wetlands in iron transport from the Amur River to the Sea of Okhotsk (Shiraiwa, 2011). The results generally showed that the forest in the middle stream and the wetland in the downstream are the major source of DFe. It also suggests that deforestation associated with urbanization in catchments might affect the coastal ecosystems by limiting the iron supply and the productivity in coastal area. In the other studies investigating DFe concentrations in stream water flowing through forests, the river water in broadleaf forests was found to contain a higher concentration of DFe compared to coniferous forests. Therefore, the vegetation type in the river basin also might influence the supply of DFe to coastal areas.

We also investigated the effect of land cover on the export rate of DFe from forested river catchments in the river basins in Kitakami Mountain, Iwate and Miyagi prefectures, northeast Japan (Endo et al., 2018). The land cover was classified into six types: broadleaf forests, coniferous forests, grasslands, paddy fields, cultivated areas, and urban areas. Water quality was monitored in winter and summer at >60 sampling points, and the relationship between the land cover and DFe load in the catchment area at each point was examined. A multiple regression analysis was applied to the area of each land cover as an explanatory variable to estimate the unit load of DFe ($g\,km^{-2}\,day^{-1}$) as a function of land cover. As a result, the unit loads of DFe were estimated to be 1.57, 139, and 439 $g\,km^{-2}\,day^{-1}$ in broadleaf forests, paddy fields and urban areas, respectively, in winter, and 61.9, 564, and 727 $g\,km^{-2}\,day^{-1}$ in grassland, paddy field and urban areas, respectively, in summer. Furthermore, the unit load was found to be 10 to 100 times higher in urban area and paddy fields than in areas of natural vegetation, indicating that human activities, including agriculture, increase the concentration of DFe in river water.

Paddy fields, which comprise a major part of the region's agriculture, possibly contain a lot of organic matter such as plant/algae residue and compost. Surface water in paddy fields is aerobic, and thus organic reactions undergo oxidative decomposition by microbial and light-mediated reactions, which produce acidic functional groups (bonding sites with iron). In anaerobic layers, such as paddy soil, iron hydroxide is reduced to form Fe(II). In the paddy, therefore, Fe(II) eluted from the soil can form a complex with the dissolved organic matter in the overlaying water, and so the concentration of DFe is likely to increase. In addition, domestic wastewater generated from urban areas generally contains dissolved organic matter and DFe at a higher concentration than river water. Those estimates of the DFe unit load implied that the anthropogenic transformation of land cover from natural forests increase the DFe load to coastal areas, although we also need to understand the reactivity and ecological role of the iron complexation in blackish zone and coastal area.

2.4.3 Relationship Between Dissolved Organic Matter and Dissolved Iron in River

In general, by separating organic matter in river water with a filter having a pore size of 0.2 to 1 μm, the substance trapped on the filter is classified as particulate organic matter (POM); organic matter passing through the filter is defined as dissolved organic matter (DOM) (Asmala et al., 2013). In aquatic systems, DOM is derived from aquatic organisms within the aqueous system (i.e., autochthonous); terrestrial ecosystems also provide DOM (i.e., allochthonous). Allochthonous organic matter generally accounts for a large portion of dissolved organic carbon (DOC) and its chemical property (e.g., bioavailability) can be influenced by vegetation and water use in river basins (Asmala et al., 2013; He et al., 2016). For example, DOM derived from forests and peatlands has a high C/N ratio and contains aromatics and humus components at higher proportion.

Organic matter that is complexed with iron in water environment is called ligand, and it includes natural organic matters such as humic acid (HA) and fulvic acid (FA), siderophore, which is produced by microorganisms such as bacteria and fungi and has extremely high affinity with Fe(III), and extracellular high polymer substances (exopolymeric substances: EPS) such as polysaccharides produced by bacteria and algae. Among these types of DOM, humic substances are dominant in various environments such as forests, rivers, and coastal zones and can be important for iron transport from a river basin to coastal area (Matsunaga, 1993). Functional groups such as carboxyl groups in humic substances form complexes with various metals including iron, thereby this DOM component greatly contribute to maintaining the insoluble metal in a dissolved state in the water environment. Therefore, humic-like DOM is a major factor determining the concentration of DFe in river water.

The authors investigated the influence of land cover and seasonality on DOM chemical composition in Kitakami Mountain watersheds, northeast Japan, to coastal areas of Sanriku Coast. Excitation-emission-matrix fluorescence combined with parallel factor analysis (EEM-PARAFAC) classified fluorescent DOM components (FDOM) into four main components (C1 to C4). This classification was consistent with the other relevant investigations about FDOM in aquatic environment. The component C1 was humic-like DOM, C2 and C3 are fulvic acid-like DOM, and C4 corresponds to protein-like DOM. The fluorescence intensities of all these components in the downstream sections were found to be about twice that in the upstream sections, while those intensities in the summer were roughly double of those in winter. In addition, PARAFAC showed significant positive correlation of DFe concentration to the peak intensity for each of the humic substances (C1 to C3), except for the upstream sections in the summer. This result implies that humic substances derived from forested basins play an important role in the DFe transport in rivers.

In addition, linear regression was applied to analyze the relation of DFe concentration to each of relevant water quality parameters. DOC concentration showed a significant positive correlation with DFe concentration in both the upstream and downstream sections, and the light absorbance at wavelength of 254 nm (A254, an indicator of humic substances) showed a significant positive correlation with DFe, except for the upstream sections in the summer. Furthermore, the regression slopes between DFe and DOC concentrations (and A254 as well) showed a significant difference between upstream and downstream sections (ANCOVA, $p < 5\%$), showing an increasing trend in the slope toward the downstream. This result implies that DOC-bound DFe concentration increased with increasing DOC concentration. Overall, a relatively large amount of DOM and humic substances is supplied from farmland (largely paddy fields) and urban areas in the downstream sections in Kitakami Mountain, which probably results in higher DFe concentration (Endo et al., 2018).

2.4.4 Flocculation and Precipitation of Dissolved Iron in Brackish Water

A large portion of DFe supplied from river water precipitate and settle in the estuary/coastal zone because of the increase in salinity, which occurs because the negatively charged DOM is neutralized by the cations (for example, Mg^{2+}, Ca^{2+}, Na^+ and K^+) present at a high concentration in the seawater. Because of such precipitation, typically, 40% to 99% of DFe in river water is removed in the estuary/coastal zone. DFe precipitation has been observed in estuarine mixing zone in the world, even in the low salinity range from 0 to 5 psu. Among major DOM components, aromatic molecules are apt to coagulate and precipitate with iron, and thus the chemical composition of DOM is also an important factor in determining DFe transport to coastal area.

The authors experimentally tested the coagulation-sedimentation process by mixing the riverine water from Kitakami Mountain region with artificial seawater (the final salinity 30). We then quantified how much DFe is transported to coastal waters as transport ratio that is defined as a ratio of DFe concentration in saline water to that in the river water. As a result, the arrival rate ranged from 1.0% to 83% in the laboratory-based experiment. Interestingly, DFe sampled in the upstream sections in the summer showed a significantly higher transport ratio than those in the downstream sections and in both the upstream and downstream sections in winter.

We also applied multiple regression analysis to investigate the relationship between the major fluorescent components of DOM and DFe transport ratio. The transport ratio is reasonably explained by the relative peak intensity of C1 to C3 ($R^2 = 0.72$). These results suggest that land cover information in a river basin allows us to estimate DFe load (refer to Section 2.4.2) as well as its transport ratio where salinity shifts, implying the importance of chemical structure of organic ligands in DFe transport.

2.4.5 Redox Reaction of Iron and Iron Intake by Algae

The bioavailability of DFe in natural waters depends largely on Fe redox transformation kinetics and resulting chemical speciation, which are affected by physicochemical factors such as light, pH, reactive oxygen species (ROS), dissolved oxygen (O_2), and organic and inorganic ligands (Rose and Waite, 2005). Therefore, it is important to investigate the redox reaction of iron to understand the role of DFe in biogeochemical cycle and primary production in coastal ecosystem. The oxidation kinetics of Fe(II) has been investigated by numerous researchers using natural water and controlled solutions. Consequently, the reaction mechanism, rate constants for individual chemical species, and detailed kinetic models have been well developed (Shaked, 2008). For example, it is well known that Fe(II) is oxidized rapidly to thermodynamically stable ferric iron (Fe(III)) by oxidants such as O_2 and hydrogen peroxide (H_2O_2) under circumneutral pH and air-saturated conditions (e.g., euphotic zones of oceans and lakes) (Lee et al., 2016, 2017).

In oxygenated surface water, thermodynamically unstable inorganic Fe(II) is oxidized to Fe(III) with a half-life within a few minutes in seawater (i.e., pH ~8.0 and temperature 25°C) (Santana-Casiano et al., 2005, 2006). Fe(II) is oxidized instantaneously under aerobic conditions, while unstable Fe(II) can be produced temporarily by various reduction actions. These reactions include light-induced ligand-to-metal charge transfer (LMCT) reactions and superoxide- and microbially mediated processes, and the thermal (dark) and photochemical reduction of organically complexed Fe(III) via reactions with redox-active moieties (e.g., quinone-hydroquinone) of HS.

Our research group investigated the ratio of steady state Fe(II) to DFe in the oxidation-reduction reaction in the presence of humic substances. The oxidation-reduction rate for iron was found to depend on the origin and nature of humic substances. Fe(II) oxidation rate showed a significant positive correlation with the aliphatic content of humic substances, suggesting that Fe(II) complexation by aliphatic components accelerates Fe(II) oxidation. In addition, the reduction rate and the steady state Fe(II) fractions in the presence of sunlight showed relatively strong positive correlations with the free radical content in humic substances. This means that radicals semiquinone in humic substances are related to the reduction properties of Fe(III). These results indicate that the chemical properties of humic substances affect the redox reaction of iron and the resulting Fe(II) formation.

Furthermore, the authors investigated the influence of DOM characteristics on Fe(II) oxidation in river basins flowing into Shizugawa bay and its coastal waters (Lee et al., 2017). The oxidation rate of Fe(II) in coastal seawater was found to be lower than the oxidation rate in freshwater (i.e., rivers). The oxidation rate constant of Fe(II) substantially decreased after the removal of humic-type DOM, suggesting that the hydrophobic DOM is a key factor that accelerates the Fe(II) oxidation in the freshwater samples. In addition, oxidation rate decreased when the hydrophobic organic substances produced outside the seawater environment (allochthonous origin such as humic substances) were removed by the nonionic exchange resin.

The oxidation rate of Fe(II) in seawater of Shizugawa Bay, however, was slower (Fig. 2.10). Therefore, the relatively slow oxidation rate that was observed in coastal seawater might be related to the autochthonous organic matter produced in the seawater environment (e.g., derived from microorganisms). The observed lower oxidation rate for coastal seawater, compared with freshwater and organic ligand-free seawater, likely was associated with microbially derived autochthonous DOM, and the variation of Fe(II) oxidation at a fixed pH was best described by fluorescence index that represents the proportion of autochthonous and allochthonous DOM in natural waters.

To clarify the relationship between DFe and the growth of coastal algae (i.e., iron bioavailability), we also investigated the iron uptake kinetics by coastal microalgae and macroalgae using river and coastal water sampled in northeast part of Japan (refer the Section 2.4.2 for site details). Microalgae *Skeletonema marinoi-dornii* and macroalgae *Eisenia bicyclis* were collected in the Shizugawa Bay and their iron intake processes was quantitatively investigated using iron radioisotope. The river water samples, which had DFe concentration ranging in 6.0–2.1 μM, was diluted with Aquil* seawater medium to decrease DFe concentration to 1.0–120 nM, which is comparable to DFe in seawater samples (i.e., 1.1–41 nM).

As a result, we observed the trend that Fe uptake rate by both of *S. marinoi-dohrnii* (Fig. 2.11) and *E. bicyclis* increase with DFe concentration. Fe uptake rates for both algae were significantly greater in the medium derived from coastal water than those from river water. Considering that the bioavailable form of Fe is normally limited to inorganic soluble

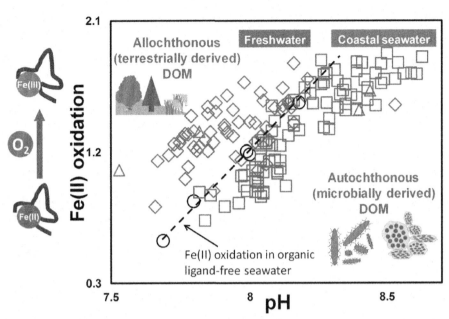

FIG. 2.10 Relationship between pH and Fe(II) oxidation rate (partial modification of Lee et al., 2017). Although dissolved organic matter (DOM) in coastal seawaters (squares and circles) is characterized mainly by microbial origin, river water (diamond and triangles) contains terrestrial DOM at higher proportion. This difference of DOM origin and characteristics can affect the rate of Fe(II) oxidation at fixed pH (e.g., terrestrial DOM can accelerate Fe(II) oxidation). The dashed line is a statistical distribution boundary between river and coastal waters.

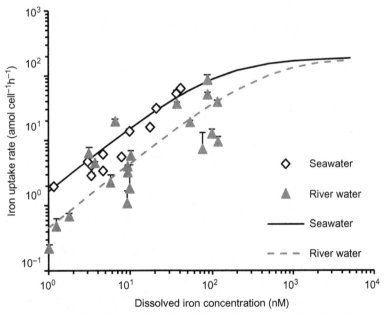

FIG. 2.11 Iron uptake rate (mean ± standard deviation) as a function of dissolved iron concentration in microalgae (*S. marinoi-dohrnii*) cultured in river water (triangles) and seawater with salinity of 30 (diamonds). Solid and dashed lines are curves representing the Michaelis-Menten equation fitted to the river water and seawater systems, respectively.

species, the difference in Fe uptake efficiency from DFe pool between river and coastal water likely is caused by the different complexation characteristics of DOM.

DOM in coastal waters, which are a mixture of oceanic, terrestrial, and indigenous organics, might have a potential to serve higher concentrations of inorganic soluble iron than that in river water (i.e., largely terrestrial origin) at a same concentration of DFe. In fact, three-dimension fluorescence PARAFAC analysis showed that the concentrations of protein components in the coastal water were higher than in river water, while the concentration of humic substances was lower in the coastal water. Considering the relationship between DOM and DFe oxidation rate, the differences in the DOM composition likely affect the oxidation-reduction rate of Fe and its bioavailability for algae (Fig. 2.10).

2.4.6 Distribution of DFe and DOM in Shizugawa Bay

We understand that high productivity in coastal waters is maintained by major nutrients supplied from river basins and open ocean, but much fewer studies have been undertaken for micronutrients, in particular DFe supply and its spatiotemporal distribution in coastal waters. We investigated the dynamics of DOM and DFe in Shizugawa Bay over the 3 years from 2014 to 2017. The results revealed that DFe concentration in the bay was approximately two orders of magnitudes lower than that for the river water, while the DOC concentration in the bay tended to be slightly higher than that of the river.

Fluorescence index (FI) was 1.6 on average in the upstream river sections, possibly because of the presence of organic matters derived from land at higher proportion, whereas FI in the bay was 1.8 on average, suggesting dominant organic matter derived from aquatic organisms. In addition, FI slightly increased to 1.7 on average as water flowed down to the river months. We also found that FI of wastewater effluent was 2.1 or higher, showing that the wastewater input to the rivers is also a factor that can cause the increase in FI in the downstream sections. DOM derived from river basins with high aromaticity flows into the bay at a high proportion especially in the spring (March to April) as suggested by the variation in DOC, fluorescence index, and aromatic index SUVA254 (A254/DOC).

The concentration of DFe in riverwater ranged from the below the detection limit to several hundred nanomoles (nM) and was higher in the downstream sections than that for the upstream sections. DFe concentration in the rivers showed the similar seasonal patterns in the upstream and downstream sections, particularly in spring. DFe concentration in the bay was 4.8 nM in the surface layer and 9.4 nM at the depth of 10 m on average. DFe concentration in the surface of the river mouth seasonally fluctuated, and it tended to be high (average 9.3 nM) in summer and winter, and decreased in spring and autumn (average 2.2 nM). DFe transported by rivers is precipitated and diluted to some extent in the brackish water of estuaries, as described in the Section 2.5.4. The high DFe concentration in the surface layer in the bay could be attributed, at least partially, to the DFe supplied from the rivers to the bay, although DFe supplied seasonally from Pacific Ocean and the bottom layer in the bay also might influence DFe distribution in the bay.

2.4.7 Contribution of DFe Transport From Rivers to Primary Production in the Bay

Based on the previously mentioned survey and experimental results, we estimated the balance of DFe in the Shizugawa Bay. Although it might not be possible to reveal all the inflow and outflow fluxes of DFe, we quantified several major processes to estimate its flux and budget in the bay for each of the four seasons, including inflow from river, coagulation-sedimentation in brackish water of estuaries, seawater exchange with Pacific Ocean, uptake by microalgae (phytoplankton), and uptake by macroalgae. In this manner, we attempted to address the contribution of DFe derived from river basins in the DFe budget and primary production in the bay.

DFe influx from rivers was calculated by multiplying the monitored time-series data of DFe concentration and the river discharge estimated from rainfall amounts and evapotranspiration in each basin. The exchanges with the ocean were estimated based on the monitored concentration and the seawater exchange volume in the upper and lower layers in the way described in the Section 2.3.6. We also applied the results of coagulation-sedimentation experiments to estimate flocculation-sedimentation flux in the brackish zone (refer the Section 2.4.4). In addition, the dominant microalgae in Shizugawa Bay were isolated and cultured in the laboratory, and culture experiments were carried out with varying concentrations of DFe to delineate the relationship between DFe concentration and phytoplankton growth rate (see Section 2.5.5). This relationship then was used to estimate Fe intake flux by coastal phytoplankton.

The result from estimating the DFe budget in each season shows that, DFe flowing from the land to the estuary was 0.22–8.5 kg day^{-1}, and the input through brackish zone to the bay was in the range of 0.02–0.71 kg day^{-1} (Fig. 2.12). Regarding the exchange with the open ocean, it was showed that the flow into the bay ranged from 0.67–14.28 kg day^{-1} and the flow out of the bay ranged from 0.01–2.54 kg day^{-1}. This result suggests that the amount of DFe supplied from the river to the bay is smaller than the amount transported via seawater exchanges in Shizugawa Bay, meaning the exchange with seawater has a great influence on the amount of DFe in the bay.

The amount of DFe supplied from the river (0.02–0.71 kg day^{-1}) corresponded to 2.3% to 51% of the iron intake rate of the microalgae in the bay, which implies that the influx from the river provides a certain contribution to the primary production. In addition, the content of iron and the intake rate of DFe in the microalgal community in the bay were

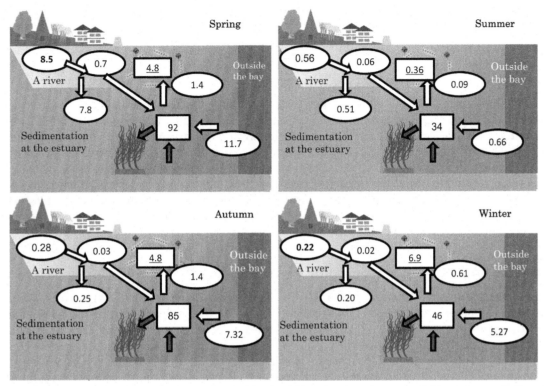

FIG. 2.12 Budget of dissolved iron (DFe) in Shizugawa Bay in each season. The values in each arrow and ellipse represent the flux of DFe per day ($kg\,day^{-1}$), the values within rectangles represent the existing quantity (kg) of DFe in the bay seawater, and the underlined numbers in the rectangle are the existing quantity (kg) of DFe as phytoplankton in the bay. In addition, uptake of DFe by macroalgae and elution from the sea floor are not estimated *(gray arrow)*.

estimated to be 0.36–6.9 kg and 0.09–1.4 kg day^{-1}, respectively. Therefore, the iron uptake rate of microalgae was about 0.1% to 1.5% of the amount of DFe (34–92 kg) in the bay, implying the microalgal uptake does not significantly affect the DFe storage in the bay (Fig. 2.12).

The above estimates are based on all the available, albeit limited, data sets in the research project, and it is important to improve the estimation accuracy of each target process by conducting continuous monitoring. In addition, some missing pathways, such as the exchange with groundwater and sediment and uptake by macroalgae, should be warranted in further investigation.

2.4.8 Conclusion

In this section, we outlined the contribution of DFe from the river basin to primary production in the coastal zone, including our research outcomes, and provide quantitative assessment on every major process on iron dynamics in the integrated system of land, river, and coastal waters. The land cover in the river basin was found to be a major factor determining DFe load to coastal area and the chemical property of organic ligand, which serves as an iron carrier. Its chemical property is closely related to flocculation and precipitation of DFe, as well as redox reaction and bioavailability of Fe in the estuary.

In addition to the recent findings in this section, it is also essential to link the topographical and hydrological characteristics to the biogeochemical processes of trace metals, including iron. For example, we assume the relative importance of riverine input is also determined by the size of river basins and topography of coastal area or bay, as they primarily determine the mixing process of river and sea water. Such outcomes will help us develop practical methods and policy to implement the environmental management integrating river basin and coastal waters. Essential trace elements, such as iron, are ubiquitous in the environment, and therefore they tend to be out of our scope of general environmental management or standards. As summarized in this section, however, essential trace elements directly influence the primary production, adding to the influence of major nutrients such as nitrogen and phosphorus. Therefore, understanding and management of coastal ecosystems can be improved if trace elements are monitored and controlled in such ecosystems.

2.5　IS THE FOREST A LOVER OF THE SEA? PARTICULATE ORGANIC MATTER?

Takashi Sakamaki and Osamu Nishimura

2.5.1　Forests, Rivers, and Oceans Linked Through Particulate Organic Matter

Particulate organic matter (POM) of various types, including those generated from forests, primary production in rivers, and human activity, is supplied to coastal waters via rivers. Both the quality and quantity of POM are influenced strongly by environmental factors, such as the terrain, vegetation, and land use in a given catchment. The response of a given ecosystem of coastal zone to materials supplied from river basins depends on its organisms, which vary in terms of nutritional demands. Consequently, the role of river-supplied POM in coastal waters is not universal, as demonstrated by previous researches about the impact of river-supplied POM on secondary production in coastal areas, which have yielded diverse results and interpretations. In this section, in addition to reviewing literature about the impacts of river-supplied POM on consumers in coastal areas via the food chain, we also present the results of our research conducted in Shizugawa Bay, Minamisanriku Town.

2.5.2　Particulate Organic Matter Supplied by Rivers to Coastal Waters

Numerous studies have used stable isotopes (δ^{13}C) or C:N ratios of organic matter (OM) to investigate the relative contributions of river- and marine-derived POM to coastal waters. The δ^{13}C value of land-derived OM generally is less than −25‰, indicating a lower ^{13}C content than that of marine-derived phytoplankton (typically −22‰ or greater) or benthic microalgae (typically −20‰). In many instances, δ^{13}C source information has provided evidence for the decreasing contribution of river-derived OM and increasing contribution of marine-derived OM in estuaries, when moving downstream. Comparative analyses of estuaries have revealed that the distribution and contribution of river-derived OM vary widely from site to site. In some cases, river-derived OM remains the dominant source over an entire estuary; in other cases, marine-derived OM increases dramatically near the river mouth because of the intrusion and mixing of seawater.

Using δ^{13}C and C:N ratios, Sakamaki et al. (2010) investigated the contribution of river-derived OM to the sediment of 20 estuarine tidal flats along the northwest coast of North America with catchments ranging in area from 7 km^2 to 8000 km^2. In estuarine tidal flats containing sediment OM with relatively low C:N ratios (11 or lower), the contribution of river-derived OM (the share of river-derived OM in sediment OM) was positively associated with catchment area of feeder rivers. In estuarine tidal flats with higher C:N ratios, however, the contribution of OM believed to have been transported by rivers was substantial even in estuaries with small catchment areas. Such estuarine tidal flats containing sediment OM with high C:N ratios are all connected to mountain streams and might have accumulated forest-derived non-labile OM brought by rivers during flooding.

The C:N ratio of marine-derived suspended OM, which consists primarily of phytoplankton, is generally low. In contrast, the C:N ratio of river-supplied OM, which is derived from terrestrial plants, microalgae in rivers, wastewater from human activities, and other sources, varies widely. The C:N ratio of OM can affect its degradability and availability to organisms. For this reason, differences in the composition of river-supplied OM might be related to differences in the contribution of river-derived OM to the estuary biomass and food web. In a study of the relationship between land use and estuarine sediment based on a comparison of 15 river basins on Okinawa Island, Morita et al. (2017) observed a significant relationship between the C:N ratio or δ^{13}C and the proportion of the catchment area covered by forests. These findings indicate that the composition of OM supplied by rivers varies over space and time depending on the vegetation, land use, and inflow of anthropogenic waste into the catchment.

2.5.3　Contribution of Catchment-Derived Particulate OM to Secondary Production in Coastal Waters

Numerous studies, many of which have employed stable isotopes of carbon, have investigated the contribution of catchment-derived OM supplied by rivers to secondary production in coastal waters. Some of these studies have shown that catchment-derived OM is assimilated by secondary producers in estuaries and coastal waters, suggesting that river-supplied OM is an important contributor to estuarine and coastal food webs. A large number of studies, however, have reported that benthic microalgae and marine phytoplankton produced in estuarine and coastal waters are the dominant contributors to secondary production in such environments and that the contribution of catchment-derived OM is small. As is evident from these disparate results, the contribution of catchment-derived OM to estuarine and coastal waters depends on the specific estuary and coastal zone (Sakamaki, 2011).

The differences among studies about the importance of river-supplied OM to estuarine and coastal waters can be explained, in part, by differences in the composition of secondary producers or feeding habits of specific secondary producers among estuaries and coastal zones. Sakamaki et al. (2010) analyzed relationships the $\delta^{13}C$ of sediment OM and the $\delta^{13}C$ values of each of six benthic species (one amphipod, four bivalves, and one bristle worm species) that are distributed widely across the 20 tidal flats included in the study. The $\delta^{13}C$ values of three bivalves (*Venerupis philippinarum, Mya arenaria,* and *Nuttallia obscurata*) were unrelated to the $\delta^{13}C$ of sediment OM. In all tidal flats, the $\delta^{13}C$ values of these organisms were similar to those of marine-derived OM and benthic microalgae, and little variance was observed among flats. These three species primarily feed on suspended OM. Because they selectively consume and assimilate this OM, they rarely use sediment OM, even in areas of substantial accumulation of catchment-derived OM. The $\delta^{13}C$ values of three other species (the bivalve *Nereis limnicola,* amphipod *Macoma balthica,* and bristle worm *Corophium spinicorne*) positively correlated with the $\delta^{13}C$ of sediment OM. This is likely because the composition of OM assimilated by these organisms changes with the OM composition of the sediment. The three species are believed to feed primarily on deposited materials. Because they consume the OM in sediment nonselectively, they will feed on and assimilate river-supplied OM when such materials exist in their habitat, resulting in a decline in $\delta^{13}C$. As illustrated by previous example, differences in the feeding habits of benthic organisms can have strong impacts on the use of river-derived OM.

As explained in Section 2.5.2, both the quality and quantity of OM supplied by rivers vary among estuaries and coastal zones. These factors can have substantial impacts on the use of river-derived OM by secondary producers in such waters. The larger the catchment and discharge, the greater the amount of catchment-derived OM supplied to an estuary or coastal zone and the greater the potential for consumption by organisms living in these waters. The quality of OM accumulated in estuarine and coastal waters, however, can vary substantially depending on the share of recalcitrant OM derived from terrestrial vegetation. Therefore, the contribution of river-supplied OM to secondary production in estuarine and coastal waters depends on the quantity and quality of the OM, as well as ecological characteristics of resident secondary producers.

Of the POM carried by rivers, the share of coarse-grained particles, which generally are defined as particles with a diameter of 1 mm or greater, is typically small. Fallen leaves, which are the main component of coarse-grained POM, however, differ substantially from fine-grained POM in both size and shape. In the fall, leaf litter accumulates in estuaries of rivers whose catchments include deciduous forests. In addition to being food for secondary producers in estuaries, this leaf litter serves a special ecological function, providing hiding places and habitats for macroorganisms. In an experiment where mesh bags containing either leaf litter from deciduous trees or pieces of nylon of roughly the same size and shape as leaf litter were placed in estuarine tidal flats, no significant difference was observed in the number of attached organisms, suggesting that leaf litter serves primarily as a habitat rather than as a food source for secondary consumers (Sakamaki and Richardson, 2008). In an experiment conducted in Hokkaido, however, leaf litter patches that accumulated in estuaries were infested with swarms of amphipods, which serve as an important food source for the righteye flounder (Sakurai and Yanai, 2006). This example illustrates that catchment-supplied POM, which varies in both size and quality, not only is consumed directly as food but also can serve diverse ecological functions in estuarine and coastal waters.

2.5.4 Role of Catchment-Derived POM in Shizugawa Bay

2.5.4.1 Concentration of POM in Feeder Rivers

In our June 2014–June 2015 investigation of three main rivers (Hachiman, Mizushiri, and Oritate rivers) that feed into Shizugawa Bay, the concentration of particulate organic carbon (POC)—defined as particles with diameters ranging from 0.7 to 250 µm—did not exhibit any seasonality and did not differ substantially between rivers (Fig. 2.13). In contrast, a positive correlation was observed between POC and the stable isotope ratio ($\delta^{13}C$), suggesting that POC increases as OM derived from aquatic algae and other primary producers increases relative to higher terrestrial plant-derived OM in rivers.

In our investigation of water samples collected over the entire Sanriku coastal zone, including Minamisanriku Town, the POC of rivers varied particularly widely depending on the degree of forest cover and human land use in the catchment. POC was negatively associated with the percentage of forest cover and positively associated with the percentage of farmland or urban area (Fig. 2.14). Generally, although leaf litter and other forest vegetation-derived POM is supplied to small streams in the upper reaches of rivers, primary production in the streams themselves is low because of shading by the tree canopy. Primary production increases in the intermediate and lower reaches of rivers as the shading effect of the tree canopy decreases because of river widening, and the influx of nutrients from surrounding farmland and urban areas increases. These factors are believed to explain the observed variation in POC associated

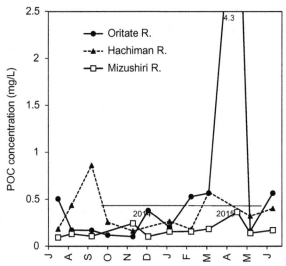

FIG. 2.13 Particulate organic carbon (POC) concentrations of stream waters collected from downstream stations of three major streams entering Shizugawa Bay.

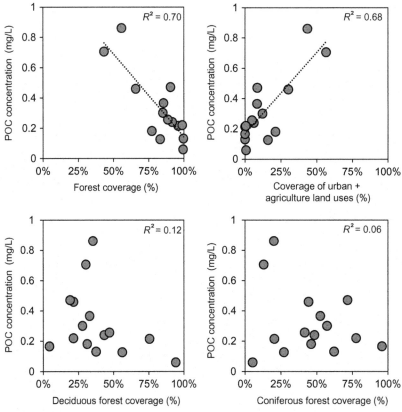

FIG. 2.14 The relationships of POC concentration in stream water with the coverages of forests and urbanized land use in watersheds (based on a data set obtained from 15 stations of Sanriku Region in summer 2017).

with landcover. No difference was observed in the relationship between POC in rivers and percentage of deciduous forest and the relationship between POC in rivers and percentage of coniferous forest.

2.5.4.2 Concentration and Origin of Particulate Organic Matter in Shizugawa Bay

We sampled the POC of surface water (depth, 0 to 10 m) on a monthly basis near the head of Shizugawa Bay, near the middle of the bay, and at the bay mouth. We detected greater POC differences between months than between

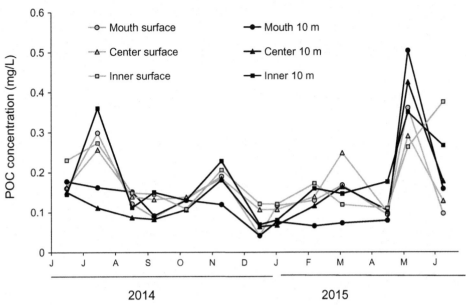

FIG. 2.15 POC concentrations in the seawaters at three stations in Shizugawa Bay. The waters were collected at surface and 10-m depth layers at each station.

sampling locations; POC concentration at different sampling locations, for the most part, moved in tandem (Fig. 2.15). The levels of variation in marine and river POC were of similar orders of magnitude. Furthermore, marine POC and δ^{13}C were positively correlated and exhibited similar trends over time. Specifically, POC concentration and δ^{13}C both decreased in November and December. We speculate that this was caused by a decline in primary production because of the decreasing water temperatures, downward transport of surface phytoplankton caused by the formation of a mixing layer, and unfavorable light conditions in the winter. In contrast, bay water POC increased dramatically in warmer months from March to October, as did δ^{13}C. We detected higher levels of algal-derived fatty acids in the warmer months than the winter months, suggesting that the increase in POC was caused by increased phytoplankton production. Given that annual phytoplankton blooms occur in other bays along the Sanriku Coast (Ofunato Bay, Ogatsu Bay, etc.), the dramatic increase in POC in March 2015 observed in this investigation is believed to be the result of an annual bloom. The steep increases in POC observed in July and November could be explained by a fall phytoplankton bloom resulting from discharges of nutrient-rich river water during the approach of Typhoon No. 8 in late July and subsequent favorable weather.

Based on the results of our investigation of Shizugawa Bay, we evaluated the relationships between the daily flux in each season of various processes (primary production and respiration, influx from rivers, sedimentation, feeding by aquaculture oysters, and oyster harvest) and the fate of POM in the inner bay, covering an area of about 14 km². The analysis indicated that primary production is the dominant source of POM in most seasons (Fig. 2.16). Furthermore, the flux of primary production was about two orders of magnitude greater than the food required by aquaculture oysters cultivated in the bay. The quantity of POM supplied by rivers was about two orders of magnitude less than that generated by primary production; thus, river-supplied POM is believed to have a relatively small impact on POM in bay waters. That said, these fluxes were estimated based on short-term measurements of about 1 day and did not account for river discharge events nor do they represent a mass balance.

2.5.4.3 Food Sources of Consumers in the Bay

In September 2015, we collected macrobenthos, benthos, and sediment samples from tidal flats and shallow waters near the head of the bay and performed carbon and nitrogen stable isotope analyses (Fig. 2.17). For suspended POM in rivers and seawater, we evaluated monthly measurements obtained over a year starting in June 2014. The carbon and nitrogen stable isotope ratios for all taxa investigated were close to those of benthic microalgae, suggesting that microalgae were assimilated as the primary food source. The carbon and nitrogen stable isotope ratios of the sediment OM of tidal flats and shallow waters near the head of the bay were similar to those of the river-derived POM. Based on these results, we believe that although the river-supplied OM accumulates in sediment, its assimilation by benthic organisms is limited and many benthic organisms selectively feed on microalgae in the sediment. Carbon and nitrogen stable isotope ratios in the soft tissues of oysters cultured by the suspension method in the bay clearly differ from those

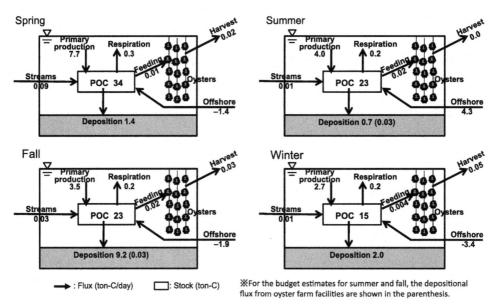

FIG. 2.16   Estimated fluxes of particulate organic carbon (POC) in the 14-km² area of the inner part of Shizugawa Bay under an assumption of calm weather condition.

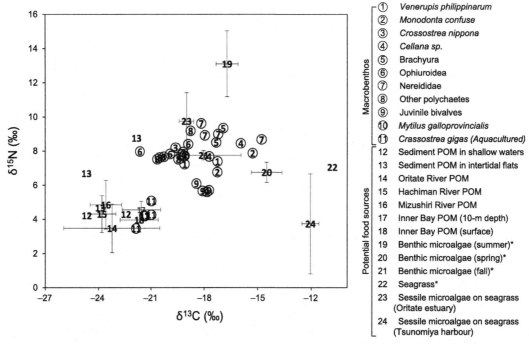

FIG. 2.17   Carbon and nitrogen stable isotope ratios of macrobenthos and their potential food sources in the inner part of Shizugawa Bay (sampled in September 2015). Taking account of stable isotopic fractionations through food chains, the stable isotopic ratios for macrobethos were plotted after subtracting 1% and 3% for δ^{13}C and δ^{15}N, respectively. For POM of seawater and stream waters, their stable isotopic ratios were determined in monthly samplings of 2014 and 2015. For the data with *, the values were determined in other sites of a previous study (Sakamaki and Richardson, 2008).

of benthic organisms, and the range of isotopic ratios overlaps that of suspended POM. Based on these two observations, we believe that suspended POM is the primary food source for the cultured oysters. Comparing their stable isotope ratios, it is apparent that the contribution of the river-supplied POM as a food source for the cultured oysters is negligible, if not zero.

We also performed a fatty acid analysis of cultured oysters and suspended POM, which we assumed to be the primary food source of the oysters. Among the essential fatty acids that cannot be produced or produced in sufficient

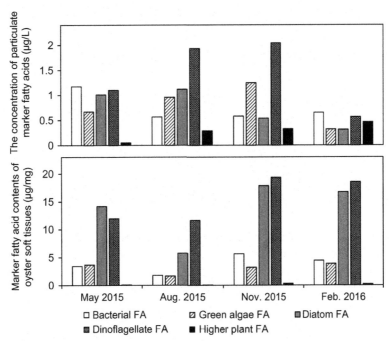

FIG. 2.18 Marker fatty acid concentrations in seawater and oyster soft tissue.

quantities by consumers, some can be generated only by certain producers. These fatty acids serve as useful markers for food web analyses. The fatty acid composition of suspended POM varied substantially among seasons (Fig. 2.18). In all seasons, suspended POM contained long-chain fatty acid markers believed to be derived from higher plants mixed with OM derived from a wide range of organisms. The fatty acid composition of the cultured oysters included relatively high quantities of dinoflagellate- and diatom-derived fatty acid markers, irrespective of sampling time or location. The fatty acid composition of POM included fatty acids derived from a wide range of organisms and changed with the season; it is likely that the cultured oysters selectively consume and assimilate fatty acids from dinoflagellates and diatoms. These results for the stable isotope ratio and fatty acid composition analysis indicate that in Shizugawa Bay, higher plant-derived OM supplied by rivers is not assimilated directly by oysters.

2.5.5 Growth and Environmental Impact of Cultured Oysters in Shizugawa Bay

2.5.5.1 Locational Variation in Oyster Growth

Starting in October 2014, we compared the growth rate of seed oysters that were collected in the summer of 2010 and suspended in culture nets in four different locations of Shizugawa Bay. To eliminate the impact of attached organisms, the oysters and culture net cages were cleaned on a monthly or bimonthly basis. Continuous monitoring of oyster growth in the four locations revealed that growth is faster in waters off the coast of the Togura District on the south side of the bay than at other locations (Fig. 2.19), which might be explained by the lower aquaculture density in the Togura zone. In the Togura zone, the horizontal distance between longlines for suspending oysters ranges from approximately 30 m to nearly 50 m. This interline spacing results in substantially sparser culture conditions than the 20- to 30-m spacing used in the Shizugawa zone near the head of the bay (Fig. 2.22).

Prior to the earthquake in 2011, the density of oyster aquaculture operations in the Togura zone was much higher than it is today. According to local oyster farmers, the growth rate of oysters was much lower then. In response to concerns that were raised during the restoration of aquaculture facilities after the earthquake, a decision was made to substantially reduce the aquaculture density to approximately one-third of that before the earthquake (see Section 2.6.1). According to interviews with fishermen involved in oyster aquaculture, the time required from seeding to shipping in the Togura zone was 18 months or more before the earthquake; it was reduced to about 10 months after the earthquake as a consequence of the reduced density.

Based on a comparison of oyster growth rates among locations and interviews with fishermen, we conclude that the oyster growth rate in Shizugawa Bay depends on the aquaculture density. As indicated in Section 2.5.4, primary production over the entire area near the head of Shizugawa Bay is about two orders of magnitude greater than the food

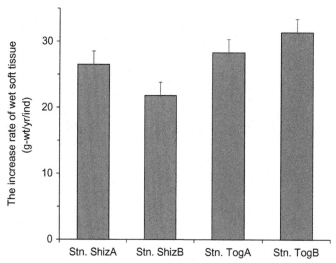

FIG. 2.19 Oyster growth rates at four stations of Shizugawa Bay in a 1-year growth experiment. ShizA and ShizB indicate two stations in Shizugawa zone on the north side of the bay, and TogA and TogB indicate those in Togura zone on the south side of the bay.

requirement of the oysters, and therefore supplies sufficient food. It was demonstrated that oyster growth depends on and is limited by the local aquaculture density. Accordingly, the food environment at spatial scales ranging from the entire (inner) bay to the spacing of individual culture rafts are important determinants of the growth of cultured oysters.

2.5.5.2 Sedimentation and Oxygen Consumption of POM From Oyster Farms

Seasonal investigations of sediment traps in Shizugawa Bay revealed that the deposition of POM at a depth of 10 m is 7 to 17 times greater directly below oyster aquaculture fields than elsewhere (Fig. 2.20; Kawahata et al., 2018). In addition to oyster feces, the additional POM deposited in oyster fields is thought to comprise OM of various origins, including materials previously attached to aquaculture equipment.

Accordingly, we conducted a mesocosm experiment using oysters of three ages (4, 16, and 28 months) from which we collected and analyzed the chemical composition and biological oxygen demand of oyster biodeposits. The organic carbon content of feces ranged from 0.05 to 0.07 gC gDW-feces^{-1} and was higher in older oysters than in younger oysters (Fig. 2.21), suggesting that assimilation rates decline with oyster age. The oxygen consumption rate for oyster feces ranged from 0.09 to 0.55 mgO$_2$ gDW-feces^{-1} h^{-1}, and was higher in older oysters. These results indicate that leaving old oysters in aquaculture fields will lead to the sedimentation of OM with higher biological oxygen demand, which in turn will lead to the environmental deterioration of oyster culture ground by lowering the dissolved oxygen at the seafloor. The resource use efficiency also declines with oyster age. From the standpoint of sustainable oyster aquaculture and the

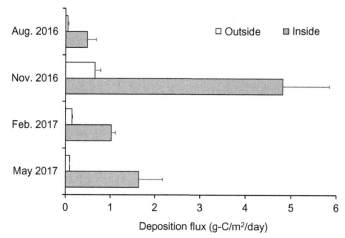

FIG. 2.20 The deposition flux of particulate organic carbon inside and outside of oyster farms. The flux was determined using sampling traps placed in the 10-m depth layer.

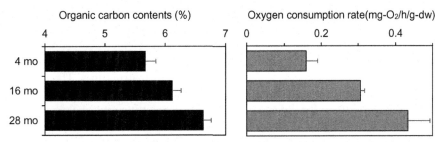

FIG. 2.21 Organic carbon contents and oxygen consumption rates of biodeposits from oysters of ages 4, 16 and 28 months.

preservation of the benthic environment, it is important to reduce the density of oyster rafts and to adopt appropriate management practices that promote growth and reduce the burden on the seafloor environment.

2.5.6 Summary and Conclusion

Regarding the title of this section "Is the forest a lover of the sea? Particulate organic matter" our analyses of the direct assimilation of river-supplied POM by secondary producers in Shizugawa Bay do not support the existence of such a relationship. As mentioned in Sections 2.5.2 and 2.5.3, some studies of other coastal regions have reported that catchment-derived POM contributes to the food supply of secondary producers. In Shizugawa Bay, which is an open bay typical of the Sanriku Coast, however, the amount of water supplied to the bay by rivers is very small compared to the contribution from the open ocean, which is large because of the openness of the bay. In such a bay, the contribution of POM derived from forests is believed to be small. To draw general conclusions regarding the contribution of forest-derived POM to oceans, it is necessary to compare Shizugawa Bay with other bays. In addition, catchments supply not only OM but also various substances, including nutrients, to coastal regions. The relationships between catchment and coastal ecosystems must be evaluated comprehensively.

Our investigations also revealed that, although phytoplankton appears to be the dominant food source for oysters in Shizugawa Bay, oyster growth is affected substantially by the local food resource supply, which varies according to aquaculture density. Furthermore, our studies showed that OM flux in oyster fields differs substantially from that at other locations in terms of POM sedimentation and the like. These results demonstrate that an understanding of POM dynamics in coastal waters and aquaculture fields is essential for evaluating their productivity as fishing grounds and their health and sustainability as environments. Further research is needed to deepen our understanding of POM dynamics in coastal waters, including linkages to catchments.

2.6 MANAGEMENT OF THE AQUACULTURES AND MARINE ENVIRONMENTS IN AN OPEN INNER BAY

Teruhisa Komatsu, Hiroki Murata, Shuhei Sawayama and Shuji Sasa

2.6.1 Aquacultures in Shizugawa Bay

In Shizugawa Bay, the aquacultures target wakame, kombu, oysters, coho salmon, scallops, and ascidians. Except for coho salmon, the target species are cultured without feeding. According to the Miyagi Prefecture Fishery Cooperative, an aquaculture of wakame spread from Onagawa Bay to Shizugawa Bay from approximately 1955–60. A raft culture method for the aquaculture of oysters was developed suitable for waters with a deep bottom in 1930 and applied to Kesennuma Bay and Okatsu Bay. In 1952, a rope-and-buoy method consisting of long horizontal surface ropes with buoys and vertical ropes with oyster clusters was developed and used in waters with deeper bottoms. This method is a current oyster culture method. In 1975, a coho salmon culture started in Japan for the first time in Shizugawa Bay.

The load from the coho salmon farming in Shizugawa Bay reached 2000 tons per year in chemical oxygen demand (COD), which was equivalent to about 90% of the total load of the bay. Nomura et al. (1996) studied the dissolved oxygen (DO) in the bay in the early 1990s and reported that the DO consumption rate in the bottom layer was $0.679\,\mathrm{mg\,L^{-1}\,day^{-1}}$, which was equivalent to the organic load that formed an oxygen deficit in water. In 1994, the number of coho salmon culture rafts decreased because coho salmon was imported to Japan at a low price, and the feed price soared. The mean

CODs of the coho salmon culture area and the entire Shizugawa Bay were 1.28 and 0.94 mg L^{-1}, respectively, at the peak period of coho salmon culture between 1989 and 1992. They decreased to 0.69 and 0.57 mg L^{-1} during the declining period, respectively (Nomura et al., 1998). Dense oyster cultures, however, have been conducted in the bay.

On March 11, 2011, a large tsunami hit the Sanriku Coast near the epicenter of the Great East Japan Earthquake in the Pacific Ocean and greatly damaged the coast's social and blue infrastructures. The first author of this section believed it was a necessity to support local fishermen by providing information on recovered distributions of macrophyte beds that were blue infrastructures for the coastal fisheries. The University of Tokyo, to which the first author belonged, and the Northwest Pacific Region Environment Cooperation Center decided to create distribution maps of the macrophyte beds before and after the tsunami in Shizugawa Bay with financial support from The Mitsui & Co. Environment Fund (Komatsu et al., 2002).

In September 2011, we visited the Shizugawa Branch of the Miyagi Prefecture Fishery Cooperative with late Yasutsugu Yokohama, professor emeritus and director of the Nature Center of Minamisanriku Town, to discuss a plan to create the macrophyte distribution map and the fisheries in Shizugawa Bay with late Norio Sasaki, chairman of the Steering Committee of the Shizugawa Bay Fishery. According to Sasaki, before the tsunami, it took 3 years to land oysters after the deployment of seed oysters in the sea because of the overcrowding of the oyster culture rafts in the inner bay area. Before the tsunami, Yokohama noted that the excessive rafts deteriorated the marine environments and that a decrease in planktonic prey for oysters made the oysters grow slowly, and that such aquaculture is not sustainable. Sasaki said that, after the tsunami, his office discussed a drastic reduction of the aquaculture rafts in the bay to change the tsunami disaster into an opportunity. The first author showed a map of the oyster culture raft distributions in Yamada Bay of Iwate Prefecture (Komatsu et al., 2002). Sasaki asked the author to create a map of the aquaculture raft distributions in Shizugawa Bay.

In Shizugawa Bay, fishermen deployed the seed oysters remaining after the tsunami in the sea. The seed oysters grew rapidly and reached the size for selling in only 7 to 10 months after the tsunami, while it took 3 years to grow them to the size for selling before the tsunami. This experience promoted the discussion of the reduction of aquaculture rafts among the fishermen. Togura Branch and Shizugawa Branch in the inner Shizugawa Bay aimed at reducing the number of oyster rafts to about 30% and 50% of the numbers in the Togura and Shizugawa zones designated as common fishery right areas for aquaculture by Miyagi Prefectural Governor before the tsunami, respectively (Fig. 2.22) (Komatsu et al., 2018b). Fig. 2.22 visualizes the distributions of the oyster and wakame culture rafts before and after the tsunami obtained by analyzing the satellite images. Our observations of the dissolved oxygen content in the bottom layer verified that the bottom environments were improved by reducing the oyster rafts. The flushing out of deteriorated bottom sediments by the tsunami and the reductions in rafts contributed to the maintenance of the sound bottom environment.

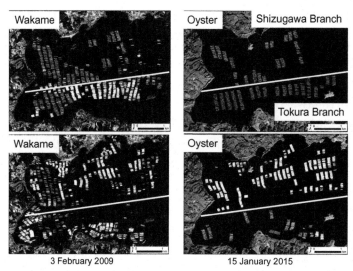

FIG. 2.22   Distributions of the aquaculture facilities mapped with satellite remote sensing in the inner area of Shizugawa Bay on February 3, 2009 before the tsunami in 2011 *(left panels)* and on January 15, 2014 after the tsunami *(right panels)*. The aquaculture facilities consisted of rope and buoy types for oyster cultures *(upper panels)* and wakame *(lower panels)*. The Togura and Shizugawa zones of the Miyagi Fishery Cooperative are located above and below the straight line.

The growth rate of the seed oysters suspended in culture nets in four different locations of Shizugawa Bay were measured continually after the tsunami (see Section 2.5.5). The growth of oysters cultured in the Togura Branch area on the south side of the bay was faster than in the other locations (Fig. 2.19), which might be explained by the lower aquaculture density there than in the other areas of the bay. The organic carbon contents of the solid excrements were higher in the older oysters than in the younger oysters (Fig. 2.21), showing that the assimilation rate of oysters declines with aging. The oxygen consumption rate of the solid excrements of the older oysters was higher than that of the younger oysters. These results indicate that culturing oysters at a lower raft density in the bay will lead to the rapid growth of oysters, less sedimentation of solid excrements, and less oxygen demand to decompose the organic matters of excrements at the seafloor. Prey resources are used efficiently by young oysters, which grow to the size of selling in a year. From the standpoint of a sustainable oyster aquaculture and the preservation of the benthic environment, it is important to reduce the density of oyster rafts and to adopt appropriate management practices that promote growth and reduce the burden on the seafloor environment. This is the Satoumi activity that changes local fishermen's intervention with nature to create sustainable and prosperous fisheries and sound marine environments.

Another point of Satoumi activity is that oyster rafts provide substrates for the diatoms in the euphotic zone an artificial bottom (see Section 2.3.9). Oysters immediately feed on the detached diatoms from surfaces oysters and rafts in the euphotic zone and excrete urea and feces that are converted to nutrients, which foster the diatoms attached to the surfaces of oysters and rafts. This is a gardening effect of the oyster culture. The energy and materials from the primary producer were transferred to the oysters as the first consumers on the rafts. When the number of oyster rafts is within the appropriate numbers mentioned, the material flow becomes fat, long, smooth and quick. In Shizugawa Bay, Satoumi is realized by the change of aquaculture methods by local fishermen.

2.6.2 Council for the Future Environment of Shizugawa Bay

We organized the Council for the Future Environment of Shizugawa Bay, with the Steering Committee of the Shizugawa Branch of the Miyagi Prefecture Fishery Cooperative, Minamisanriku Town, Miyagi Prefecture, the Ministry of the Environment of Japan, and WWF Japan to share scientific research results and discuss the future environment, aquaculture, and fisheries in Shizugawa Bay (Komatsu et al., 2018b). The council was held on April 30, 2015 (Fig. 2.23), May 16, 2016, January 16, 2017, October 11, 2017, July 31, 2018 and March 11, 2019. The ecological modeling study led by Yanagi

FIG. 2.23 Picture showing the first council for the future Shizugawa Bay held at the Hiraiso Civic Center of Minamisanriku Town from 14:30 to 17:00 on April 30, 2015. The participants were 19 fishermen, 3 officers of Miyagi Prefecture, 1 journalist of a TV channel, 2 officers of Minamisanriku Town, 1 officer of WWF, and 7 members of this study. The participants discussed how to realize sustainable aquacultures and sound marine environments based on the results of the field surveys presented in this study.

predicts the future environment and aquaculture catches (refer to Section 6.2). Therefore, the council discussed not only the results in Sections 2.2 to 2.6, but also those in Section 6.2. The following subsections present the contents of the discussions.

2.6.2.1 Is the Forest a Lover of the Sea?

Phytoplankton absorb carbon, nitrogen, and phosphate at a mol ratio of 106:16:1, respectively, and maintain the body at the same ratio, which is called the Redfield ratio. When a mol ratio of nutrient contents is similar to the Redfield ratio, phytoplankton can absorb nutrients efficiently to grow smoothly. When a nutrient's content is lower than the requirements of the phytoplankton present in the water, it controls the reproduction of phytoplankton. Such a nutrient control level is half-saturation constant K of the Michaelis-Menten equation, one of the best-known models of the enzyme kinetics relating substrate concentration and reaction velocity. Therefore, even if the seawater has nutrients similar to the Redfield ratio, phytoplankton growth is controlled by a limiting nutrient that is less than its half-saturation constant of phytoplankton growth. We assume that the half-saturation constants of dissolved iron and nitrogen for a diatom in Shizugawa Bay are 3.5 nM (see Section 2.4.5) and 1.6 μM, respectively. We compared the dissolved iron and dissolved nitrogen concentrations in Shizugawa Bay with the half-saturation constants of dissolved iron (3.5 nM) and dissolved nitrogen (1.6 μM), respectively, and DIN/DIP with the Redfield ratio (Fig. 2.24).

We plotted the observed values on a graph with horizontal and vertical axes represented by dissolved iron and DIN/DIP, respectively (Fig. 2.24). In summer, the nitrogen concentration levels were less than their half-saturation constants for the diatom. The DIN/DIP values were <16 in summer and winter. Even in autumn, DIN/DIP was <16 in the bottom layer of the central area and the bay head area. In summer and autumn, dissolved iron shortages occurred in the surface layers of the central bay area and the bay mouth area but not in the surface layer of the bay head area or the bottom layer of the bay mouth area.

The nutrients, dissolved iron, and particulate organic matter transported from the rivers to the bay are not sufficient for the oyster and wakame cultures by the enrichment of nutrients, as described in Sections 2.3, 2.4 and 2.5. These results were presented at the council on May 16, 2016. After the presentations, one local fisherman commented that it is better to use "the forest is a friend of the sea" instead of "the forest is a lover of the sea." This comment is an exquisite expression because the forest does not play a definitive role in the material flow of the iron and nitrogen in Shizugawa Bay and its watershed. The forest, however, favors oysters and other organisms in the bay with good environments through the prevention of sediment runoff (Kitahara, 1998) and water storage effects to alleviate the influx of freshwater into the

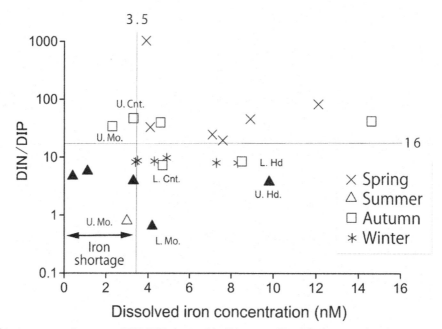

FIG. 2.24 Dissolved iron concentration versus DIN/DIP observed in Shizugawa Bay. The horizontal and vertical axes are the dissolved iron concentration and DIN/DIP, respectively. The half-saturation constant of a diatom in Shizugawa Bay was 3.5 nM, which the vertical line passes through. The DIN/DIP Redfield ratio is 16, which the horizontal line passes through. The observed values of nitrogen below 1.6 nM are indicated with closed symbols. In the graph, the letters U, L, Mo, Cnt, and Hd indicate the upper layer, lower layer, bay mouth area, central bay area and bay head area, respectively. For example, a symbol with "U. Mo." represents a value measured in the upper layer of the bay mouth area.

bay (Fujieda, 2007) during heavy rainfall. During heavy rainfall, the soil on the surface layer is washed out, and landslides occur because of the maldeveloped root systems of the trees in the planted coniferous forests that have not been managed by thinning, pruning, and lower grass cutting. In a summer-green deciduous forest, however, light penetrates the forest floor to develop the root systems and lower grasses. Therefore, it is better to plant summer-green deciduous trees than coniferous trees when one cannot manage a forest. The Ministry of Environment of Japan needs to promote the afforestation of summer-green deciduous trees based on scientific results obtained by this research, namely the multifunction of the forest.

2.6.2.2 Model Calculation of Aquaculture Rafts and the Discussion

Miyagi Prefecture is the greatest production prefecture of ascidians in Japan. Since September 2013, South Korea has banned the import of all fishery products of 8 prefectures near Fukushima Diaiichi Nuclear Plant in Japan because of radioactive contaminants diffused by the accident at the nuclear plant. The landing of the ascidians produced by aquaculture was about 13,000 tons in 2016, of which 7600 tons was not consumed in Japan. The Tokyo Electric Power Co. Ltd. bought the excessive production, burned it, and compensated the fishermen for the deficit from the Korean ban. The Miyagi Prefecture Fishery Cooperative decided to reduce the ascidian aquaculture in 2017. The scallops produced in Miyagi Prefecture have been contaminated with paralytic shell toxin and banned since April 2018. Because this event likely will continue for a long time, the fishermen of Miyagi Prefecture discern that the scallop culture is finished. In the innermost area of Shizugawa Bay, wakame loses its green color because of a lack of nitrogen, lowering its commercial value.

In this situation, the fishermen of Shizugawa Bay started to discuss the conversion of the scallop and ascidian culture rafts to wakame ones in the central part of the bay and the reduction of the wakame culture rafts to appropriate numbers in the innermost area of the bay (Fig. 2.25). They needed a scientific foundation, which we provided, for discussing the conversion and reduction of rafts. We codesigned four scenarios of converting and reducing aquaculture rafts with local fishermen. Based on the scenarios, the ecological modeling research group developed an ecological model of Shizugawa Bay to calculate the marine environments and aquaculture productions depending on the scenarios (refer to Section 6.2).

On October 11, 2017, the council discussed the results of the model calculation (Yamamoto et al., 2018). Scenarios were as follows:

Case 1: Current situation of the number of rafts of each cultured species.
Case 2: Reduction of the wakame rafts to 75% of their current number in the innermost area and the conversion of 25% of the current ascidian and scallop rafts to wakame rafts in the central area.
Case 3: Reduction of the wakame rafts to 50% of their current number in the innermost area and the conversion of 50% of the current ascidian and scallop rafts to wakame rafts in the central area.
Case 4: Reduction of the wakame rafts to 25% of their current number in the innermost area and the conversion of 75% of the current ascidian and scallop rafts to wakame rafts in the central area.

The results of the ecological model calculations indicated that the production of oyster cultures did not change in the four scenarios. Depending on the percentage of the reduced number of rafts, the yield of wakame per raft increased, while the total yield and total benefit of the wakame culture in the innermost area decreased (Yamamoto et al., 2018) (Fig. 2.26). It is necessary to take into account the cost per raft to evaluate the final benefit. The total benefit of the wakame culture in the innermost area is obtained from the following equation:

$$\text{Total benefit} = \sum_{k=0}^{n} P_k - nC, \tag{2.1}$$

where n, P_k and C are the number of wakame culture rafts in the innermost area (depending on the scenario), the wakame yield of a raft k deployed in the innermost area (in cash, depending on its location and number of rafts) and the cost per raft, which is constant, respectively. Based on Eq. (2.1), the reduction according to the Case 2 scenario brought the greatest benefit among the four scenarios (Yamamoto et al., 2018) (Fig. 2.26). The dissolved oxygen contents in the bottom layer in Cases 2, 3 and 4 were better than in Case 1 because the excrements of the ascidians and scallops decreased (Yamamoto et al., 2018).

We examined whether wakame competes with phytoplankton as prey of oysters for nutrient uptake. Oysters actively feed on prey, namely, phytoplankton and attached microalgae (see 2.3.9), from autumn to early winter, while wakame needs nutrients during its growth season from late winter to early spring. Thus, phytoplankton and attached microalgae are not influenced by the number of wakame cultures because of the differences in timing. Oysters consume phytoplankton and attached microalgae to convert urea and solid excrements that bacteria decompose to

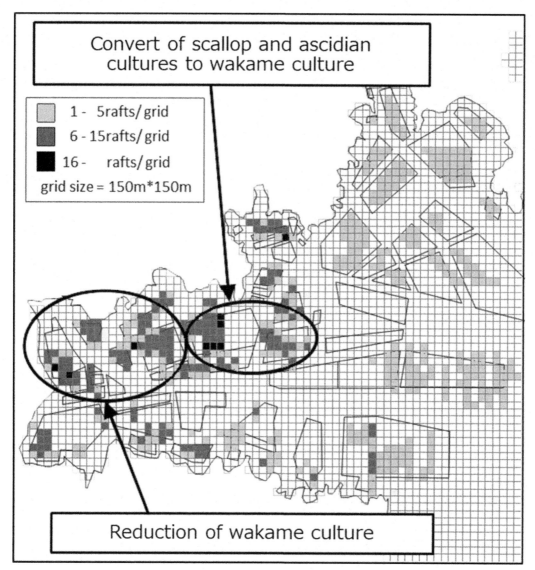

FIG. 2.25   Map shows the densities of the aquaculture rafts of wakame, oyster, and scallop cultures plotted on 150 × 150 m grids for the calculation of the ecological model in the inner Shizugawa Bay. The darker grids show a high density of rafts. The polygons indicate common fishery right areas of the aquacultures. Left and right ellipses correspond to the areas where the wakame aquaculture rafts are reduced in the innermost area of the bay head, and where a portion of the scallop and ascidian culture rafts are converted to wakame culture rafts, respectively, depending on the scenarios.

nutrients. The aquacultures of scallops and ascidians, which are filter feeders, compete with those of oysters. When the aquacultures of scallops and ascidians are converted to wakame, the competition of oysters disappeared. One 100 m long horizontal rope suspends vertical ropes at 1–2 m intervals to which clusters of scallops and ascidians are attached. Because the drag coefficients of the scallop and ascidian clusters are greater than that of the wakame, the conversion of the scallop and ascidian cultures to wakame cultures increased the water exchange between the bay head and the open ocean, which is the most important source of nutrients. Therefore, the increased water exchange in the innermost area benefits wakame, phytoplankton, the marine environments, and eventually oysters.

We presented these results at the council on July 31, 2018. The local fishermen agreed with the results based on their experiences. The Shizugawa Branch of the Miyagi Fishery Cooperative took into account the results to decide the target species of aquacultures at the next update of the common fishery rights of the aquacultures in Shizugawa Bay in 2018 fiscal year. In this way, the council serves to promote sustainable aquacultures, keeping fishermen's incomes in harmony with the marine environments using scientific data. Discussion at the council among stakeholders and scientists is one of the Satoumi approaches for realizing sustainable aquaculture in an open-type inner bay. It is possible to make Shizugawa Bay prosperous and sustainable by continuing the council regularly in the future. Our Satoumi experiences in Shizugawa Bay can be applicable to other rias-type bays in Japan and perhaps in the world.

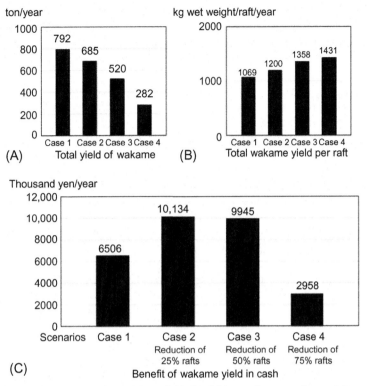

FIG. 2.26   Total yield of the wakame culture per year (*upper left panel*, A), the yield per raft per year (*upper right panel*, B) and the total annual benefit of the wakame culture per year, in cash (*lower panel*, C), in the inner area of Shizugawa Bay.

2.6.3 ASC Certificate of the Oysters in the Togura Branch

The Aquaculture Stewardship Council and the Marine Stewardship Council are independent, international non-profit organizations that manage certification and labeling programs to create beneficial aquacultures and fisheries that do not create negative environmental impacts and are sustainable. Only food materials with internationally certificated ecolabels, such as ASC and MSC, were used for the London and Rio Olympic Games in 2012 and 2016. In 2020, the Tokyo Olympic Games will use food materials only with ecolabels. Only four marine products had an ecolabel from the Marine Stewardship Council until 2016. On March 30, 2016, the oysters cultured in the Togura zone of Shizugawa Bay acquired ASC certification, which succeeded in reducing about 70% of the oyster culture rafts, with the supports of WWF Japan (Maekawa, 2016) and this study on marine environments, including a map of the distributions of the aquaculture rafts (Fig. 2.22) (Komatsu et al., 2018b). Matsuda (2007) noted that fisheries and aquacultures are important for Satoumi to realize sustainable fisheries and marine environments through human interventions with nature. The oyster cultures in the Togura zone are a sound aquaculture aligned with Matsuda's point. The marine environments in Shizugawa Bay, however, cannot be managed by only the Togura zone. Because it is necessary to cooperate with the other zone in Shizugawa Bay to maintain sound marine environments and sustainable aquacultures, the council will play an important role in discussing them among the two zones. In the future, it is needed to extend the council to Utastu zone located near the bay mouth of Shizugawa Bay. It is important to visualize the current situations of the marine environments, monitored with remote sensing and observations, and the future aquaculture productions and marine environments, predicted by ecological modeling according to the scenarios that were codesigned with the stakeholders and scientists.

2.6.4 Goal of Minamisanriku Town's Recovery and the Ramsar Convention

In 2012, after the tsunami, Minamisanriku Town created a recovery plan as follows (Minamisanriku Town, 2012): "We have lived here on areas given by the sea and protected by the mountains. However, we have sometimes suffered from the wrath of nature. We will build our town in harmony with nature, esteeming blessings from the mountains and the sea, succeeding in place and culture and not forgetting the reverence for and the awesomeness of nature."

The words *sato*, *chi*, and *yama* in Japanese mean village, area, and mountain in English, respectively. *Sato-chi* indicates an area consisting of a village, paddy fields, and mountains surrounding the village. These mountains are *Sato-yama*. They are located between a pristine nature setting and a village with agriculture fields, and are reservoirs for agriculture and grasslands. Human activities, accompanied with agriculture and forestry, have formed and maintained specific environments of Satoyama in harmony with nature. The environments provide foods, timbers, beautiful landscapes, and cultural spirits. Because the flat area of a rias-type bay is narrow because of the mountains in the hinterland, the villages along the bay closely depend on the mountains by using the forests as second-growth for the production of wood charcoal and firewood. The local people have thinned trees, pruned branches, and cut the lower grasses on the forest floor to obtain energy, wood, and fertilizers. These activities called Satoyama make tree trunks fat, roots deep, trees sound, and the biodiversity of forest floors rich because of increases in light availability. The Forest Cooperative of Minamisanriku Town obtained FSC certification from the Forest Stewardship Council in October 2015 (Maruoka and Yasumatsu, 2016). The FSC's stated mission is to "promote environmentally appropriate, socially beneficial, and economically viable management of the world's forests." One of the principles of FSC-certified forests is the maintenance of the ecological functions and integrity of the forest. This is a concept of Satoyama activity, which uses forests near villages in a sustainable way. The forests of Minamisanriku Town certified by the FSC connect to the oyster cultures in Shizugawa Bay that are certified by the ASC. This means that Satoyama connects with Satoumi by international ecolabels.

On the coastal ecotone, Minamisanriku Town presented the registration of the macrophyte beds consisting of seagrass and seaweeds shallower than 6 m at the lowest low tide in the wetlands of Shizugawa Bay to the Ramsar Convention, which is the intergovernmental treaty that provides the framework for the conservation and wise use of wetlands and their resources. A characteristic of the seaweed beds in Shizugawa Bay is the coexistence of *Eisenia bicyclis* in temperate waters and *Saccharina japonica* in boreal waters. Shizugawa Bay is the northern limit of *E. bicyclis* and the southern limit of *S. japonica* in Japan (Abe, 2017). The wetlands in Shizugawa Bay are wintering areas for Brant Brent geese (*Branta bernicla*), which are a national natural treasure and designated an endangered species by the Ministry of the Environment of Japan. In winter, 100 to 200 birds pass through Shizugawa Bay because there are seagrasses and seaweed beds that supply prey and rocks above the sea surface near the shore that supply resting places. These characteristics made Shizugawa Bay a wetland of the Ramsar Convention at the 13th meeting of the Conference of the Contracting Parties to the Ramsar Convention on Wetlands (COP13) in Dubai on October 18, 2018. The registration of Shizugawa Bay as a wetland of Ramsar Convention places a link between the forest certified by FSC and the oysters certified by ASC. The link from the mountains to the sea through the coast is completed from a point of view of the sustainable use of nature. Minamisanriku Town approaches the goal of recovery with the ideas that the local community lives in harmony with nature through appropriate human interventions, namely, Satoumi and Satoyama activities, such as sustainable forestry and fisheries that maintain a sound material flow (refer 2.1.2).

2.6.5 Toward the Future

In 2012, the cabinet of the Government of Japan decided five national strategies for promoting biodiversity: penetration of biodiversity into societies; reconstruction of the relation between human and nature in a community through revision; ensuring a link among forests, villages, rivers, and the sea; behaving with a global perspective; and strengthening the scientific bases of biodiversity and connecting them to the policies (Ministry of the Environment of Japan, 2012). The Law for the Promotion of Nature Restoration enacted in 2003 aims to restore ecosystems and natural environments, such as rivers, wetlands, tidal flats, macrophyte beds, and coral reefs lost by past human activities and operations through the participation of various local stakeholders to conserve, restore, and create these ecosystems, and to maintain those already conserved, recovered, or created (Ministry of the Environment of Japan, 2002). The Government of Japan is working to realize SDGs after the decision of the General Assembly of United Nations on Sustainable Development Goals (SDGs) on September 15, 2015 (Ministry of Foreign Affairs of Japan, 2016). One of the goals is SDG14 to conserve marine resources for sustainable use.

The tsunami on March 11, 2011 recovered wetlands and tidal flats lost by reclamation works on the Sanriku Coast during the high economic growth and succeeding periods from 1960 to the 1990s. Wetlands produce rich corrosive substances, such as fluvic acids combined with dissolved iron, to form complexes in a reductive environment. Dissolved iron increases the primary productivity of the sea. In the Amur River, dissolved iron is entrained from back marshes into the river during floods to maintain a high concentration of dissolved iron in the river water (Shiraiwa, 2011). Research conducted in Furen Lake on Hokkaido Island, Japan, showed a positive relationship between the concentration of the dissolved organic carbon and dissolved iron (Shiraiwa, 2017). Their significant correlations were found in meadowlands ($r = -0.61$), natural forests ($r = 0.47$), wetlands and marshlands ($r = 0.73$). Considering the National Strategies for Promoting Biodiversity of Japan, Law for the Promotion of Nature Restoration, and SDG 14, it was possible not to reclaim the saltmarshes restored by the tsunami but to conserve the saltmarshes supplying dissolved iron for the primary production of the sea. Discussions about future marine environments after the tsunami were needed among the residents, fishermen, government officers, and scientists based on national strategies, the act, and SDG14. The Ministry of Land Construction and Transport of Japan, however, immediately decided on the construction of large seawalls and the reclamation of saltmarshes and tidal flats, justifying the decision for the reason to restore current coastal areas to those before the tsunami.

The Satoumi method fits the biodiversity strategy of Japan (Ministry of the Environment of Japan, 2012), the Law for the Promotion of Nature Restoration (Ministry of the Environment of Japan, 2014), and SDG 14 (Ministry of Foreign Affairs of Japan, 2016). The Satoumi method is suitable for securing fishery production, biodiversity and sound marine environments through using the full ecological functions of macrophyte beds and tidal flats by the appropriate human interventions. Sea urchin fisheries are indispensable for maintaining seaweed forests. Thus, it is defined as "wise use of wetland" that Ramsar Convention requests. The appropriate number of oyster cultures increases the cycling speed of nutrients by a gardening effect to provide a substrate for the attached microalgae. The ecological modeling of Shizugawa Bay presented in Section 6.2 shows that an appropriate number of oyster culture rafts that does not drop the primary production in the inner bay increases the cycling rate of nutrients. The ecological modeling compared the speed in the inner bay in 2020 under the Case 2 scenario with that in 2009 before the tsunami, when too many rafts were deployed. The model calculated a primary production rate and cycling rate of nitrogen defined as a primary production rate of phytoplankton and wakame divided by the DIN of the standing stock in the inner bay, respectively. The primary production rate and cycling rate of nitrogen in 2020 were 1.4- and 4.5-fold of those in 2009, respectively, suggesting the efficient reuse of nutrients in 2020 but not in 2009. By stopping the over-deployment of oyster culture rafts and reducing the rafts to an appropriate number, the growth of oysters to a size for selling shortens to 1 year through the efficient use of prey and amelioration of the bottom environment by the decrease in the organic loads of excrements (refer Section 2.5). Such appropriate human interventions, namely satoumi activities, sustainably enable an increase in the production rates of marine organisms and sound marine environments.

This study developed a management method using the Satoumi concept for the sustainable development of Shizugawa Bay, which is representative of open inner bays. The developed methods here are universally applicable to a rias-type bay where aquacultures are conducted. We would like to expect that the methods will be spread widely as a new approach of the Ministry of the Environment of Japan in the future and in other countries.

References

Abe, T., 2017. Toward the registration of Shizugawa Bay to wetlands of RAMSAR convention. Tsuchioto 59, 3 (in Japanese). http://www.reconstruction.go.jp/topics/main-cat6/sub-cat6-1/20171222_tutioto.pdf.

Akashige, S., Hirata, Y., Takayama, K., Soramoto, K., 2005. Seasonal changes in oxygen consumption rates and filtration rates of the cultured Pacific oyster *Crasssotrea gigas*. Nippon Suisan Gakkaishi 71 (5), 762–767 (in Japanese with Englshi abstract and figure captions).

Asmala, E., Autio, R., Kaartokallio, H., Pitkänen, L., Stedmon, C.A., Thomas, D.N., 2013. Bioavailability of riverine dissolved organic matter in three Baltic Sea estuaries and the effect of catchment land use. Biogeosciences 10 (11), 6969–6986.

Bundy, R.M., Abdulla, H.A., Hatcher, P.G., Biller, D.V., Buck, K.N., Barbeau, K.A., 2015. Iron-binding ligands and humic substances in the San Francisco Bay estuary and estuarine-influenced shelf regions of coastal California. Mar. Chem. 173, 183–194.

Endo, Y., Natsuike, M., Miyamoto, M., Yoshimura, C., Fujii, M., 2018. Effects of land cover on dissolved iron loading and its load unit in the Kitakami mountains in winter. J. Jpn. Soc. Civ. Eng. Ser. B1 74 (4), 535–540 (in Japanese with English abstract).

Estes, J.A., Palmisano, J.F., 1974. Sea otters: their role in structuring nearshore communities. Science 185, 1058–1060.

European Commission, 2013. Building a green infrastructure for Europe. European Union, Brussel, p. 23.

Fujieda, M., 2007. Water-holding capacity and basin storage at forest catchments in Japan. Bull. Forestry For. Prod. Res. Inst. 403, 101–110 (in Japanese with English abstract).

Fukushima, K., Tomita, R., Yokoyama, K., 2016. Spatial distribution of dissolved organic matter concentration and characteristics in river basins of the Kesennuma Bay. J. Jpn Soc. Civ. Eng. Ser. G 72, 165–172 (in Japanese with English abstract).

Hatakeyama, S., 2000. Afforestation by Fishermen. The Forest Is a Lover of the Sea. Kodansha, Tokyo, p. 176 (in Japanese).

He, W., Chen, M., Schlautman, M.A., Hur, J., 2016. Dynamic exchanges between DOM and POM pools in coastal and inland aquatic ecosystems: a review. Sci. Total Environ. 551, 415–428.

Kawahata, T., Fujibayashi, M., Yugami, Y., Nishimura, O., Sakamaki, T., 2018. Influence of oyster farming on chemical composition of particulate organic matter and oxygen consumption rate in seawater. J. Jpn. Soc. Civ. Eng. Ser. G 74(7), 63–71 (in Japanese with English abstract).

Kitahara, Y., 1998. Forests prevents a surface erosion. Shinrin Kagaku (Forest Sci.) 22, 16–22 (in Japanese).

Komatsu, T., Takahashi, M., Ishida, K., Suzuki, T., Hiraishi, T., Tameishi, H., 2002. Mapping of aquaculture facilities in Yamada Bay in Sanriku Coast, Japan, by IKONOS satellite imagery. Fish. Sci. 68 (Suppl 1), 584–587.

Komatsu, T., Sagawa, T., Sawayama, S., Tanoue, H., Mohri, A., Sakanishi, Y., 2012. Mapping is a key for sustainable development of coastal waters: examples of seagrass beds and aquaculture facilities in Japan with use of ALOS images. In: Ghenai, C. (Ed.), Sustainable Development-Education,

Business and Management-Architecture and Building Construction-Agriculture and Food Security. In Tech Publishing Co., Rijeka, Croatia, pp. 145–160

Komatsu, T., Terauchi, G., Dazai, A., Aoki, M., Nakura, Y., Sasaki, H., Tsujimoto, R., Sasa, S., Sakamoto, S.X., Yanagi, T., 2014. Restoration of seagrass and seaweed beds in Shizugawa Bay towards the recovery of coastal fisheries from damage due to the Great East Japan Earthquake. Bull. Coast. Oceangr. 52 (1), 103–110 (in Japanese with English abstract and figure captions).

Komatsu, T., Ohtaki, T., Sakamoto, S., Sawayama, S., Hamana, Y., Shibata, M., Shibata, K., Sasa, S., 2015. Impact of the 2011 tsunami on seagrass and seaweed beds in Otsuchi Bay, Sanriku coast, Japan. In: Ceccaldi, H.J., Hénocque, Y., Koike, Y., Komatsu, T., Stora, G., Tusseau-Vuillemin, M.-H. (Eds.), Marine Productivity: Perturbations and Resilience of Socio-Ecosystems. Springer, Cham, pp. 43–53.

Komatsu, T., Otaki, T., Sasa, S., Sawayama, S., Sakamoto, S.X., Gonzalvo Maro, S., Asada, M., Hamana, M., Murata, H., Tanaka, K., Tsujimoto, R., Terauchi, G., 2017. Restoration process of blue infrastructures supporting coastal fisheries along the Sanriku Coast after the huge tsunami. Bull. Coast. Oceangr. 54 (2), 117–127 (in Japanese with English abstract and figure captions).

Komatsu, T., Sasa, S., Hamana, M., Sakamoto, S., Asada, M., Terauchi, G., Tsujimoto, R., Yanagi, T., 2018a. Temporal and spatial changes in a coastal ecotone in Shizugawa Bay, Sanriku coast due to the impacts of the tsunami on 11 March 2011 and the following artificial impacts. In: Santiago-Fandiño, V., Sato, S., Maki, N., Iuchi, K. (Eds.), The 2011 Japan Earthquake and Tsunami: Reconstruction and Restoration. Springer, Cham, pp. 265–278.

Komatsu, T., Sasa, S., Montani, S., Yoshimura, C., Fujii, M., Natsuike, M., Nishimura, O., Sakamaki, T., Yanagi, T., 2018b. Studies on a coastal environment management method for an open-type bay: the case of Shizugawa Bay in southern Sanriku Coast. Bull. Coast. Oceanogr. 56 (1), 21–29 (in Japanese with English abstract and figure captions).

Lee, Y.P., Fujii, M., Terao, K., Kikuchi, T., Yoshimura, C., 2016. Effect of dissolved organic matter on Fe(II) oxidation in natural and engineered waters. Water Res. 103, 160–169.

Lee, Y.P., Fujii, M., Kikuchi, T., Natsuike, M., Ito, H., Watanabe, T., Yoshimura, C., 2017. Importance of allochthonous and autochthonous dissolved organic matter in Fe(II) oxidation: a case study in Shizugawa Bay watershed, Japan. Chemosphere 180, 221–228.

Maekawa, S. (2016) Regeneration of the sea and aquaculture certification—disaster and fisheries of Minamisanriku Town. In: Omoto, R., Sato, S., Naito, D. (Eds.), International Certification for Stock Management. Ecolabels Linking Local to Global Activities, University of Tokyo Press, Tokyo, 66–83. (in Japanese).

Magni, P., Montani, S., Takada, C., Tsutsumi, H., 2000. Temporal scaling and relevance of bivalve nutrient excretion on a tidal flat of the Seto Inland Sea, Japan. Mar. Ecol. Prog. Ser. 198, 139–155.

Martin, J.H., Coale, K.H., Johnson, K.S., Fitzwater, S.E., Gordon, R.M., Tanner, S.J., Hunter, C.N., Elrod, V.A., Nowicki, J.L., Coley, T.L., Barber, R.T., 1994. Testing the iron hypothesis in ecosystems of the equatorial Pacific Ocean. Nature 371 (6493), 123–138.

Maruoka, Y., Yasumatsu, N., 2016. Possibilities of "reconstruction tourism" in the disaster areas of the great East Japan Earthquake: based on the case of Minamisanriku town, Miyagi prefecture. Bull. Soc. Sea Water Sci. 70, 231–238 (in Japanese).

Matsuda, O., 2007. Chapter 4, taking advantage of multi-functions of fisheries to regenerate marine environments. In: Matsuda, O., Muraoka, H., Kobayashi, E., Yanagi, T. (Eds.), Converting Seto Inland Sea to Satoumi: Regeneration Policy from a New Point of View. Kosekisha-Koseikaku, Tokyo, pp. 29–38 (in Japanese).

Matsunaga, K., 1993. The Sea Will Die if the Forest Disappears. Kodansha, Tokyo, p. 190 (in Japanese).

Minamisanriku Town, 2012. Basic Plan of Minamisanriku Town for Recovery From the Disasters (draft). Minamisanriku Town Office, Minamisanriku Town, p. 14 (in Japanese).

Ministry of Foreign Affairs of Japan, 2016. The SDGs Implementation Guiding Principles. Ministry of Foreign Affairs of Japan, Tokyo, p. 39. https://www.mofa.go.jp/mofaj/gaiko/oda/sdgs/pdf/000252819.pdf.

Ministry of the Environment of Japan, 2002. Law for the Promotion of Nature Restoration. Ministry of the Environment of Japan, Tokyo, p. 8. http://www.env.go.jp/nature/saisei/law-saisei/law_e.pdf.

Ministry of the Environment of Japan, 2012. Living Harmony with Nature: The National Biodiversity Strategy of Japan. Ministry of Environment of Japan, Tokyo, p. 3. http://www.biodic.go.jp/biodiversity/about/library/files/nbsap2012-2020/nbsap2012-2020_cop11ver_EN.pdf.

Ministry of the Environment of Japan, 2014. Basic Policy for Nature Restoration. Ministry of the Environment of Japan, Tokyo, p. 11.

Montani, S., Tada, K., Okaichi, T., 1988. Purine and pyrimidine bases in marine particles in the Seto Inland Sea, Japan. Mar. Chem. 25, 359–371.

Morita, A., Touyama, S., Kuwae, T.O., Nishimura, O., Sakamaki, T., 2017. Effects of watershed land-cover on the biogeochemical properties of estuarine tidal flat sediments: a test in a densely-populated subtropical island. Estuar. Coast. Shelf Sci. 184, 207–213.

Nomura, M., Chiba, N., Xu, K.-Q., Sudo, R., 1996. The formation of anoxic water mass in Shizugawa Bay. Bull. Coast. Oceanogr. 33 (2), 203–210 (in Japanese with English abstract and figure captions).

Nomura, M., Chiba, N., Xu, K.-Q., Sudo, R., 1998. The effect of pollutant loading from the fishery cultivation on water quality in inner bay. J. Jpn. Soc. Water Environ. 21 (11), 719–726 (in Japanese with English abstract and figure captions).

Rose, A.L., Waite, T.D., 2005. Reduction of organically complexed ferric iron by superoxide in a simulated natural water. Environ. Sci. Technol. 39 (8), 2645–2650.

Sakamaki, T., 2011. Dynamics and biological use of riverine organic matter in estuarine tidal flats. Jpn. J. Ecol. 61, 63–69 (in Japanese).

Sakamaki, T., Richardson, J.S., 2008. Retention, breakdown and biological utilisation of broadleaf deciduous tree leaves in an estuarine tidal flat of southwestern British Columbia, Canada. Can. J. Fish. Aquat. Sci. 65, 38–46.

Sakamaki, T., Shum, J.Y.T., Richardson, J.S., 2010. Watershed effects on chemical properties of sediment and primary consumption in estuarine tidal flats: importance of watershed size and food selectivity by macrobenthos. Ecosystems 13, 328–337.

Sakurai, I., Yanai, S., 2006. Ecological significance of leaf litter that accumulates in a river mouth as a feeding spot for young crest head flounder (*Pleuronectes schrenki*). Bull. Jpn. Soc. Fish. Oceanogr. 70, 105–113.

Santana-Casiano, J.M., González-Dávila, M., Millero, F.J., 2005. Oxidation of nanomolar levels of Fe(II) with oxygen in natural waters. Environ. Sci. Technol. 39 (7), 2073–2079.

Santana-Casiano, J.M., González-Dávila, M., Millero, F.J., 2006. The role of Fe(II) species on the oxidation of Fe(II) in natural waters in the presence of O2 and H2O2. Mar. Chem. 99 (1–4), 70–82.

Shaked, Y., 2008. Iron redox dynamics in the surface waters of the Gulf of Aqaba, Red Sea. Geochim. Cosmochim. Acta 72 (6), 1540–1554.

Shiraiwa, T., 2011. Earth Environmental Studies on Forests Accompanying Fish: Am River Fostering Oyashio and Okhotsk Sea. Showado, Kyoto, p. 226 (in Japanese).

Shiraiwa, T., 2017. Dissolved iron transport from land to river/estuary: land-use and land-cover impacts on dissolved iron discharges. Nippon Suisan Gakkaishi 83, 1011 (in Japanese).

Yamamoto, Y., Yoshiki, K., Komatsu, T., Sasa, S., Yanagi, T., 2018. Analysis of optimal aquaculture quantity for realizing sustainable marine environments with use of a model integrating land and sea areas in Shizugawa Bay. J. Jpn. Soc. Civ. Eng. Ser. B2 74 (2), I_1279–I_1284 (in Japanese with English abstract).

Yanagi, T., 2013. Japanese Commons in the Coastal Seas: How the Satoumi Concept Harmonizes Human Activity in Coastal Seas with High Productivity and Diversity. Springer Japan, Tokyo, p. 113.

Nutrient Management

Wataru Nishijima*, Kuninao Tada†, K. Ichimi†, T. Asahi‡, Takeshi Tomiyama§, J. Shibata§, and Yoichi Sakai§

*Environmental Research and Management Center, Hiroshima University, Higashi-Hiroshima, Japan
†Seto Inland Sea Regional Research Center, Kagawa University, Kagawa, Japan
‡Faculty of Agriculture, Kagawa University, Kagawa, Japan
§Graduate School of Biosphere Science, Hiroshima University, Higashi-Hiroshima, Japan

Many enclosed and semi-enclosed seas with densely populated and industrialized watersheds are subject to excess eutrophication and deterioration of the quality of water and sediment because of human activities. Conversely, part of the production in the sea is supported by the nutrient loading from land. A healthier marine environment will be achieved with fewer anthropogenic effects. Too little nutrient loading from land, however, carries the risk of reducing the productivity of higher trophic levels, such as fish production, through a production decline in the lower trophic levels, although excess nutrient loading from land induces eutrophication and the deterioration of the quality of water and sediment.

Nutrient management in excessively eutrophic enclosed seas is simple: reduce nutrient loading from land to prevent various adverse effects, such as red tides, the deterioration of sediment, and the accompanying excessive phytoplankton growth. Seas that have overcome an environmentally problematic condition to some extent achieve a healthy environment in many areas, whereas environmentally problematic conditions remain in some areas and seasons. In this situation, The methodology adopted to improve excessive eutrophication and deterioration of the environment in the

© 2019 Elsevier Inc. All rights reserved.

environmentally problematic areas is not adequate and will induce the reduction of primary and secondary production, resulting in the decline of useful aquatic species. We also need to know that the condition without nutrient loading by human activities, however, will be inherent in the area even though the production will be much lower in the future.

The coastal area has been the site for human inhabitation since ancient times, and our lives and the adjacent coastal area have long affected each other. If productivity in the adjacent seas can remain higher than that of the inherent condition, and do so without environmental problems, we can set a goal for our nutrient management of the sea that will have high productivity and healthy environmental conditions under controlled human stress.

In this third chapter, we consider the nutrient management in which adverse effects of nutrient loading from human activities are avoided, and healthy environmental conditions and high productivity are maintained.

3.1 PRODUCTIVITY AND TRANSFER OF ENERGY IN A LOWER TROPHIC ECOSYSTEM

Wataru Nishijima

As primary producers, marine phytoplankton and their photosynthesis support secondary production and consequently higher trophic level species, including commercial fish. The primary production in enclosed and semi-enclosed seas is much higher than that in the open sea and is supported by nutrients, mainly from land and adjacent seas. The distribution of nutrients from land and adjacent seas varies depending on the area of the sea. Therefore, the effects of the change in nutrient loading from land on productivity also vary depending on the area of the sea. Phytoplankton and their production exhibit high spatial and seasonal variability not only because of nutrient concentration, but also physical, chemical, and biological parameters, such as the horizontal and vertical gradient of salinity and temperature, light availability, the formation of stratification, and predator-prey species distributions. Zooplankton and their production, called secondary producers, are more directly important to many fish as food than phytoplankton and primary production. Therefore, we should see an energy transfer from primary production to secondary production, and then to higher trophic level production.

3.1.1 Light Conditions

Light availability is an essential factor in determining the growth of phytoplankton. The photic zone is much shallower in the coastal area than in open sea because of high light attenuation with a high concentration of suspended particles and chromophoric dissolved organic matter (CDOM). Light available for photosynthesis at a depth is determined by the flux of photosynthetically active radiation (PAR) defined as light with wavelengths between 400 and 700 nm (visible light). The light attenuation coefficient (K_d) for PAR has been used as an optical property for water. The Secchi depth developed by Secchi is closely related to K_d and has been recorded for >150 years. Secchi depth has been monitored in Japanese coastal waters since the 1970s, when coastal waters, including enclosed and semi-enclosed seas, were exposed to severe eutrophication and pollution. Secchi depth has been monitored worldwide for a long time and is a good tool to understand the inherent optical conditions and its historical changes with maritime development, urbanization, and industrialization. Moreover, it also has been used to evaluate the effectiveness of marine management, and nutrient and pollution control.

The Lambert-Beer equation estimates K_d from vertical profiles of down-welling irradiance:

$$I_Z = I_0 \exp\left(-K_d Z\right) \tag{3.1}$$

where I_Z is light at depth z ($\mu mol\, s^{-1}\, m^{-2}$), I_0 is light at a surface ($\mu mol\, s^{-1}\, m^{-2}$), K_d is the light attenuation coefficient (m^{-1}), and Z is depth (m).

Light is attenuated by the absorption and scattering properties of the water, CDOM, phytoplankton, and tripton. K_d is expressed:

$$K_d = K_W + K_{CDOM} + K_{PHY} + K_{TRI} \tag{3.2}$$

where K_W is the specific light attenuation coefficient of water (m^{-1}), K_{CDOM} is the specific light attenuation coefficient for CDOM (m^{-1}), K_{PHY} is the specific light attenuation coefficient for phytoplankton (m^{-1}), and K_{TRI} is the specific light attenuation coefficient for tripton (m^{-1}).

K_d also is expressed as a function of Secchi depth:

$$K_d = b/SD \tag{3.3}$$

where b is a coefficient, and SD is Secchi depth (m^{-1}).

3.1.1.1 Spatial and Temporal Distributions of Phytoplankton and Secchi Depth

K_d varies according to variations in absorption and the scattering properties of the water, CDOM, phytoplankton, and tripton, as shown in Eq. (3.2). A high phytoplankton concentration in eutrophic waters is an important factor in the determination of K_d. Spatial distributions of mean chlorophyll a (Chl.a) concentration and mean Secchi depth in summers from 2005 to 2014 in Seto Inland Sea is shown in Fig. 3.1 (Nishijima et al., 2016). More than 10 μm L^{-1} of Chl.a concentration was observed in the northeastern part of Osaka Bay, the northern part of Hiroshima Bay, and the western part of Suo Nada, which are highly enclosed, whereas Chl.a concentration was <2 μm L^{-1} in wide areas, including Bungo Channel, Kii Channel, and many offshore areas. Secchi depth, however, showed a tendency opposite to Chl.a. <5 m of Secchi depth was observed in many areas of the northeastern part of Osaka Bay, the northern part of Hiroshima Bay, and the western part of Suo Nada, whereas >10 m of Secchi depth was observed in many areas of the Bungo Channel and Kii Channel.

Secchi depth apparently is controlled by phytoplankton, and if we could decrease the Chl.a concentration to the level of that in Bungo Channel and Kii Channel, >10 m of Secchi depth would be expected even within these bays. We examined this hypothesis in the shallow Suo Nada, where water depth increases rapidly toward the connecting Iyo Nada (Nishijima et al., 2016). Fig. 3.2 shows the 5-year-mean Chl.a concentration at each depth and 5 year-mean Secchi depth in summer in Suo Nada. The evaluation was conducted by dividing Suo Nada into four areas by water depth, and these areas were: <10 m, 10–20 m, 20–40 m, and >40 m.

The results in areas shallower than 10 m that were located near the coast, and the areas with depths of 20–40 m are shown. Chl.a concentration in areas shallower than 10 m decreased from 6.0 μg L^{-1} in the 1980s to 2.5 μg L^{-1} in late 2000s, whereas Secchi depth was approximately 3 m in both periods and did not change. Chl.a concentration in areas with depths of 20–40 m, however, decreased from 2.5 μg L^{-1} in the early 1980s to 1.0 μg L^{-1} in the late 1990s and then

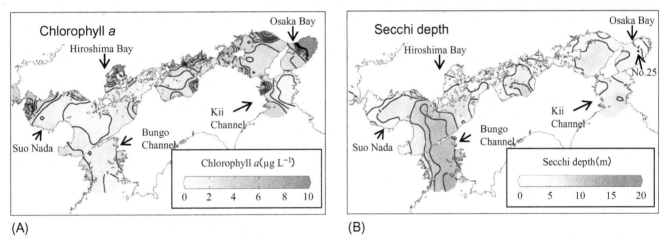

FIG. 3.1 Spatial distributions of (A) mean chlorophyll a concentration and (B) mean Secchi depth in summer from 2005 to 2014 (Nishijima, 2016 was modified).

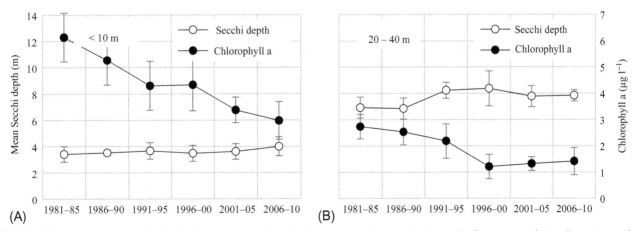

FIG. 3.2 Time course of mean Secchi depths and mean chlorophyll a concentrations averaged over the four seasons during 5-year intervals in areas with depths (A) <10 m and (B) 20–40 m in the Suo Nada during 1981–2010. Error bars show standard deviations.

maintained that level. In these areas, Secchi depth increased with the decreased Chl.*a* concentration and was approximately 8 m. Chl.*a* concentration in the areas shallower than 10 m in the late 2000s and in the areas with depths of 20–40 m in the late 1980s was the same (approximately 2.5 μg L^{-1}) even if mean Secchi depth was quite different, being 3.4 and 7.7 m, respectively, suggesting that factors other than phytoplankton were determining light attenuation differently in these areas. Moreover, the contribution of phytoplankton to light attenuation in the areas shallower than 10 m was low because the decrease in Chl.*a* concentration did not induce an increase in Secchi depth even when Chl.*a* concentration was high.

Although Secchi depth seems to be controlled by phytoplankton in Fig. 3.1, these results indicate that a decrease in phytoplankton, by measures such as the reduction of nutrient loading from land, does not necessarily lead to an improved Secchi depth.

3.1.1.2 Region-Specific Background Secchi Depth

The scenario for improving the optical conditions through nutrient control is that a reduction in nutrient loading from the land causes a decrease in phytoplankton concentration, and then specific light attenuation to phytoplankton decreases. The contribution of phytoplankton or factors other than phytoplankton on light attenuation, however, is different from region to region as shown in Section 3.1.1.1. We need to manage the sea with knowledge of the contribution of phytoplankton to light attenuation and the limits of nutrient management on the improvement of the optical conditions. The effect on the optical condition of nutrient management is the variation of specific light attenuation to phytoplankton. Therefore, it is meaningful to know the Secchi depth excluding the contribution of phytoplankton, which is called region-specific background Secchi depth (BSD) (Nishijima et al., 2018).

BSD can be estimated by an accumulated dataset of Chl.*a* concentrations and Secchi depths obtained at the same time and the same place. For example, long-time monitoring data will be suitable because of the high variation in the data. BSD has been estimated based on seasonal monitoring data from 1981 to 2014 by the Ministry of the Environment (MOE). Fig. 3.3 is an example of the determination of BSD at site No. 25, located in the center of Osaka Bay. We plotted Chl.*a* concentration against the natural logarithm of the reciprocal of Secchi depth (ln(1/SD)) for each season using monitoring data from 1981 to 2014 (Nishijima et al., 2018). BSD was determined by extrapolating the regression line to a Chl.*a* concentration of zero. BSD is based on the premise that light attenuation factors other than phytoplankton are relatively stable. Therefore, it is expected that remarkable seasonal changes, such as river inflow and resuspension of bottom sediment, are necessary to determine BSD by season.

The distributions of spring and summer BSD in Seto Inland Sea are shown in Fig. 3.4. In Seto Inland Sea, BSD is related to salinity and water depth. As a result, BSD was high in the Bungo Channel and Kii Channel that connect to the open sea, whereas low BSD was observed in coastal areas, especially the inner parts of Osaka Bay and Hiroshima Bay, with a large river mouth. The causes of low BSD in these areas would be high suspended solids and CDOM supplied through rivers and resuspension of loose sediment because of shallow water.

Because freshwater through rivers and groundwater mainly supplies nutrients to the sea, the Chl.*a* concentration will be high in the areas with low BSD. In the coastal areas, it is likely that the Secchi depth is low because of high Chl.*a* concentration, which is not always right. We should evaluate the contribution of light attenuation by phytoplankton and other factors separately and understand the limits of the improvement of optical conditions by nutrient management. Based on this knowledge, we should conduct comprehensive environmental management, including nutrient control.

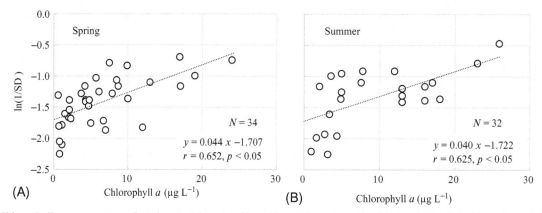

FIG. 3.3   Chlorophyll *a* concentrations plotted against the natural logarithm of the reciprocals of the Secchi depth for (A) spring and (B) summer at monitoring site No. 25 for the period from 1981 to 2014 (Nishijima et al., 2018).

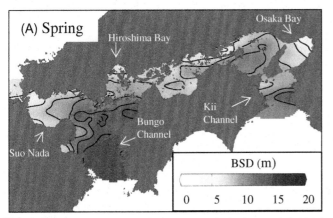 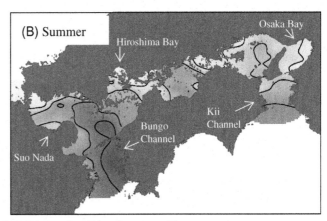

FIG. 3.4 Spatial distribution of region-specific background Secchi depth (BSD) in spring (A) and summer (B).

3.1.2 Spatial and Temporal Distributions of Primary Production

Primary production is basic production supporting the entire ecosystem and, therefore, is essential information about nutrients and fishery management for ecosystem conservation. It will be difficult, however, to monitor primary production widely and periodically because the direct measurement of primary production requires much labor and effort. Therefore, primary production is estimated using basic water monitoring parameters, such as Chl.a concentration and remote sensing imagery. Primary production can be estimated roughly by phytoplankton concentration, water temperature, and optical condition. Currently, primary production in the sea is estimated on a global scale by remote sensing imagery. It is still difficult to estimate primary production in coastal waters, however, because of a high level of suspended particles and high CDOM concentrations. Conversely, we can estimate primary production using water quality data about the Chl.a concentration, water temperature, and Secchi depth, which are measured directly and periodically and accumulate for long-time monitoring.

In Seto Inland Sea, primary production varies spatially from region to region because of various environments connected by waters and varies temporally because of the reduction in anthropogenic nutrient loading. We measured primary production directly in Hiroshima Bay and established an equation to estimate primary production using the Chl.a concentration and Secchi depth to elucidate the spatial and temporal distribution of primary production in Seto Inland Sea. Estimation of primary production was conducted in autumn when nutrients were sufficient and would not inhibit phytoplankton growth (Nakai et al., 2018):

$$K_d = (0.15SD + 0.68)/SD \tag{3.4}$$

$$PP = 12 \int [2.23\text{Chl}.a(z) \times \exp(-K_d z)]dz \tag{3.5}$$

where PP is primary production (mgC m^{-2} day^{-1}), Chl.a is chlorophyll a concentration at depth Z (mg m^{-3}).

Assimilation number was estimated as 2.23 mgC mgChl.a^{-2} h^{-1} based on primary production and the measured Chl.a concentration. Primary production was calculated at ranges of depths and times; range varies from the surface to where the irradiance was 1% of that at the surface and time ranged from sunrise to sunset. The measurement of Chl.a concentration in the periodic monitoring generally is conducted at the surface. Thus, we used Chl.a concentration at the surface for all depths.

Distribution of Chl.a concentration in autumn in the early 1980s and late 2000s in Seto Inland Sea is shown in Fig. 3.5. The areas with higher than 10 µg L^{-1} of Chl.a concentration were located in Osaka Bay and the northern part of Hiroshima Bay. The monitoring sites were classified into five types based on Chl.a concentration in 1981–85. Fig. 3.6 (Nishijima, 2018) shows the historical change in primary production in each group. Primary production in Seto Inland Sea was estimated as 13,271 tC day^{-1} and decreased to 10,653 tC day^{-1} (20% reduction) in the late 2000s. The degrees of decrease in primary production varied among groups. The decreased rate was much higher in the group with a >10 µg L^{-1} of Chl.a concentration in 1981–85 than those in other groups and achieved approximately 45%. The rate was 14% in the groups with a Chl.a concentration lower than 10 µg L^{-1} in 1981–85, which covered 94.4% of the total area of Seto Inland Sea. Moreover, it was found that the rate was only 2% in the group with a Chl.a concentration lower than 2 µg L^{-1} in 1981–85. Primary production decreased almost linearly in this period. The lower rates calculated by linear approximation were 1.48%/year, 1.36%/year, 0.82%/year, 0.45%/year, and 0.04%/year in the groups with a Chl.a concentration higher than 20, 10–20, 5–10, 2–5 µg L^{-1}, and lower than 2 µg L^{-1} of Chl.a concentration in 1981–85, respectively. These results indicated that the decrease in primary production since the 1980s did not occur uniformly in

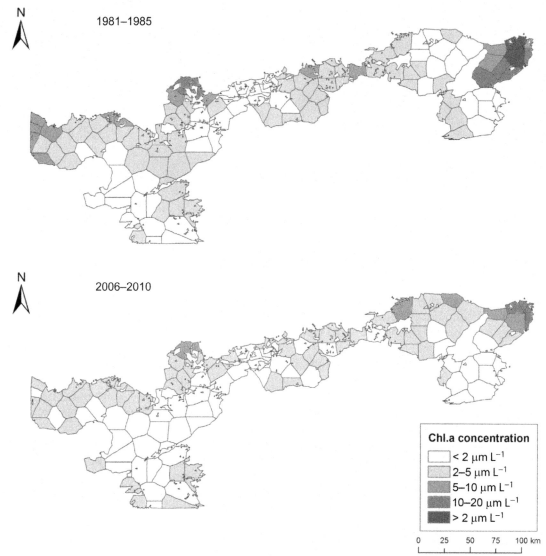

FIG. 3.5 Spatial distribution of mean chlorophyll *a* concentration in autumn in 1980–1985 and 2006–2010.

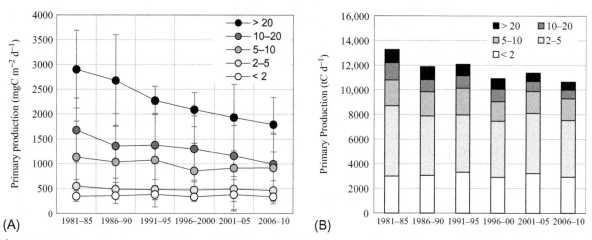

FIG. 3.6 Historical change in primary production of (A) each group classified by mean chlorophyll *a* concentration from 1981 to 1985 and of the whole area (B) in Seto Inland Sea.

Seto Inland Sea. The greater decrease was observed in the areas with a large urban watershed with high loading of nutrients. Osaka Bay and Hiroshima Bay were typical, followed by the western and northern areas of Suo Nada and coastal areas of Kure and Fukuyama cities. Conversely, a small decrease was observed in the offshore areas with a low loading of nutrients. These results, however, were derived only from the analysis of autumn data, and we need further annual-based analyses including periods with nutrient depletion or lowered nutrients.

Products produced by primary production are used by zooplankton and higher trophic-level species. Excess products, however, will not be used and will be transferred to sediment, causing the deterioration of sediment, oxygen depletion near the sediment, and the generation of bottom hypoxic water. We should consider the primary production from the viewpoint of secondary consumers if primary production is excessive in an area.

3.1.3 Transfer of Primary Production to Primary Consumers

Main secondary consumers are zooplankton, such as copepods, in a pelagic ecosystem. In the pelagic ecosystem controlled by bottom-up regulation, zooplankton biomass is controlled by phytoplankton biomass. It is also known, however, that growth and reproduction of zooplankton promoted by an increase in phytoplankton biomass as food will reach a maximum at some biomass level of phytoplankton that is comparative to or less than that observed in eutrophic estuaries, semi-enclosed, and enclosed seas. In the case of cultivation of *Calanus pacificus* under various diatom (*Thalassiosira fluviatilis*) concentrations, the growth rate of *C. pacificus* reached a maximum at 10 ppm of diatom concentration corresponding to approximately $10 \, \mu g \, L^{-1}$ of Chl.a concentration (Vidal, 1980). In other words, approximately $10 \, \mu g \, L^{-1}$ of Chl.a concentration was sufficient food concentration for *C. pacificus*, which would not be expected to grow even if food concentration increased, and the risk of deteriorating sediment quality by transferring unused food to sediment increases. The egg production rate of *Calanus sinicus* larger than $5 \, \mu m$ was best estimated by the concentration of Chl.a, and 90% maximum fecundity was achieved at $2.26 \, \mu g \, L^{-1}$ in April and $1.72 \, \mu g \, L^{-1}$ in June (Uye and Murase, 1997).

We calculated the transfer efficiency of primary production to copepods as secondary consumers based on primary production and secondary production by copepods in July to October 2014–17 in Osaka Bay (Fig. 3.7). Primary and secondary production were calculated based on Eq. (3.5) and the methods of Uye, (Uye and Shimazu, 1997) respectively. Primary production rates were $519 \pm 555, 700 \pm 819, 1210 \pm 913 \, mgC \, m^{-2} \, day^{-1}$, and secondary production rates were $246 \pm 138, 200 \pm 128, 206 \pm 118 \, mgC \, m^{-2} \, day^{-1}$ in three sites from inner to outer parts, respectively. The variation in primary production rate among the three sites was high, whereas that of secondary production was low. Extremely high transfer efficiency was sometimes observed by high variation in Chl.a concentration, but as the Chl.a concentration increased, the transfer efficiency decreased. Although primary production supports secondary consumers, this analysis clearly shows excess primary production does not enhance secondary production. At a minimum, the reduction of the nutrient supply to the areas with a Chl.a concentration higher than $10 \, \mu g \, L^{-1}$ is expected to improve the problems produced by excessive growth of phytoplankton without the reduction of secondary production even if primary production declines.

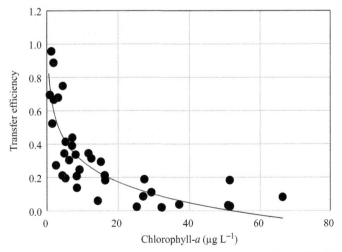

FIG. 3.7   Transfer efficiency from primary production to secondary production in copepods during July–October in 2014–2017 in Osaka Bay.

3.2 NITROGEN AND PHOSPHOROUS CYCLES IN TIDAL FLATS AND SEAGRASS BEDS

Kuninao Tada, K. Ichimi, and T. Asahi

Shallow water areas, such as tidal flats and seagrass beds are very important for the functioning of the coastal eco-system and acceleration of the material cycle. The area of tidal flats and seagrass beds, however, has been decreasing. In Seto Inland Sea, the biggest enclosed sea in Japan, surveys of the area began in 1898 at the tidal flats and in 1960 at the seagrass beds. Both areas have decreased to less than half of their original areas (Sato-umi Net, n.d.). Although it was known qualitatively that tidal flats and seagrass beds are important for various organisms and the material cycle, few studies have been conducted to quantitatively measure the material balance of biophilic elements in the world (e.g., Valiela and Teal, 1979; Wolvaver and Spurrier, 1988) and in Japan (Sasaki, 1989; Ichimi et al., 2011, and Kokubun et al., 2009). It follows, therefore that the material cycle in the tidal flats and seagrass beds is not fully understood. We do not know how we can conserve tidal flats and seagrass beds because we do not know what happens in the tidal flats and seagrass beds nor what function they have. This means that we have no policy to conserve and regreen the tidal flats and seagrass beds, and we have no policy for their maintenance and restoration.

The authors, a Kagawa University research team, have been collecting data to quantify the nitrogen (N) and phosphorous (P) inflow and outflow in a tidal flat and seagrass bed. In this section, we will focus on that, and the discussion will be based on the information obtained during a previous study.

3.2.1 Material Inflow and Outflow in Tidal Flats and Tidal Flat Function in the Breakdown of Materials Into Inorganic Compounds

The difference in the tidal level occurs because of the gravity of moon and sun in the sea, and we have one or two tidal cycles in one day (more properly, 24 h and 50 min). The tidal flat inclines gently from the land to the sea and is the specific environment that is submerged or emerged depending on the difference in tidal levels. This place, called the tidal flat, is divided into estuary, intertidal flat, and lagoon based on their formation (Fig. 3.8. The intertidal flat has sandiness and a wide area in the sea with shoals. In the estuary, the bottom sediment is affected and is sand and clay, which is a continuous input from the river. The lagoon tidal flat is formed at the convoluted area from the estuary or the sea shore. As with other tidal flats, we can see the sandbank in the offshore. Generally, an area that is >100 m wide and >1 ha in size at low tide is called a tidal flat. In the tidal flat at the boundary between the land and sea, various benthos organisms, such as large and small algae, and bivalves are observed, and they accelerate the speed of the material cycle. The decomposition rate by bacteria is extremely high because of the abundance of oxygen.

In general, the tidal flat located at the estuarine region is eutrophic and has an abundance of organisms, but the intertidal flat is oligotrophic and has a poor number of organisms. Our research team has been conducting observations to quantify the nitrogen (N) and phosphorous (P) inflow and outflow during 1 year at the Shinkawa-Kasugagawa River Estuary (72 ha) in Takamatsu City and the intertidal flat (1 ha) located west of the Ohgushi peninsula in Sanuki City, Kagawa Prefecture Fig. 3.9. The measurement of N and P concentrations in the seawater was obtained at the boundary between offshore and the tidal flat during one tide. We estimated the water volume of inflow from offshore to the tidal flat and outflow from the tidal flat to offshore based on the difference in the tidal level. N and P inflows and outflows were computed by multiplying the N and P concentrations by the water volumes of inflow and outflow under the flowing tide and falling tide. If N and P inflows under the flowing tide are greater than their outflows, the differences in N and P are maintained in the tidal flat; however, if they are lower, the differences in N and P have

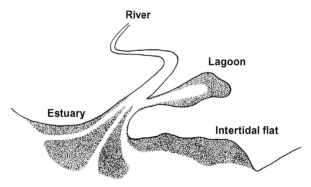

FIG. 3.8   Various tidal flats: Estuary, Intertidal flat, and Lagoon and their formation.

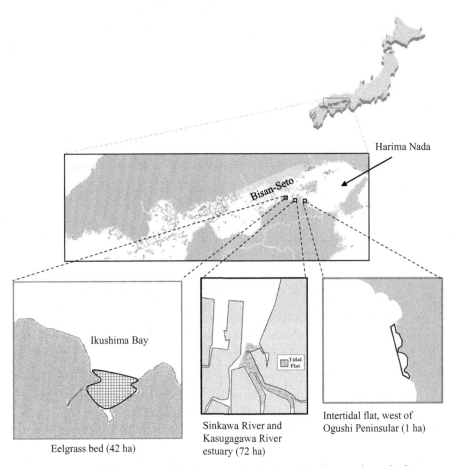

FIG. 3.9 The study fields of tidal flats and eelgrass bed (Hatching area in Ikushima Bay shows eelgrass bed).

disappeared. Additionally, we estimated N and P loading from the river from the inflow water volume and N and P concentrations of the river water (Ichimi et al., 2011). The results of the observations from 2006 to 2010 showed that the estimated N outflow from tidal flat to offshore (39.7 t year^{-1}) was lower than the estimated N inflow from the river to the tidal flat (68.7 t year^{-1}). We suspected that part of the N was released to the air by denitrification. Moreover, N was composed of approximately 60% particulate organic nitrogen (PON) in the N inflow from the river to the tidal flat, but the PON contribution decreased to only 10% in the N outflow. P behavior, however, was different, and P loading from the river to the tidal flat (10.6 t year^{-1}) was similar to P outflow (11.9 t year^{-1}). Similar to N, greater than half of the P was composed of particulate phosphorous (PP) in the P inflow from the river to the tidal flat, but the PP contribution decreased to only approximately 10% in the P outflow. Because the tidal flats are the place where organic compounds are decomposed, it has been shown quantitatively that the tidal flat highly contributes to water quality purification and is a natural wastewater treatment system (Fig. 3.10, Ichimi et al., 2011).

The authors conducted the same observations at the same tidal flat in 2017. Comparing the data sets from 2006 to 2010, the biomass of benthic micro algae (Chlorophyll *a* contents) in the surface sediments decreased in 2017, the benthos biomass, such as the short-necked clam and *Musculista senhousia* apparently decreased (Ichimi et al., 2011, Ichimi personal com.). From these results, it is thought that the nutrient condition of this tidal flat has been decreasing. Therefore, we will suppose that the tidal flat from 2006 to 2010 was in the high nutrient condition, and it was in the low condition in 2017. N outflow from the tidal flat to offshore (45.3 t year^{-1}) in 2017 was slightly higher than N loading from the river to the tidal flat (33.5 t year^{-1}). N was composed of about 40% PON in the N inflow from the river to the tidal flat, but the PON contribution decreased to only 10% in N outflow. Also, P outflow from the tidal flat to offshore (6.0 t year^{-1}) was slightly higher than P loading from the river to the tidal flat (4.9 t year^{-1}). Similar to N, more than half of P was composed of PP in the P inflow from the river to the tidal flat, but the PP contribution decreased to only about 30% in P outflow.

In addition, we conducted observations for inflow and outflow water during the flowing tide and falling tide at the intertidal flat located west of the Ohgushi peninsula, where it was thought to be in a lower nutrient condition. N inflow (0.162 t year^{-1}) and N outflow (0.133 t year^{-1}) were similar. The particulate nitrogen composed 67% of total N in the

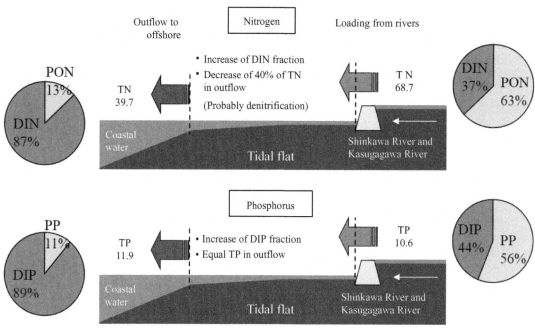

FIG. 3.10   Nitrogen and phosphorous inflow and outflow in the tidal flat. *Adapted from Ichimi, K., 2016. Mass balance of nitrogen and phosphorous in Shinkawa-Kasugagawa River estuary in the eastern Seto Inland Sea, Japan. J. Japan Soc. Water Environ. 39, 125–129 (in Japanese).*

inflow water, and it slightly decreased to 58% in the outflow water. It was thought that the decomposition function in the intertidal flat was lower compared with that of the tidal flat located in the estuary.

As described previously, the composition of nitrogen compounds (the rate of PON and DIN; DIN: dissolve inorganic nitrogen) was different in inflow and outflow water in the tidal flat, and also the change in the composition of nitrogen compounds with N difference between inflow and outflow differed depending on the nutrient condition level (Table 3.1). In the tidal flat with a relatively high nutrient condition level, organic decomposition was progressive. In the tidal flat, the composition rate of PON in outflow nitrogen apparently was lower than that in the inflow nitrogen, and the rate of DIN in outflow nitrogen apparently was higher than in inflow nitrogen. In the tidal flat where there is a relatively low nutrient condition level, however, the composition rates of PON and DIN were not noticeable, although there was a tendency for PON to decrease and DIN to increase on the tidal flat. It was thought that the level of organic material decomposition was low.

From the results we obtained in this study for the estuary and intertidal flat, we estimated the biodegradative capability of all the tidal flats in Kagawa Prefecture (estuary: 225 ha, intertidal flat: 752 ha), and it was estimated that 38–69 t N of the particulate organic nitrogen was decomposed to inorganic nitrogen in 1 year. There were 16 sewage plants in Kagawa Prefecture, however, and their degradation capability was calculated 87–174 t N year^{-1}. It was found that the tidal flats existing in the natural environment have 22%–79% of the degradation capability of the sewage plants.

TABLE 3.1   Nitrogen Inflow and Outflow, and Changes of Nitrogen Form in Tidal Flats

	Inflow		Outflow to offshore	
Shinkawa and Kasugagawa River Estuary	68.7 t year^{-1}		68.7 t year^{-1}	
2006–2010 (high nutrient)	PON	63%	PON	13%
	DIN	37%	DIN	87%
Shinkawa and Kasugagawa River Estuary	33.5 t year^{-1}		45.3 t year^{-1}	
2017 (low nutrient; decreasing from 2006 to 2010)	PON	45%	PON	11%
	DIN	55%	DIN	89%
Intertidal flat, west of Ohgushi Peninsula	0.162 t year^{-1}		0.133 t year^{-1}	
2016 (further lower nutrient)	PON	67%	PON	58%
	DIN	33%	DIN	42%

3.2.2 Material Inflow and Outflow in the Seagrass Bed and Its Function

The seagrass bed has been noted as an egg-laying site for fish and a growth site for juvenile fish. Few studies have been conducted, however, regarding the function of the seagrass bed for the material cycle of biophilic elements, such as nitrogen and phosphorous., A quarter of all seagrass beds in Japan exists in Seto Inland Sea (Ministry of Environment (Sato-umi Net, n.d.)). Most of the seagrass beds were located in the central part of the sea, with the largest portion being the area of eelgrass bed (Fig. 3.11; Nature Conservation Bureau, Environment Agency and Marine Parks Center of Japan, 1994). In this section, we will focus on nitrogen inflow and outflow in the eelgrass bed in Ikushima Bay.

The seaweed communities of eelgrass (*Zostera marina*) grow in shallow water areas with calm waves 5 to 10 m under the low-tide line with a sand and mud bottom. Eelgrass is a flowering plant, and the length of its leaf is generally 60 to 100 cm. The eelgrass is composed of a leaf sheath, which rises from a rhizome with a root in the sand and mud bottom. Most eelgrass is distributed in the temperate and tropical regions, and it reproduces in two ways: by seed and by a horizontal expansion of the root in the sand mud bottom to establish a new leaf sheath. In Seto Inland Sea, it reproduces from the winter to spring, and the biomass reaches the maximum at early summer. After that, it stops growing, most of the algal bodies are dead and settled or flow out as floating seaweed. The eelgrass plays a great role in material cycles in the coastal environment because it photosynthesizes by taking the nutrients from the seawater, and, after its death, it delivers the organic matter to the surrounding area. Although we do not directly eat it as brown seaweed, the eelgrass bed is a habitat for many organisms. For example, juvenile rock fish and red seabream live there and bigfin reef squid lay eggs in the eelgrass bed. Fishermen have carefully protected the eelgrass as a fish habitat for a long time.

The authors have been conducting observations in the eelgrass bed at Ikushima Bay adjacent the Bisan-Seto, central part of Seto Inland Sea. Ikushima Bay is small bay of 42 ha, and almost all the bay is covered by a seagrass bed from spring to summer (Fig.3.9). In this bay, the biomass of eelgrass reached the maximum ($364 \, g \, dw \, m^{-2}$), decreasing when most of the eelgrass is depleted from summer to fall, and increasing biomass the next April. No river flows into the bay; the water mass is exchanged in and out of the bay by tidal current.

We conducted observations of the nitrogen concentrations of inflow and outflow water during the flowing tide and falling tide, and the nitrogen inflow and outflow in the bay were estimated as the tidal flat previously described (Fig.3.12). As a result, the DIN inflow into the bay during the eelgrass growth period was estimated at $8–52 \, mg \, N \, m^{-2} \, day^{-1}$.

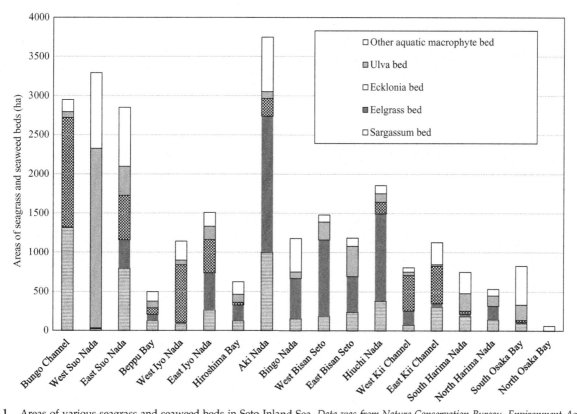

FIG. 3.11 Areas of various seagrass and seaweed beds in Seto Inland Sea. *Data was from Nature Conservation Bureau, Environment Agency and Marine Parks Center of Japan, 1994 and was modified.*

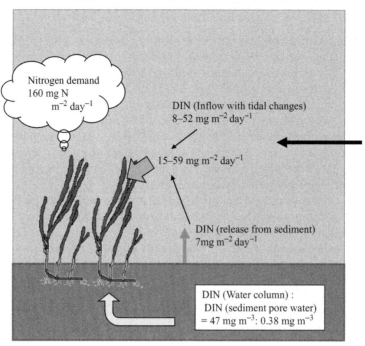

FIG. 3.12 Nitrogen balance in the eelgrass beds.

Additionally, DIN release from the bottom sediment to the water column of the bay was estimated as $7\,mg\,m^{-2}\,day^{-1}$ by a core incubation experiment. Nitrogen demand of eelgrass during its growth period in the entire bay, however, was estimated as $160\,mg\,N\,m^{-2}\,day^{-1}$, so it was unsatisfied with the DIN inflow and release from the bottom sediment. Eelgrass can uptake nutrients from both the leaf sheath and root. At the 10 cm depth of sediment where the rhizome of the eelgrass exists, DIN concentrations in the pore water were lower (average of $0.706\,mg\,N\,L^{-1}$) during the eelgrass growth period and higher (maximum of $0.937\,mg\,N\,L^{-1}$) during its senescent period (Fig.3.13). The DIN concentration in the pore water was higher than that of the seawater in the water column, but the standing stock of DIN in the pore water (the integrated value of 0 to 10 cm depth: $0.38\,mg\,N\,m^{-2}$) was 1/120 of that of the seawater in the water column ($47\,mg\,N\,m^{-2}$), because the pore water volume was much lower than that of seawater in the water column. We know the DIN concentration in the pore water, but we cannot estimate the DIN contribution from the root, because the DIN uptake rate from the root is unknown. Based on previous reports, about 50% of the annual nitrogen uptake is drawn from the water column and the rest is taken from the sediment (e.g., Pedersen and Borum, 1993). Additionally, it was shown that leaves and roots are equally important in nutrient acquisition based on a numerical model (Zimmerman et al., 1987).

As just described, our results demonstrated that the DIN supply ($15–59\,mg\,N\,m^{-2}\,day^{-1}$) from the DIN inflow into the bay by the tidal current and DIN release from the bottom sediment were only 10% to 40% of the DIN demand ($160\,mg\,N\,m^{-2}\,day^{-1}$) of the eelgrass during its growth period. It is expected that DIN uptake from the root plays a substantial role for the residual N demand.

When it rained, it was observed that the salinity decreased and DIN concentration increased rapidly. At that time, the nitrogen content of the leaf sheath doubled, from $16.3\,mg\,N\,g^{-1}$ to $35.1\,mg\,N\,g^{-1}$ over 6 days, and the C/N ratio decreased from 22.6 to 14.2. The C/N ratio of eelgrass's leaf sheath was reported as about 22 (Duarte, 1990). This

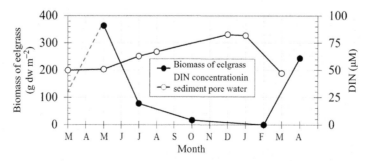

FIG. 3.13 Monthly changes of eelgrass biomass and DIN concentrations of the pore water (average value from 0 to 10 cm depth).

finding indicates that nitrogen was taken up excessively by eelgrass during the short term. Our results suggest that the eelgrass can uptake efficiently the DIN that supplies the bay water intermittently and can accumulate nitrogen in its tissues. We think there are a number of nitrogen absorption mechanisms.

Our study focused on nitrogen inflow and outflow in the eelgrass bed, and we learned that the great biomass of the eelgrass covers the entire the bay during its growth period, and after summer, it is depleted and floats out as the floating seaweed offshore. During the eelgrass growth period, its nitrogen demand is extremely large ($160\,\mathrm{mg\,N\,m^{-2}}$-$\mathrm{day^{-1}}$), and its biomass is also large ($364\,\mathrm{g\,N\,m^{-2}}$), making it the nitrogen stock in the bay. Although we did not measure the flowing out amount of the eelgrass in this study, we observed that only a small amount of eelgrass was deposited inside the bay and that most of it went offshore. The floating eelgrass will be decomposed and returned to seawater as DIN, making the eelgrass a large nitrogen stock in the coastal zone from the spring to summer, after which, it flows out, transporting the nitrogen to the offshore area where it is decomposed to DIN. For all of these reasons, the eelgrass contributes greatly to the nitrogen cycle of the coastal water.

3.3 FISH PRODUCTION AND PRESERVATION IN SETO INLAND SEA

Takeshi Tomiyama, J. Shibata, Yoichi Sakai

Seto Inland Sea is a highly productive water body, and the catch per unit area in coastal fisheries is extremely high. The annual catch, however, has declined after the peak of 485 thousand tons in 1985 to 150 thousand tons in 2015. The catch of fish, excluding crustaceans and shellfish, also has continued to decrease to 135 thousand tons in 2014 from the peak of 351 thousand tons in 1982 (Fig. 3.14).

A characteristic of the fisheries in Seto Inland Sea is that a large portion of the catch was anchovy and sand lance, which are regarded as planktivorous fish. Fluctuations in the annual total catch of fish are well correlated with the catch of planktivorous fish (Fig. 3.14). The most important target species is the Japanese anchovy *Engraulis japonicus* in Seto Inland Sea, followed by the western sand lance *Ammodytes japonicus*. The catch of both species in 2012–16 exceeded 50% of the total catch in Seto Inland Sea and 60% of fish catches (Annual Statistics of Fishery and Fish Culture). Therefore, both the anchovy and sand lance are important as target species and play an important role as a food for higher trophic piscivorous fishes. In this section, we describe the production and conservation of these planktivorous and the piscivorous fishes.

3.3.1 Stock Characteristics and Conservation of Anchovy

Japanese anchovy of Seto Inland Sea group is a target for stock assessment by the Fisheries Agency and the Fisheries Research Agency of Japan, and stock levels and trends have been evaluated. This stock contains not only groups in Seto Inland Sea, but also groups in the Pacific Ocean (Kono and Takahashi, 2018). The spawning season of this species is from May to October in Seto Inland Sea. The number of spawned eggs, as estimated based on the monthly surveys by each prefecture, is large in most areas, suggesting no specific spawning ground formation. It has been pointed out that the variation in the survival rate before recruitment is large because there is no significant relationship between the number of eggs spawned and recruits (Kono and Takahashi, 2018). Many unidentified factors affect the survival of the

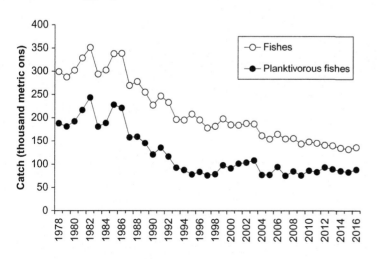

FIG. 3.14 Changes in the annual catch of fish and planktivorous fish in Seto Inland Sea. The sum of the catch of Japanese sardine (*Sardinops melanostictus*), round herring (*Etrumeus teres*), Japanese anchovy (*Engraulis japonicus*), larval anchovy and sardine, and western sand lance (*Ammodytes japonicus*). Data were obtained from Annual Statistics of Fishery and Fish Culture.

anchovy. No information is available on the composition ratio of the group originating from the Inland Sea and the group from the Pacific Ocean.

In Seto Inland Sea, the proportion of anchovies in the catch is very high, but in the 1990s the stock of Japanese sardine *Sardinops melanostictus* was also large. Sardines, however, were almost absent in Seto Inland Sea in the mid-2000s. On the Pacific coast, declines in anchovies and increases in sardines have continued since 2010, and attention must be paid to the stock dynamics of both species in Seto Inland Sea.

3.3.2 Resource Characteristics and Preservation of Sand Lance

Western sand lance is a particularly important fish species in Seto Inland Sea. In the eastern region, especially in Hyogo Prefecture, boiled juvenile sand lance (Shinko) is a spring feature. This species joins in the anchovy as a target of stock assessment in the eastern region. The annual catch of the sand lance in Seto Inland Sea was the lowest ever in 2017, and the serious reduction in this stock has become a concern.

Sand lances burrow into the sand, estivating without feeding during the high-water temperature period of summer to autumn. After estivation, they spawn demersal eggs on the bottom sediment, making a suitable sandy environment extremely important as a habitat, a ground for estivation, and a spawning ground for this species. The extraction of gravel and sand, which began in the 1960s, however, has caused serious impacts on the sand lance in Seto Inland Sea. In the Hyogo Prefecture, the extraction of sand was terminated in 1965, and many sandy grounds preferable for the sand lance remain. In other parts, especially in central and western Seto Inland Sea, the sand extraction continued until the 2000s, and the stock of sand lance collapsed to an extremely low level. In these areas, fisheries for sand lance are locally performed at small scales. To conserve the sand lance stocks, it is essential we elucidate habitats and spawning grounds currently being preserved and to understand the environment necessary for the sand lance.

To ascertain the reproductive status of the sand lance, we collected larvae after hatching over a wide area of Seto Inland Sea (Shigematsu et al., 2017). A net with a 1.3-m mouth opening was used for collection. Of the collected larvae, 81% were sand lance. The sand lance was more abundant in the eastern area (Bisan Strait to Harima-Nada) than the central to western Seto Inland Sea (Aki-Nada to western Hiuchi-Nada). Longitudinal variation also was found in body size of larvae between the east and west, and the total length of the larvae tended to be greater in the eastern area (Fig. 3.15). Furthermore, this tendency for larger size in the eastern area also was observed in rockfish, such as the *Sebastiscus marmoratus* and *Sebastes inermis* species complex that were collected during the same survey. Because the water temperature in winter is lower in the eastern part of the sea, the longitudinal variation in the larval body size is thought to be caused by the difference in the timing of spawning (parturition in rockfish).

The dynamics of the catch of sand lance varied greatly between the east of Bisan Strait and the west of Bingo-Nada. Therefore, it was suggested that there are two regional groups of sand lance in the Seto Inland Sea, a western group mainly inhabiting the area around the coast of the Hiroshima Prefecture and an eastern group mainly inhabiting the Harima-Nada. Scientific evidence supporting the existence of multiple regional groups of sand lance in Seto Inland Sea, however, does not exist. To confirm the population structure of sand lance in Seto Inland Sea, we analyzed genetic population structure using microsatellite DNA, which is useful for elucidating the genetic population structure of wild animals, and was newly developed for sand lance (Table 3.2) in advance of the analysis (Shibata et al., 2018). The results showed that the groups of sand lance collected from the western part of Seto Inland Sea (Hiroshima-Ehime

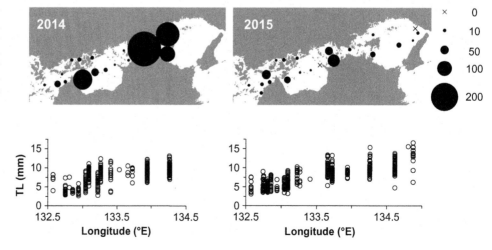

FIG. 3.15 Distribution and body size of larval sand lance in Seto Inland Sea. The upper panel shows the number of collected individuals per 5- min tow of ring net in January 2014 and January 2015, and the lower panel shows the relationship between longitude and total length of collected larvae. Shigematsu et al. (2017) was modified.

TABLE 3.2 Characteristics of Novel Developed 11 Microsatellite DNA Loci for Sand Lance, *Ammodytes japonicus* (Shibata et al., 2018)

Loci ID	GeneBank Accession No.	Repeat motif	Primer sequence (5′–3′) F	R	Annealing temperature (°C)	Number of alleles	Allele size (bp)
Amm001	LC075600	$(AG)_{10}$	CGGTTGCCTCTTGTCTCATC	CAGGGCGACTCATTTCCATTC	59	6	115–125
Amm008	LC075605	$(ACAG)_{14}$	ACCTACACACAGTGAAAGGTACCTTGC	GCTCTTGGCTGATCTGCATGCAAC	59	14	269–337
Amm016	LC075613	$(GT)_{10}$	GCATGTGCAATAGTGTGAATAAGTGTGG	TGAGCCATGGGTGTAGGCAGTG	59	8	119–133
Amm017	LC075614	$(GT)_{15}$	TGTCCTCTCCATCTGCTCTGTGTCTC	TACCATTGAGCATGAGCAGGCCG	59	18	148–212
Amm021	LC075618	$(GT)_{15}$	GGAAACCCAATTTGCTGCCAACAGTC	CTCCTGTCCATAATGGCCTTGAAGACC	59	22	155–211
Amm022	LC075619	$(AGAT)_{11}$	CAGCCATCACATGCTTCTGCTGC	GCTGCAGAAACAGATGATTGTCAAGAGG	59	18	131–207
Amm024	LC075621	$(GGCT)_{12}$	CCCACGCCAAATGCAGCACATATC	TGGAAGTTGCTGACTCACCTCTGC	59	16	154–218
Amm025	LC075622	$(AAAC)_6$	AGGACTAGGAGGAATAAAGACCGAGGC	TTTAATGCCCGCCAGTGGTGTTTGTC	59	5	184–200
Amm027	LC075624	$(ATCC)_6$	TCGGAGAAATGGCTGGAGGC	TCGCTCACTTAGTCGCAGATTGAC	59	7	169–201
Amm028	LC075625	$(AG)_9$	GGGAGTCACACAAGACACAGCCTTAAC	TTCATTGGGACCATCTTTCTCCCGC	59	6	202–212
Amm029	LC075626	$(AG)_9$	TGTTTGATCTTTATGATGGTGGCAGAGC	AACGTGACGCTATAGTGCCGTC	59	8	187–205

coast) and eastern part (Hyogo-Osaka coast) were not genetically different. Evaluation results used the fixation index F_{ST}, an index of genetic differentiation between the two groups. An F_{ST} of 1 indicates that the two groups are completely genetically differentiated and that of 0 indicates they are genetically identical. The F_{ST} value between the eastern and western groups of sand lance was a very small value of 0.005, which was statistically not different from 0, that is, it confirmed that sand lance inhabiting the eastern and western regions of the Seto Inland Sea were genetically identical. Even if there was a regional difference in the dynamics of catches of sand lance in the Seto Inland Sea, they belong to a single group, in which individuals migrate between the eastern and western parts of the sea to some extent. This supports the expectation of the natural recovery of the sand lance population by improving environmental conditions of the habitat, even in the western area where the catch of the sand lance has reduced drastically. Therefore, locating the main spawning areas of sand lance and the elucidation of suitable characteristics of sand lance habitat and spawning grounds will greatly contribute to the conservation and recovery of sand lance stocks in the Seto Inland Sea. Furthermore, because microsatellite DNA markers can be used for assessment of population size as well as for understanding the genetic population structure, the microsatellite DNA markers also are expected to provide useful indices for monitoring the change in the sand lance stocks following future stock management.

Sand lance are caught mainly in the eastern Seto Inland Sea, and a relatively large-scale estivation and subsequent spawning ground remain there. Little is known, however, about the scale of estivation and spawning grounds remaining in the central to the western Seto Inland Sea, where sand lance stock has collapsed because of sand extraction. We estimated the possible spawning grounds of the sand lance using particle tracking simulation and the information from sand lance larvae collected in the larval surveys. In this simulation, the number of days after hatching was determined from the body size of the sand lance larvae, and the starting point of the collected larvae at each sampling location was assigned as the spawning ground. It was proposed that several small-scale spawning grounds have remained around Mihara Strait in the Aki-Nada in the central area (Fig. 3.16).

To evaluate the current conditions of the sandy grounds of the Mihara Strait as potential estivation grounds of the sand lance, we conducted sediment samplings at 11 survey points using a square-shaped dredge towed by the training vessel Toyoshio-maru of Hiroshima University in 2014 and 2015. We subsequently analyzed the physical characteristics of the sediment samples. Sand lance individuals were observed in the substratum sediments at a five survey points, with three on the western side and two on the eastern side of the Mihara Strait (Fig. 3.17). Sediments at these five survey points commonly were composed of sandy elements of 0.25–4.0 mm diameter, which constituted >80% of the weight ratio of the sediment samples, being similar to the conditions of estivation grounds in Harima-Nada, eastern Seto Inland Sea (Sakai et al., 2018). Sediments from the other survey points of Mihara Strait and especially zones

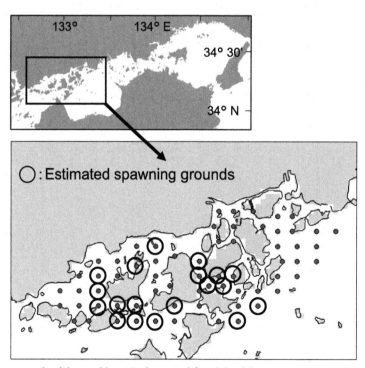

FIG. 3.16   Estimated spawning grounds of the sand lance in the central Seto Inland Sea.

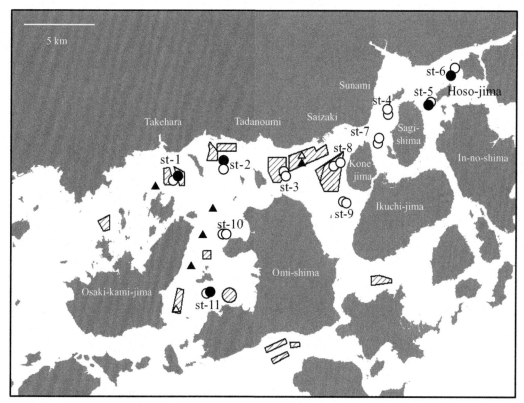

FIG. 3.17 Dredge survey points for estivation grounds of the sand lance in the Mihara Strait (circles: st-1 to st-11), mid-western Seto Inland Sea, Japan. Shaded areas indicate bottom sands in quarrying areas in the past (prohibited by 1997 in Hiroshima Prefecture). Solid circles indicate the points where the sand lance individuals appeared in the sampling sediments. Triangles indicate dredge survey points in 2015 by Hiroshima Prefecture (solid triangles indicate the sand lance collection points) (Hiroshima Prefecture, 2017). *Redrawn after Sakai, Y., Endo, A., Iwasaki, N., Tomiyama, T., Shibata, J., Yamaguchi, S., Nakaguchi, K., 2018. Estivation grounds of the sand lance* Ammodytes japonicus *(Ammoditidae) in the Mihara Strait, mid-western Seto Inland Sea, Japan. Bull. Hiroshima Univ. Museum 10, 19–27 (in Japanese with English abstract).*

around Tadanoumi and Sunami conspicuously included pebble elements and had a significantly lower weight ratio of sandy elements, suggesting the unsuitability of the ground conditions of these points for burrowing by the sand lance (Sakai et al., 2018). Dredge surveys by Hiroshima Prefecture's research team also confirmed the sand lance occurrence in sandy sediments in the western side of the Mihara Strait (Fig. 3.17) (Hiroshima Prefecture, 2017), which, together with our results, emphasized the importance of the zone of the habitat for the sand lance population. Therefore, it is suggested that sandy grounds available for estivation of the sand lance currently are limited geographically in the Mihara Strait, which might restrict population recovery of the sand lance in these waters.

The sand lance in Seto Inland Sea is known to burrow into the ground for rest (or escape) and for long-term estivation over 120 days during the high-water temperature season in the fine sandy grounds. It is crucial to understand the suitable and preferable sandy conditions for the recovery of the sand lance populations, usually fine sandy grounds of 0.5–2.0 mm grain diameter (Tanda, 1998). However, no data are available on the preference of physical properties of bottom sediments by the sand lance in the western Seto Inland Sea populations. To test the habitat preference of the sand lance, we conducted aquarium experiments that exhibited natural bottom substrates in the Mihara Strait (Endo et al., 2019).

We obtained sand lance individuals of the 1+ age class commercially captured off Hojo, Matsuyama, and Ehime Prefecture in southern Aki-Nada, for the experiment. We laid 20 natural sediment samples in glass tanks filled with seawater [26 × 15 × 17 cm (H)] to a depth of 8 cm. The water temperature in the tanks was controlled at the range of 15–17°C to prevent estivation. We randomly chose eight individuals and transferred each to an experimental tank; the individuals that were burrowing were counted 30 min later. The procedure was repeated six times for each sediment sample, and a mean value of the burrowing rate was calculated. We measured the following properties of the sediment: grain component factors, i.e. median grain size and sorting, shell content rates, and share strength value. Share strength of bottom sediments represents the penetration resistance of the sediment and has been applied recently in evaluation of habitat conditions of sand burrowing benthic animals.

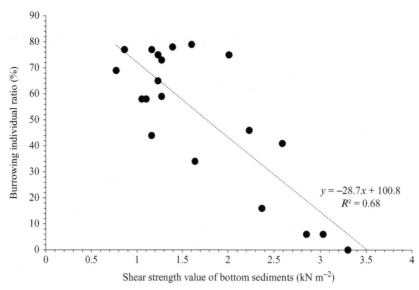

FIG. 3.18 Relationship between the burrowing individual ratio of the sand lance and the shear strength values of bottom sediments in the aquarium experiment (after Endo et al., 2019). A linear regression line representing the best-fit model by the stepwise multiple regression analysis is shown by a dotted line.

Using a stepwise multiple regression analysis, it was confirmed that the share strength was the most important property for explaining variation in the burrowing ratio. The burrowing rate showed a peak in sediments with low share strength values (Fig. 3.18), suggesting that ease of burrowing penetration seriously affected habitat preference of the sand lance. Because the sand lance dives into the sand by rushing headfirst into sediment using swimming force, a small penetration resistance of sediment severely affects the ease of burrowing. The median grain size of sandy substrates previously has been thought to be the most important factor in sediment preference of the sand lance, but the share strength value is a new aspect that contributes greatly to our understanding of the suitability and availability of sandy grounds for estivation and burrowing by the sand lance.

3.3.3 Production of Piscivorous Fish

The sand lance has its own economic value as an important fishery resource in Seto Inland Sea and is an important prey species for many fish. In Sendai Bay (Togashi et al., 2015), Ise Bay (Uzaki et al., 2014), and Seto Inland Sea, consumption of sand lance has been observed by at least 21 fish of 16 families in seven orders (Table 3.3). These predators include important target species in coastal fisheries in Seto Inland Sea, indicating the contribution of the sand lance to the production of these predatory fishes. Therefore, protecting sand lance resources in Seto Inland Sea contributes not only to the sand lance itself, but also to raising and maintaining the production of many fishery fish.

In the Seto Inland Sea, many fishery fish, such as Japanese Spanish mackerel *Scomberomorus niphonius*, Japanese amberjack *Seriola quinqueradiata*, Japanese flounder *Paralichthys olivaceus*, and daggertooth pike conger *Muraenesox cinereus* are important catch target species. The decrease in the catch of all fish that occurred in Seto Inland Sea could be explained by the reduction of planktivorous fish at least until the mid-2000s. In spite of the relatively constant catch of planktivorous fish after the latter half of the 2000s, however, the catch of all fish has decreased (Fig. 3.14). This reduction is because of the decrease in fish at higher trophic levels than that of the planktivorous fish. In particular, the annual catch of the cutlassfish *Trichiurus japonicus* has declined drastically from >10,000 t in 2007 to <4000 t after 2015. The catch of the cutlassfish in Seto Inland Sea accounts for about half of its catch nationwide, and therefore we should learn the factor causing this decline for the sustainable fisheries of the cutlassfish.

Analysis of the stomach contents of the cutlassfish caught by bottom trawls from Hiuchi-Nada, the central sea area, showed that the anchovy were the most prevalent food, with the sand lance and the shrimp *Leptochela gracilis* also of high importance (Fig. 3.19) (Niino et al., 2017). The importance of sand lance and shrimp was shown first in this study, although feeding habits of the cutlassfish have been reported from several areas, including around Iyo-Nada, in the western sea. Because all three species support the production of the cutlassfish, it is necessary to consider their production as well as the production of higher trophic predators.

TABLE 3.3 Predatory Fishes for Sand Lance Around Japanese Coastal Area

Order	Sub-order	Family	Species	Sendai Bay[Togashi et al., 2015]	Ise Bay[Uzaki et al., 2014]	Seto Inland Sea (● Niino et al., 2017, ○ Shibata et al. private communication)
Auguilliformes		Cogridae	*Conger myriaster*			○
Aulopiformes	Aulopoidei	Synodontidae	*Trachinocephalus myops*		●	
			Saurida elongata		●	○
Gadiformes		Gadidae	*Gadus macrocephalus*	●		
Syngnathiformes	Syngnathoidei	Fistulariidae	*Fistularia petimba*		●	
Perciformes	Scorpaenoidei	Scorpaenidae	*Inimicus japonicus*			○
			Sebastes spp.[a]			○
		Triglidae	*Lepidotrigla microptera*	●		
			Chelidonichthys spinosus		●	
	Percoidei	Lateolabracidae	*Lateolabrax japonicus*	●	●	○
		Carangidae	*Seriola quinqueradiata*			○
		Haemulidae	*Plectorhinchus cinctus*		●	
		Sparidae	*Acanthopagrus schlegelli*		●	
		Sciaenidae	*Pennahia argentata*		●	
	Cottoidei	Hexagrammidae	*Hexagrammos otakii*	●		
	Scombroidei	Trichiuridae	*Trichiurus japonicus*			●
Pleuronectiformes		Paralichthidae	*Paralichthys olivaceus*	●	●	
		Pleuronectidae	*Platichthys bicoloratus*	●		
			Pseudopleuronectes herzensteini	●		
			Pseudopleuronectes yokohamae	●		
Tetraodontiformes	Tetraodontoidei	Tetraodontidae	*Lagocephalus* spp.[a]		●	

[a] *Lagocephalus* spp. and *Sebastes* spp. ,which were not identified in species level, were counted as one taxon, respectively.

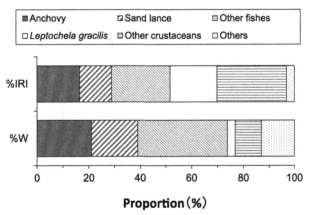

FIG. 3.19 Feeding habits of the cutlassfish *Trichiurus japonicus* in the central part of Seto Inland Sea. A total of 465 individuals were used for analysis. Data are shown by the weight proportion and the index of relative importance (index combining occurrence frequency, proportion in number, and proportion in weight). Niino et al. (2017) was modified.

3.4 NUTRIENT MANAGEMENT IN SETO INLAND SEA

Wataru Nishijima

We should consider the nutrient management in the sea from two viewpoints: environmental conservation to reduce various adverse effects from eutrophication and maintenance productivity of the fisheries. From the viewpoint of environmental conservation, we should solve various environmental problems, such as red tides and the deterioration of sediment, which occur in the warm season primarily because of the excessive growth of phytoplankton. From the viewpoint of productivity, aquacultures, such as oyster and clam farming and seaweed culturing, should be considered, as well as coastal fisheries. In Seto Inland Sea, sea laver (*Pyropia yezoensis*) and *Undaria pinnatifida* that directly depend on nutrients are cultured, and their growth period is the cold season. Oyster farming (*Crassostrea gigas*) is also an important fishery in Hiroshima Bay and indirectly depends on nutrients through the use of phytoplankton as food. The growth period of the oyster is also the cold season. It is known that lack of nutrients during the growth period is the cause of decreases in oyster production and discoloration of sea laver. From the viewpoint of maintaining the productivity of higher trophic level species, the productivity of food, especially zooplankton, must be maintained or enhanced.

3.4.1 Characteristics of Waters

A large difference in freshwater input nutrients is found among areas in Seto Inland Sea because of complicated geographic features. In highly enclosed areas, such as Osaka Bay and Hiroshima Bay, with an urbanized watershed, the phytoplankton concentration is high as shown in Section 3.1.2. Seto Inland Sea generally is classified as bay and nada, and there is a large difference in characteristics between coastal and offshore areas even in the same bay or nada. To control phytoplankton, we must know the characteristics of phytoplankton growth.

Expected important factors that are different among areas in Seto Inland Sea include the physical factors of water temperature, light condition, and stratification, and the chemical factors of nutrient concentration and salinity. The distribution of important physico-chemical factors, except for nutrients in summer when excess phytoplankton growth is expected, is shown in Fig. 3.20. Stratification was expressed by the simplified Brunt Väisälä frequency (N^2) calculated by the difference of density between the surface and bottom. Salinity showed a similar tendency as stratification and Secchi depth. Low salinity, low Secchi depth, and strong stratification were observed in the northeastern part of Osaka Bay, the northern part of Hiroshima Bay, and the northern coastal areas (Honshu) of Seto Inland Sea. Water temperature, however, showed a different tendency from other physico-chemical parameters. Water temperature was high in high enclosed bay and coastal areas the same as other parameters, was low from Bungo Channel to Iyo Nada and Aki Nada, and was high in Hiuchi Nada.

The frequency of red tide occurrence in the past 10 years was calculated at the monitoring sites from 2001 to 2010 based on a report by the Japan Fisheries Agency Setonaikai Fisheries Coordination Office (Fig. 3.21). Although the red tide map in the report does not indicate the accurate area, we know the approximate areas where red tides occurred. The areas where red tides occurred was frequently coincident with those with low salinity, low Secchi depth, and

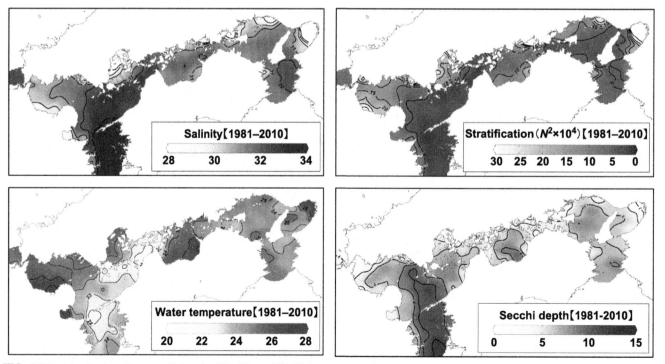

FIG. 3.20   Distribution of mean salinity, simplified Brunt Väisälä frequency, surface water temperature, and Secchi depth in summer from 1981 to 2010.

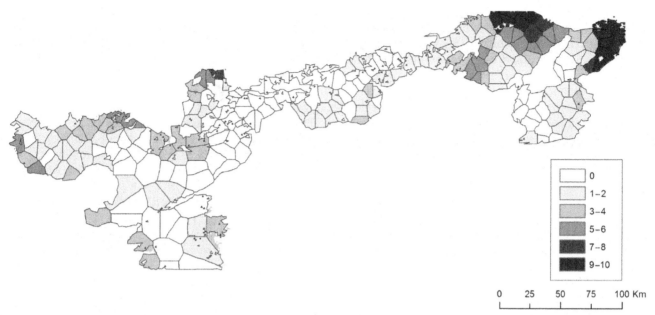

FIG. 3.21 Frequency of red tide occurrence from 2001 to 2010.

strong stratification. Because nutrients from land are supplied to the sea primarily by river water, groundwater, and wastewater treatment effluent, salinity is a good indicator. In the figure, 30.47 of salinity is the mixed ratio of 10% freshwater with seawater, and this value is the mean salinity on the bottom in the most southern three sites of Bungo Channel. The area with salinity values higher than 30.47 was the most frequent area for red tides.

Occurrence rate of red tide in Seto Inland Sea decreased from 299 in 1976 and reached <100 in 1994, and then maintained the same level. The recent 5 year-mean occurrence number of red tides was 82 (2013–17). Although the area and period of red tides were different from before, red tides still are occurring even though nutrient loading from land is decreasing. The frequent occurrence areas for red tides, however, are limited, as are the coastal areas with high freshwater input. We should take specialized measures in such areas.

3.4.2 Management of Nutrient Loading From Land

Environmental conservation is achieved through control of excess phytoplankton growth by reduction of nutrient loading from land; productivity is based on primary production affected by the nutrient conditions. Consequently, these two concepts are incongruent regarding nutrient levels. Environmental conservation and productivity maintenance or enhancement, however, can be consistent if we focus on the period and area emphasized in the management. In the period, nutrient concentration is kept low to avoid excessive growth of phytoplankton in the warm season, but it is kept high to enhance productivity. In the area, nutrient concentration is reduced in the coastal areas where nutrients are likely high, but it is increased in the offshore area where nutrients are likely insufficient. In general, it is difficult to increase nutrient concentration offshore without an increase in the coastal area. We should consider measures to reduce nutrient concentration without decreasing it offshore.

Although it is difficult to seasonally change nutrient loading from land, an operation that changes nutrient concentrations in discharged water seasonally has been tried in many sewage treatment plants. Called seasonal operation for nutrient control, it was started to improve discoloration in sea laver that was caused by a lack of nitrogen, and therefore, its objective is to discharge nitrogen from autumn to winter, the growing season for sea laver. To produce nitrogen, denitrification is restrained by operating a part of the denitrification reactors aerobically, resulting in nitrates that are discharged, or aeration is reduced in aerobic reactors to restrain nitrification resulting in ammonium nitrogen that is discharged. The nitrogen concentration increases to about twice that of the normal operation using the seasonal operation for nutrient control. The operation for the increase in phosphorus concentration in the discharged water is not conducted frequently. In general, phosphorus is removed by coagulation or biological technology. In the former case, the amount of coagulant is reduced to increase phosphorus concentration in the discharged water. In the latter case, excess phosphorus uptake in the aeration reactor or discharge in an anaerobic reactor is conducted. Currently, the seasonal operation is conducted to increase nitrogen or phosphorus concentration in the discharged water in the cold season. We should, however, consider the reduction of nutrient loading in warm season from the viewpoint of environmental conservation. It is also important to manage untreated sewer overflow and discharge into the sea during heavy rains to control nutrient load. Sewer overflow occurs frequently in the warm rainy season, and therefore, reducing the nutrient load caused by sewer overflow is a major challenge for the future.

Currently, the discharge of nitrogen and phosphorus through the sewer system accounts for 26.0% and 26.3% of the total loading from land into Seto Inland Sea, respectively (Setouchi Net, n.d.). The seasonal operation can be applied in advanced wastewater treatment plants. The rate of domestic wastewater treated in advanced wastewater treatment plants to total domestic wastewater was about 41% in 2013. The rate is not complete in the watershed of Seto Inland Sea; the rate, however, was expected to be higher than the mean value, about 41% in Japan, because of strict regulations applied in the area. The rates of nitrogen and phosphorus loadings through a sewer system to total loading to Seto Inland Sea are high and are 33.4% and 40.5% that of domestic loading, respectively. The seasonal operation of sewage treatment plants will be a measure with a certain impact.

It makes less sense, however, that the nutrient management for loading from land, including the seasonal operation, is conducted for the fertilization of seaweed aquaculture operated in a limited area. The seasonal operation for nutrient control should be implemented for environmental conservation and productivity management for the entire Seto Inland Sea. The operation should be implemented to overcome various problems mainly caused by excess phytoplankton growth during the warm season and enhancement of lower trophic level production supporting higher trophic level production. In the measure for discoloration of sea laver, direct fertilization will be more effective and will be considered. It is necessary to consider the method of fertilization for sea laver from the viewpoint of environmental conservation because nutrients that are not used by sea laver will be new loading in the sea.

3.4.3 Management of Nutrients Using an Ecological Method

It is also possible to reduce the amount of nutrients used for phytoplankton in the warm season and increase them in the cold season using an ecological method. The method is to develop habitats for seagrasses or benthic macroalgae intensively in places where a large amount of nutrients flows into the sea, such as river mouths, discharge points in sewage, and industrial wastewater treatment plants, and coastal areas with submerged groundwater discharge. The seagrasses or benthic macroalgae are expected to uptake nutrients in the warm season in coastal areas where the nutrient concentration is relatively high and prevent excessive growth of phytoplankton as a result. After the growing season, they will die and flow out widely and gradually discharge nutrients by decomposition in the cold season.

Zostera marina is one of the common seagrasses growing on sand-mud sediment in the coastal areas of Japan, including Seto Inland Sea, and is suitable for this ecological method for nutrient management. *Z. marina* is also a common species in Europe and North America and is known to grow from 15°C to 20°C and die and decline in high temperatures (Lee et al., 2007).

Monthly variation of water temperature at 5 m depth in Hiroshima Bay from 2013 to 2017 is shown in Fig. 3.22. The 5 m depth water temperature ranged from 15°C to 20°C from May to July in Hiroshima Bay, and *Z. marina* has a life cycle that starts with growth from March–April and declines beginning in August (Fig. 3.23). The start time and period of each stage will be a little different depending on the habitat, but the growth period of *Z. marina* is similar to that of

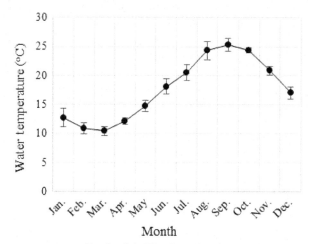

FIG. 3.22   Monthly variation of water temperature at 5 m depth in Hiroshima Bay from 2013 to 2017.

FIG. 3.23 Life cycle of *Zostera marina* in Hiroshima Bay.

phytoplankton. Consequently, Z. *marina* and phytoplankton will compete for nutrients and extension of Z. *marina* habitats is expected to prevent excessive growth of phytoplankton in the warm season.

Seagrasses, including Z. *marina*, have nutrient storage strategies, that is, the uptake of nutrients when available, and storage and use of reserves when environmental conditions become adequate for plant growth. The maximum nitrogen content in the root, rhizome, and leaf of *Zostera noltii* and *Cymodocea nodosa* were twice the minimum in an investigation in the Mediterranean Sea (Kraemer and Mazzella, 1999). Although the month with the maximum value is different, the minimum value is from June to July in all areas, indicating that nutrients stored in the cold season are used for growth in the warm season. Moreover, rapid and temporal uptake of nutrients when nutrient concentration increases in water from rain is observed in Z. *marina* as mentioned in Fig.3.3. The rapid uptake of nutrients will depend on nutrient concentration in water and nutrient content in tissues, suggesting the possibility to show strong uptake of nutrients in the growing period when nutrient content is relatively low.

Many individuals in seagrass beds will die in the declining stage, and dead leaves flow out to be carried widely. The nutrients in dead leaves will be released into the water by decomposition, and the rate is reported to be 0.26%–1.0% day^{-1} (Harrison, 1989). The dead leaves will remain for a few months based on the decomposition rate, indicating that nutrients will be released from dead leaves continuously, like a slow-release fertilizer. A part of tissues of Z. *marina*, however, is known to be stored in sediment for a long time as the so-called blue carbon.

To summarize the advantages of the method using seagrasses or benthic macroalgae:

1) It is possible to remove nutrients before diffusion to wide areas because nutrients from land are carried offshore through coastal areas that are also the habitat of seagrasses or benthic macroalgae.
2) The nutrient uptake is enhanced when the nutrient concentration in the water is high.
3) The nutrients are finally released to water slowly in wide areas during the cold season when the risk caused by excessive growth of phytoplankton is low.
4) The available nutrients for phytoplankton are reduced in the warm season and are increased in the cold season, preventing excessive growth of phytoplankton and supporting annual primary production through the increase in primary production during the cold season.

3.4.3.1 Effectiveness of Nutrient Reduction by Z. marina

Primary production of Z. *marina* during the growth period has been estimated in some areas of Japan, including Seto Inland Sea. The estimated production ranged from 2.2 to 7.8 gC m^{-2} day^{-1} (Nishijima, 2016), although the estimation method varied. I estimated reduction to nutrient loading from land by Z. *marina* during the growing period based on these reported productions.

The nitrogen and phosphorus uptake rates converted from the carbon uptake rate (primary production rate) were 129–459 mgN m^{-2} day^{-1} and 8.6–30.6 mgP m^{-2} day^{-1}, respectively, if the C:N:P ratio in tissues is assumed to be 255:15:1 (Duarte, 1990). Nitrogen and phosphorus loadings from land into Seto Inland Sea are 16.5 kgN km^{-2} day^{-1}

(374 t N day^{-1}) and 1.02 kg P km^{-2} day^{-1} (23.2 t P day^{-1}), except for Hibiki Nada in 2009. The seagrass bed area, except for Hibiki Nada, is 6381 ha based on the investigation in 1989–90. The nitrogen and phosphorus uptake rates by the seagrass bed for total loading from land were estimated to be 2.2%–7.8% and 2.4%–8.4%.

The seagrass bed has been reduced in the Seto Inland Sea from 22,635 ha, except for Hibiki Nada, in the 1960s. If the area of seagrass beds in the 1960s could be recovered, the nitrogen and phosphorus uptake rates relative to total loading from land would be an estimated 7.8%–27.8% and 8.4%–29.9%, respectively. The uptake potential of *Z. marina* for nutrients is not small, and it is practical to prevent excessive growth of phytoplankton by the increased ability of nutrient uptake by *Z. marina* in the warm season and to release the nutrients from dead leaves in the cold season.

3.4.3.2 Evaluation on Prevention of Excessive Growth of Phytoplankton by Extension of Seagrass Bed in Northern Part of Hiroshima Bay

We evaluated the degree of phytoplankton growth that was prevented by the increase in *Z. marina* beds in the northern part of Hiroshima Bay. The evaluation was conducted in May using the mathematical model for Seto Inland Sea developed in this project. The coastal area *Z. marina* beds that could be developed were defined as areas where >20% of surface photosynthetically active radiation was reached on the surface of the sediment and no other conditions, such as flow rate, were not considered. The PAR on the surface of the sediment was estimated from the light attenuation coefficient K_d calculated using Eq. (3.4) as shown in Section 3.1.2 using Secchi depth and water depth measured at 50 m intervals. We validated this model in the natural coastal area of Miyajima where *Z. marina* beds were widely developed and confirmed that the model could estimate the area of *Z. marina* beds with a 10% error rate. The current area of *Z. marina* beds is 100 ha in the northern Hiroshima Bay, and the possible habitat areas for *Z. marina* is estimated to be 370 ha, which is about 2.3% of the 16,000 ha of the entire areas of northern Hiroshima Bay. Because most of the coastal lines in the inner part of the bay are reclaimed and deep, most of the potential habitat areas for *Z. marina* is located around islands.

When *Z. marina* grows in all possible habitats and competes with phytoplankton for nutrients, Chl.*a* concentration is estimated by the mathematical model and compared to Chl.*a* concentration present in May (Fig. 3.24). We considered the difference in nutrient uptake by *Z. marina* between depths. Although little potential habitat area for *Z. marina* exists in the inner part of the bay where Chl.*a* concentration is high, Chl.*a* concentration is clearly reduced from 6.0–6.5 to

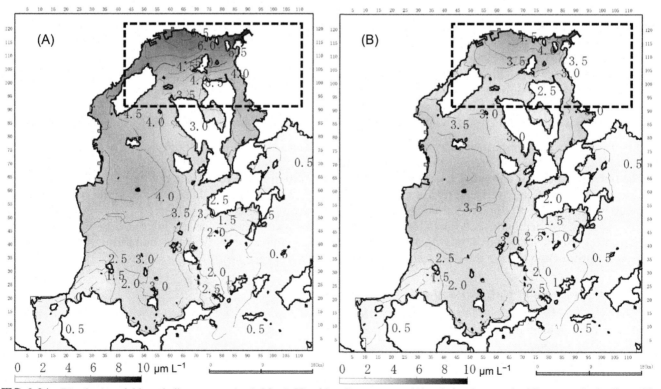

FIG. 3.24 Distribution of chlorophyll *a* concentration in May in Hiroshima Bay at present (A) and after the growth of *Zostera marina* in all possible habitats in the area surrounded by the broken line (B).

2.0 µg L^{-1} because of the high retention area. About 1.0 µg L^{-1} of reduction in Chl.a concentration is observed in the other inner part of the bay. Moreover, the extension of $Z.$ $marina$ affects Chl.a concentration in the central bay that is out of the extension area for $Z.$ $marina$, resulting in a Chl.a concentration in the area that is reduced by about 0.5 µg L^{-1}. The southern area of the bay that is connected to Aki Nada, however, would not be strongly affected by nutrient loading from land through Hiroshima Bay, showing <2.0 µg L^{-1} of Chl.a concentration after the extension of $Z.$ $marina$ beds, which is not different from the current value.

3.5 SUMMARY AND PROSPECTS

Wataru Nishijima

The entire Seto Inland Sea area was not in the state of eutrophication even in the 1980s when pollution was severe under economic and population growth. Eutrophication progressed in coastal areas, mainly enclosed waters, such as Osaka Bay and Hiroshima Bay, whereas some parts of offshore areas and waters strongly affected by the open sea were not eutrophic and maintained low phytoplankton concentrations. As a result of the reduction in nutrient loading from land, Chl.a concentration and primary production decreased remarkably in the areas where eutrophication progressed with excess phytoplankton growth, whereas little decrease in primary production was observed in the areas where Chl.a concentration was low, even in the 1980s. From the viewpoint of transfer efficiency as expressed by the rate of secondary production to primary production, it is not true that higher primary production always ensured higher secondary production. Too high primary production did not enhance secondary production, and, as a result, excess production not used by secondary consumers had negative impacts, such as red tide occurrence and sediment pollution. The reduction in nutrient loading from land could have contributed to environmental conservation through the prevention of excessive growth of phytoplankton in coastal areas.

As an indicator of higher trophic level production, fish catch has declined to approximately one-third after 2014 compared with the approximately 350,000 t year^{-1} peak year. The fishery in Seto Inland Sea is characterized by high rates of sardines, anchovies, and sand lance as planktivorous fish. The decline of the catch from the 1980s to 2000 was caused by the decrease in planktivorous fish. After 2000, the catch of planktivorous fish was maintained, and the decline in the catch of higher trophic level fish occurred. The decline in the catch of planktivorous fish from the 1980s could be explained by the remarkable decline in sardines. Sand lance, which continued to decline after 2000, has a habit of "summer sleep" in the sand during the warm season and produces eggs in the bottom sediment. The strong dependency of sand lance on the bottom sediment affects its survival, as shown by changes in the sediment environment because of sea sand collection conducted at a large scale in many sites of Seto Inland Sea. It would be difficult to explain the decline in fish catch from the 1980s by a series of changes from nutrients to primary production and secondary production. We should consider the causes of the catch decline from various aspects, such as species alternation of small planktivorous fish repeated on a decade timescale, decreases in seagrass and macroalgae meadows, and the deterioration of the sediment, and then we should implement proper measures.

It is evident that planktivorous fishes support piscivore fishes, but the food analysis of cutlassfish *Trichiurus lepturus*, a popular piscivore fish in Seto Inland Sea, revealed that *Leptochela gracilis* was an important food, as well as anchovy and sand lance. There is little knowledge about small crustaceans, such as *L. gracilis*. We expect further research about small crustaceans as the important components supporting the ecosystem in Seto Inland Sea.

The marine environment in Seto Inland Sea has been improved since the 1980s, but there are still environmental problems, caused mainly by excessive phytoplankton growth, such as red tides, and the deterioration of sediment in the coastal areas with large freshwater input. It is not appropriate to try to solve the problems occurring in limited coastal areas by the reduction in nutrient loading from land, which will affect the nutrient condition in the entire water body. We evaluated the effectiveness of a measure in which nutrients were reduced for phytoplankton growth in the warm season and were increased in the cold season by using $Z.$ $marina$. We showed that nutrient uptake by $Z.$ $marina$ increased when the nutrient concentration in water increased from rain, indicating that $Z.$ $marina$ was a suitable plant for nutrient management using the ecological method. It is not appropriate, however, to consider seagrasses and benthic macroalgae such as $Z.$ $marina$ just as a tool for nutrient management. Seagrass and macroalgae beds are important habitats providing a varied environment for marine life in Seto Inland Sea that has a relatively monotonous sediment with a sand and mud bottom. The fact that seagrass beds have been greatly reduced in recent years has possibly changed the habitat for much marine life, including fish, possibly being a direct and an indirect cause of the decline in the fish catch. Finally, I would recommend reproduction and creation of shallow areas composed of seagrass beds or macroalgae beds as an important solution to achieve the ideal Seto Inland Sea with high productivity and healthy environmental conditions (Ichimi, 2016; Harrison, 1989; Nishijima, 2016).

References

Duarte, C., 1990. Seagrass nutrient content. Mar. Ecol. Prog. Ser. 67, 201–207.

Endo, A., Iwasaki, N., Shibata, J., Tomiyama, T., Sakai, Y., 2019. The burrowing sand lance *Ammodytes japonicus* (Teleostei, Ammodytidae) prefers benthic sediments of low shear strength. J. Ethol. 37, 213–219, https://doi.org/10.1007/s10164-019-00591-9.

Harrison, P.G., 1989. Detrital processing in seagrass systems: a review of factors affecting decay rates, remineralization and detritivory. Aquat. Bot. 35 (3–4), 263–288.

Hiroshima Prefecture (2017) Hiroshima-ken Umi-jyari Saisyu Kankyo Chousa Houkoku (Hiroshima Prefectural Survey Reports on Environmental Conditions of Waters Suffered the Bottom Sands Quarrying on the Past), 72 p. (in Japanese) https://www.pref.hiroshima.lg.jp/soshiki/102/h28umijari.html.

Ichimi, K., 2016. Mass balance of nitrogen and phosphorous in Shinkawa-Kasugagawa River estuary in the eastern Seto Inland Sea, Japan. J. Japan Soc. Water Environ. 39, 125–129 (in Japanese).

Ichimi, K., Hamaguchi, K., Yamamoto, A., Tada, K., Montani, S., 2011. Retention and release system of phosphorous in Shinkawa-Kasugagawa River Estuary in the Western Part of Japan. Bull. Coast. Oceanogr. 48, 167–178 (in Japanese with English abstract).

Kokubun, H., Tsuchihashi, Y., Takayama, Y., 2009. Quantification of nutrient flux in tidal flat and *Zostera Marina* Bed under Ebb and flow in AGO Bay, Coastal Engineering Committee. J. Japan Soc. Civil Eng. B2 (Coast. Eng.) 65, 1081–1085 (in Japanese with English abstract).

Kono, N., Takahashi, M., 2018. Stock assessment and evaluation for Japanese anchovy in the Seto Inland Sea (fiscal year 2017). In: Marine Fisheries Stock Assessment and Evaluation for Japanese Waters (Fiscal Year 2017/2018). Fisheries Agency and Fisheries Research Agency of Japan, Tokyo, pp. 833–883 (in Japanese).

Kraemer, G.P., Mazzella, L., 1999. Nitrogen acquisition, storage, and use by the co-occurring Mediterranean seagrasses *Cymodocea nodosa* and *Zostera noltii*. Mar. Ecol. Prog. Ser. 183, 95–103.

Lee, K.-S., Park, S.R., Kim, Y.K., 2007. Effects of irradiance, temperature, and nutrients on growth dynamics of seagrasses: a review. J. Exp. Mar. Biol. Ecol. 350 (1–2), 144–175.

Nakai, S., Soga, Y., Sekito, S., Umehara, A., Okuda, T., Ohno, M., Nishijima, W., Asaoka, S., 2018. Historical changes in primary production in the Seto Inland Sea, Japan, after implementing regulations to control the pollutant loads. Water Policy 20, 855–870.

Nature Conservation Bureau, Environment Agency and Marine Parks Center of Japan, 1994. The Report of the Marine Biotic Environment Survey in the 4th National Survey on the Natural Environment. Vol. 2 Algal and Sea-Grass Beds (in Japanese).

Niino, Y., Shibata, J., Tomiyama, T., Sakai, Y., Hashimoto, H., 2017. Feeding habits of the cutlassfish *Trichiurus japonicus* around Hiuchi-Nada, the central Seto Inland Sea, Japan. Nippon Suisan Gakkaishi 83, 34–40 (in Japanese with English abstract).

Nishijima, W., 2016. Conservation, restoration and creation of tidal flats and seagrass beds. J. Japan Soc. Water Environ. 39A (3), 92–96.

Nishijima, W., 2018. Management of Nutrients Concentrations in the Seto Inland Sea, Bulletin on Coastal Oceanography. Vol. 56, pp. 13–19.

Nishijima, W., Umehara, A., Sekito, S., Okuda, T., Nakai, S., 2016. Spatial and temporal distributions of Secchi depths in the Suo Nada of the Seto Inland Sea, Japan, exposed to anthropogenic nutrient loading. Sci. Total Environ. 571, 543–550.

Nishijima, W., Umehara, A., Sekito, S., Okuda, T., Nakai, S., 2018. Determination and distribution of region-specific background Secchi depth based on long-term monitoring data in the Seto Inland Sea, Japan. Ecol. Indic. 84, 583–589.

Pedersen, M.F., Borum, J., 1993. As annual nitrogen budget for a seagrass Zostera marina population. Mar. Ecol. Prog. Ser. 101, 169–177.

Sakai, Y., Endo, A., Iwasaki, N., Tomiyama, T., Shibata, J., Yamaguchi, S., Nakaguchi, K., 2018. Estivation grounds of the sand lance *Ammodytes japonicus* (Ammoditidae) in the Mihara Strait, mid-western Seto Inland Sea, Japan. Bull. Hiroshima Univ. Museum 10, 19–27 (in Japanese with English abstract)

Sasaki, K., 1989. Material circulation in intertidal flat. Bull. Coast. Oceanogr. 26, 172–190 (in Japanese with English abstract).

Sato-umi Net, n.d. Sato-umi and Ecosystem, Ministry of Environment. https://www.env.go.jp/water/heisa/satoumi/02.html.

Sato-umi Net, n.d. Setouchi Net, Ministry of Environment. https://www.env.go.jp/water/heisa/heisa_net/setouchiNet/seto/g2/g2cat03/tokusohou/hasseifuka.html.

Shibata, J., Tomano, S., Umino, T., Tomiyama, T., Sakai, Y., Nakai, S., Okuda, T., Nishijima, W., 2018. Isolation, characterization and PCR multiplexing of microsatellite loci for western sand lance (*Ammodytes japonicus* Duncker and Mohr 1939). Japan Agric. Res. Quart. 52, 307–313.

Shigematsu, Y., Ochi, Y., Yamaguchi, S., Nakaguchi, K., Sakai, Y., Shibata, J., Nishijima, W., Tomiyama, T., 2017. Winter longitudinal variation in the body size of larval fishes in the Seto Inland Sea, Japan. Fish. Sci. 83, 373–382.

Tanda, M., 1998. Relationship between the habitat of sand lance and the sandy bottom in the Seto Inland Sea. In: Hirano, T. (Ed.), Marine Coastal Environment. Fuji Techno System, Tokyo, pp. 348–355 (in Japanese).

Togashi, H., Nakane, Y., Kurita, Y., 2015. The use of stable isotopes for food web analyses. Bull. Jpn. Soc. Fish. Oceanogr. 79, 369–372 (in Japanese).

Uye, S., Murase, A., 1997. Relationship of egg production rates of the planktonic copepod *Calanus sinicus* to phytoplankton availability in the Inland Sea of Japan. Plankton Biol. Ecol. 53, 529–538.

Uye, S., Shimazu, T., 1997. Geographical and seasonal variations in abundance, biomass and estimated production rates of meso- and macrozooplankton in the Inland Sea of Japan. J. Oceanogr. 44 (1/2), 3–11.

Uzaki, N., Sawada, T., Hibino, M., Kato, T., Tani, K., 2014. Investigation of Predation for Sand Lance. FY2013 Aichi Prefecural Fisheries Experiment Station Business Report, pp. 96–97 (in Japanese).

Valiela, I., Teal, J.M., 1979. The nitrogen budget of a salt marsh ecosystem. Nature 280, 652–656.

Vidal, J., 1980. Physioecology of Zooplankton. I. Effects of phytoplankton concentration, temperature, and body size on the growth rate of *Calanus pacificus* and *Pseudocalanus* sp. Mar. Biol. 56, 111–134.

Wolvaver, T.G., Spurrier, D.J., 1988. The exchange of phosphorus between a euhaline vegetated marsh and the adjacent tidal creek. Estuar. Coast. Shelf Sci. 26, 203–214.

Zimmerman, R.C., Smith, R.D., Alberte, R.S., 1987. Is growth if eelgrass nitrogen limited? A numerical simulation of the effect of light and nitrogen on the growth dynamics of *Zostera marina*. Mar. Ecol. Prog. Ser. 41, 167–176.

CHAPTER

4

International Management
of Marine Environment

Haejin Kim, Katsumi Takayama*, Naoki Hirose*, Akihiko Morimoto†,
Ryota Shibano†, Masashi Ito†, Tetsutaro Takikawa‡, Xinyu Guo†,
Yucheng Wang†, Taishi Kubota†, Naoki Yoshie†, Katazakai Saki§,
Jing Zhang§, and Takafumi Yoshida¶*

*Ocean Modeling Group, Center for Oceanic and Atmospheric Research, Research Institute for Applied Mechanics,
Kyushu University, Kasuga, Japan
†Center for Marine Environmental Studies, Ehime University, Matsuyama, Japan
‡Graduate School of Fisheries and Environmental Sciences, Nagasaki University, Nagasaki, Japan
§Graduate School of Science and Engineering, University of Toyama, Toyama, Japan
¶Regional Activity Center, Northwest Pacific Region Environmental Cooperation Center, Toyam City, Japan

OUTLINE

Integrated Coastal Management in the Japanese Satoumi
https://doi.org/10.1016/B978-0-12-813060-5.00004-3
© 2019 Elsevier Inc. All rights reserved.

The Environment Research and Technology Development Fund of the Ministry of the Environment, Japan S-13 was started from FY2014, aiming to develop a coastal environmental management methodology for sustainable coastal areas. This project selected three target sea areas, namely, the coastal area of the Sanriku region, the Seto Inland Sea and the Japan Sea to verify the developed management methodology. Among the three areas, our group focused on the Japan Sea.

The Japan Sea is a semi-enclosed international sea area which is surrounded by the Eurasia Continent and the Japanese archipelago. The Japan Sea connects to the East China Sea through the Tsushima Strait, to the Pacific Ocean through the Tsugaru Strait, and to the Okhotsk Sea through the Soya and the Mamiya Straits. The water depth of these straits is very shallow which makes the Japan Sea extremely closed. In the southern part of the Japan Sea, the Tsushima Warm Current, branched from the Kuroshio Current, flows northeastward along the Japanese coastline to the Pacific Ocean through the Tsugaru Strait. On the other hand, in the northern part of the Japan Sea, the Liman Current flows down along the Russian coastline. The average depth of the Japan Sea is 1,700 m. In winter, the surface water which is cooled in the coastal area of Russia sinks down to the deeper layer and forms cold water masses with rich dissolved oxygen, called the Japan Sea Proper Water.

In recent years, the Japan Sea faced rapid environmental changes. The first change is global warming. Rise of seawater temperature and sea level, and ocean acidification are reported in many parts of the world. It is well known that the rise of seawater temperature occurs much faster in the Japan Sea than in other sea areas. According to data from the Japan Meteorological Agency, the long-term trend of water temperature shows an increase of more than 1°C in the past 100 years. The average increase of seawater temperature in the world is 0.5°C, thus, the seawater temperature in the Japan Sea rises twice or more as fast as the world average.

The second change is rapid environmental changes in the East China Sea. As the East China Sea is located in the upstream part of the Japan Sea, seawater, heat and various substances are transported from the East China Sea to the Japan Sea by the Tsushima Warm Current. The environment of the East China Sea has changed dramatically due to rapid economic growth in the littoral countries. The East China Sea is an important spawning ground for marine life living in the Japan Sea; therefore, the environmental changes in the East China Sea may influence not only those species but also the entire biodiversity in the Japan Sea.

When we discuss the coastal management in the Japan Sea, it is necessary to identify how much influence these two changes have on the environment of the Japan Sea. Thus, our group developed new numerical ecosystem models: coupled physical-ecosystem models and a particle tracking numerical model for transport and survival of high trophic species of the Japan Sea. Using these numerical ecosystem models, we tried to understand the impacts of global warming and the environmental changes in the East China Sea on the Japan Sea as well as the mechanism of the impacts. Then, based on the outputs of this study, we considered the new coastal management of the Japan Sea, explaining it in the following sections:

- Section 4.1 is a forecast of future environmental changes in the Japan Sea using numerical ecosystem model,
- Section 4.2 is the relationship between the East China Sea and the Japan Sea,
- Section 4.3 is the response of the typical species in the Japan Sea, Japanese squid (*Todarodes pacificus*) and snow crab (*Chionoecetes opilio*), to the environmental changes and their conservation options,
- Section 4.4 is the land-sea integrated management in Toyama Bay, and
- Section 4.5 is coastal management in the Japan Sea.

4.1 NUMERICAL SIMULATION FOR THE DISSOLVED OXYGEN CONCENTRATION IN THE JAPAN SEA

Haejin Kim, Katsumi Takayama, and Naoki Hirose

4.1.1 Introduction

To investigate the physical and biogeochemical responses of the Japan Sea (JS) to various climate change scenarios, we established a coupled physical-biogeochemical model. To improve the accuracy of predictions over a century, the distributions of temperature, salinity, and materials from the surface to the deep layers must be represented as accurately as possible. In the deep layers of the JS, a water mass with almost homogeneous temperature and salinity is present, called "Japan Sea Proper Water". Deep circulation is not easily understood through analysis of its nearly homogeneous temperature and salinity; other tracers are helpful to evaluate deep-ocean circulation. In observational studies of deep circulation in the JS, circulation properties were estimated at spatial and temporal distributions of tracers such as dissolved oxygen (DO) and nutrients (e.g., Senjyu and Sudo, 1994).

In this study, we focused on the DO concentration in the JS and attempted to represent its characteristics using numerical simulation modeling. Many observational data on DO concentration exist; it is behind only temperature and salinity in this respect according to the World Ocean Database (Boyer et al., 2013). Furthermore, because the DO concentration in the ocean is determined by the physical phenomenon of air-sea gas exchange, and is increased (decreased) by the biogeochemical activity of photosynthesis (decomposition of organisms), it is a suitable tracer for evaluating the results of a coupled physical-biogeochemical model. However, simulation of DO concentration using a three-dimensional model has never been conducted in the JS, and the characteristics of DO concentration reproduced by the coupled model used in this study must be verified qualitatively and quantitatively compared with observations.

The residence time of deep waters in the JS is estimated to be around 100 years, which is short compared to a residence time of a 1000 years in the open ocean (Watanabe et al., 1991). Therefore, the model requires an integration time of more than 100 years to validate the DO concentration in the deep layer of the JS compared with observations, but such a long integration time is not realistic in terms of calculation cost. In addition, the number of observations of DO concentration in deep layers is insufficient, and it is thus difficult to confirm the variations in DO concentration simulated by the model. Therefore, to evaluate model performance, we compared DO concentrations between the model and observations, and examined whether the model can represent the seasonal variation of DO concentrations captured in measurement data from subsurface layers, where the number of observations is relatively large and the residence time is relatively short compared with the deep layer. Additionally, because the variations in DO concentration can be divided into physical and biogeochemical components in the model, we quantified the effects of biological activities on DO concentration in the JS using a coupled physical-biogeochemical model.

4.1.2 A Coupled Physical-Biogeochemical Model

We used the RIAM Ocean Model developed by Lee et al. (2003) from the Research Institute for Applied Mechanics, Kyushu University, as our physical model. The domain of the model is the northwestern Pacific from 105°E to 180°E, 15°N to 63°N. The grid components are spaced by 1/4 degree in the zonal direction and 1/5 degree in the meridional direction. The grid is divided vertically into 38 layers.

The meteorological data on surface fluxes of momentum and heat were obtained from the daily mean Japanese 25-year ReAnalysis (JRA-25). The river runoff from the Changjiang was given by the monthly mean data derived from the Chinese Ministry of Water Resources (MWR, 2011). Meanwhile, the climatological monthly mean data of the Global Runoff Data Centre (GRDC) was used for the Huang He and Amur rivers. The parameters are described in more detail in Hirose (2011).

We simulated biogeochemical processes using a nitrogen-based lower trophic ecosystem model that comprised dissolved inorganic nitrogen (DIN), phytoplankton (PHY), zooplankton (ZOO), and detritus (DET; Onitsuka et al., 2007). We also incorporated DO into our model, as described in Fig. 4.1.

All our calculations of the biogeochemical compartments were based on our physical model. The evolution of DO concentration was determined by

$$
\begin{aligned}
d(\text{DO})/dt = &\text{Advection } + \text{ Diffusion} \\
&+ \text{ Photosynthesis (PHY) } - \text{ Respiration (PHY)} \\
&- \text{ Excretion (ZOO) } - \text{ Decomposition (DET)} \\
&\pm \text{ Exchange with the atmosphere.}
\end{aligned} \tag{4.1}
$$

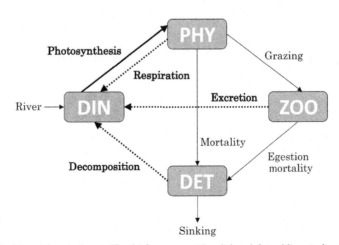

FIG. 4.1 Schematic diagram of the biogeochemical part. The thick arrows with solid and dotted lines indicate oxygen production and consumption processes, respectively.

The variation in DO concentration by the biogeochemical processes on the right side of Eq. (4.1) are defined as follows:

$$\text{Photosynthesis} = V_{\max} \times \left(\frac{\text{DIN}}{\text{DIN} + K_{DIN}}\right) \times \exp(kT)$$
$$\times \frac{I}{I_{opt}} \exp\left(1 - \frac{I}{I_{opt}}\right) \times \text{PHY}, \tag{4.2}$$

$$I = I_0 \exp\left(-\int_0^z k_e \, dz\right), \tag{4.3}$$

$$k_e = \alpha_1 + \alpha_2 \times (\text{PHY}), \tag{4.4}$$

$$\text{Respiration} = R \times \exp(k_r T) \times \text{PHY}, \tag{4.5}$$

$$\text{Excretion} = (\alpha_z - \beta_z) \times G_R \times \max\{0, 1 - \exp[\lambda(\sigma^a - \text{PHY})]\}$$
$$\times \exp(k_g T) \times \text{ZOO}, \tag{4.6}$$

$$\text{Decomposition} = V_{PN} \times \exp(k_{PN} T) \times \text{DET}, \tag{4.7}$$

where T is the water temperature, I is the light intensity, I_0 is the solar radiation at the sea surface, and k_e is the dissipation coefficient. We calculated these terms using the values in Table 4.1. The units of the biogeochemical variables are all $\text{mmolN}\,\text{m}^{-3}$ (=μM). The Redfield ratio (N:O_2=16:−138; Redfield et al. (1963) were used to obtain the oxygen concentrations. Dissolved oxygen is produced by photosynthesis, which, according to Eq. (4.2), depends on the concentrations of PHY and DIN. The light intensity (I) decreases gradually as the depth increases, with its exact decay rate depending on the concentration of PHY in the upper layer (Eqs. 4.3 and 4.4). If the light intensity is greater than or equal to the optimum light intensity (I_{opt}), the light intensity is given by the optimum light intensity ($I = I_{opt}$). The oxygen consumed by the respiration of PHY was calculated using Eq. (4.5), and the respiration of ZOO was estimated based on the excretion of ZOO to DIN (Eq. 4.6). Eq. (4.7) represents the consumption of oxygen as DET decomposes (remineralizes) into DIN.

TABLE 4.1   Parameters Used in the Equations

Symbol	Definition	Value	Unit
V_{max}	Maximum photosynthesis rate of PHY at 0°C	0.6	d^{-1}
K_{DIN}	Half saturation constant of PHY for DIN	1.5	$\text{mmolN}\,\text{m}^{-3}$
K	Temperature coefficient for the photosynthetic rate	0.0693	$°\text{C}^{-1}$
I_{opt}	Optimum light intensity	70	$\text{W}\,\text{m}^{-2}$
α_1	Light dissipation coefficient of seawater	0.05	m^{-1}
α_2	Self-shading coefficient	0.06	$\text{m}^2\,\text{mmolN}^{-1}$
R	Respiration rate of PHY at 0°C	0.03	d^{-1}
k_r	Temperature coefficient for respiration	0.0519	$°\text{C}^{-1}$
α_z	Assimilation efficiency of ZOO	0.7	
β_z	Growth efficiency of ZOO	0.3	
G_R	Maximum grazing rate of ZOO at 0°C	0.3	d^{-1}
Λ	Ivlev constant	1.4	$\text{m}^3\,\text{mmolN}^{-1}$
σ^a	Threshold value for grazing	0.043	$\text{mmolN}\,\text{m}^{-3}$
k_g	Temperature coefficient for grazing	0.0693	$°\text{C}^{-1}$
V_{PN}	Decomposition rate of DET at 0°C	0.05	d^{-1}
k_{PN}	Temperature coefficient of DET for decomposition	0.0693	$°\text{C}^{-1}$

The final term on the right side of Eq. (4.1) represents the gas exchange between the atmosphere and the surface of the ocean, which was calculated as follows:

$$\text{Exchange with the atmosphere} = K_a \cdot (DO_a - DO_{k=1}), \tag{4.8}$$

where K_a is the reaeration coefficient with a restoring time scale of 0.6 days. We obtained the saturated DO concentration (DO_a) using equation derived by Weiss (1970):

$$\begin{aligned}
\ln DO_a \left(mL\, L^{-1} \right) = \\
-173.4292 + 249.6339 \left(\frac{100}{T} \right) + 143.3483\ \ln \left(\frac{T}{100} \right) - 21.8492 \left(\frac{T}{100} \right) \\
+ S \left[-0.033096 + 0.014259 \left(\frac{T}{100} \right) - 0.00170 \left(\frac{T}{100} \right)^2 \right].
\end{aligned} \tag{4.9}$$

We used our physical model to calculate the temperature (T) of the surface water in Kelvin and salinity (S) in psu.

The sunlight for photosynthesis of PHY was given by shortwave radiation flux of JRA-25 reanalysis data. The initial condition of DIN concentration was given by the January data in World Ocean Atlas 2009 (WOA09) whereas the PHY, ZOO, and DET concentrations were uniformly distributed from the surface to 200 m layer with values of 0.1, 0.08, and 0.01 mmol m^{-3}, respectively. The DIN concentrations at the southern and eastern boundaries of the model, which are located at coordinates 15°N and 180°E, were set to the monthly mean concentration from the WOA09. The boundary conditions for the PHY, ZOO, and DET concentrations were equal to their initial conditions. The DIN flux from the Changjiang River increased linearly at a rate of 2.5 mmol m^{-3} year^{-1}, reaching 70 mmol m^{-3} by 1991 (Zhang et al., 1999). We held the DIN concentrations discharged by the other rivers constant (100 mmol m^{-3}).

The coupled physical-biogeochemical model had been integrated from 1979 to 2007. Following an initial spin-up period of 9 years, the model results were used to calculate climatological monthly means over the 20-year period from 1988 to 2007. The analysis was limited to the upper levels of the ocean, from the surface to a depth of 200 m, because the spin-up time was too short for deep ocean.

4.1.3 Effect of Biogeochemical Processes on the Seasonal Variation in DO Concentration

First of all, we describe the seasonal variation of temperature, salinity, DIN, and chlorophyll profiles averaged in the JS using the results of CTL experiment to understand their relationship (Fig. 4.2). In winter, the four variables show vertically homogeneous structure in the upper ocean by winter mixing. The surface temperature gradually increases from spring and has maximum value in summer (Fig. 4.2A). The surface mixing of the upper JS is typically initiated in autumn. The salinity near the surface is low from summer to autumn together with high precipitation, and relatively high saline condition occurs from winter to spring due to inflow of Kuroshio water (Fig. 4.2B). The reliability of these physical conditions has been demonstrated by Hirose (2011).

The enhanced winter mixing results in supply of DIN from deeper layers, thereby the surface DIN shows maximum concentration in winter (Fig. 4.2C). In spring, the stronger solar radiation and the development of stratification lead to spring bloom with maximum chlorophyll concentration more than 1.1 µg L^{-1} at the surface (Fig. 4.2D). The DIN within the upper stratified layer initiates to be depleted at the onset of spring bloom. Due to the consumption of DIN in the upper layer, subsurface chlorophyll maximum (SCM) locates at 30 m in summer and it tends to slightly deepen in autumn. The development of SCM layer around 30–50 m was in good agreement with the observational study of Rho et al. (2012). In addition, Onitsuka et al. (2007) confirmed that the seasonal variations of the biogeochemical variables, which were reproduced using an earlier version of the coupled model used in this study, were in good agreement with the observations.

We assessed the contributions of biogeochemical processes to the variation in DO concentration by conducting two comparative experiments: the CTL experiment and the no-ecosystem (NE) experiment. In the CTL experiment, we calculated the DO concentration by summing the contributions of all the processes defined in Eq. (4.1), such as biological production and consumption as well as the gas exchange with the atmosphere. But the NE experiment neglected the biogeochemical processes and thus simulated the DO concentration response to the air-sea gas exchange only.

(1) Vertical profile

To assess the performance of the two comparative experiments on DO variation, we compare the model results and WOA09 observation data. Fig. 4.3A–D show vertical DO profiles from the surface to a depth of 200 m averaged over

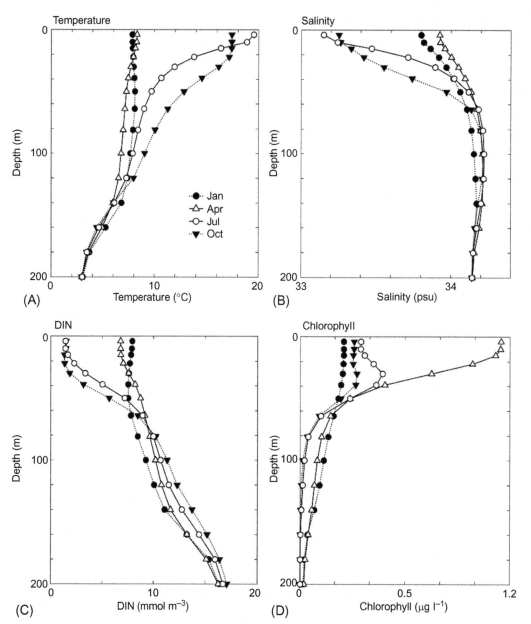

FIG. 4.2 Seasonal variations of vertical structure in (A) potential temperature (°C), (B) salinity (psu), (C) dissolved inorganic nitrogen (DIN) concentration (mmol m⁻³), and (D) chlorophyll concentration (μg L⁻¹) averaged over the Japan Sea. The figures represent the results of control (CTL) experiment in January (dotted line with filled circle), April (solid line with open triangle), July (solid line with open circle), and October (dotted line with filled upside-down triangle).

the JS for each season. The CTL experiment showed the monotonic pattern of DO concentration with depth similar to the WOA09 in winter (Fig. 4.3A). The DO maximum in the CTL experiment occurred at the surface in March by beginning of spring bloom. The depth of peak DO concentration descended through the subsurface layer from spring to autumn (Fig. 4.3B–D), but the depth remained stable at 30 m during summer. These results are in good agreement with the observation data. Although intensive photosynthesis occurred on the surface during springtime (Fig. 4.2D), the peak DO concentration was below the surface (20–30 m) between April and May. It appears that the low oxygen solubility in the warming water at the surface limited the DO concentration. More oxygen could dissolve in the cooler water just below the surface. In summer, the subsurface DO maximum occurred at the same depth as the peak chlorophyll concentration (Figs. 4.2D and 4.3C). As the surface mixing initiates, the subsurface DO maximum goes deeper level.

The NE model did not represent the evolution of the subsurface DO maximum between spring and autumn. Also, the DO concentrations in the NE experiment were gradually increased below the winter mixed layer (about 120 m

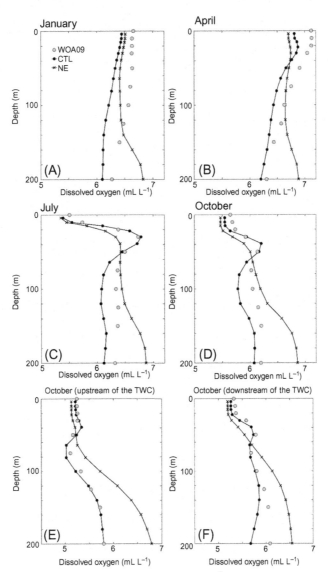

FIG. 4.3 Vertical profiles of average dissolved oxygen (DO) concentration (mL L^{-1}) in the Japan Sea in (A) January, (B) April, (C) July, and (D) October; (E) in the upstream region of the Tsushima Warm Current (TWC; 130°–133°E, 35.8°–37.8°N); (F) in the downstream region of the TWC (135°–138°E, 37.8°–39.8°N) in October. Dots represent observational data, and solid lines with dots and crosses indicate values calculated using the control (CTL) and no-ecosystem (NE) experiments, respectively.

averaged in the JS) to 200 m in all seasons, contrary to the observation or the CTL experiment. We attribute this result to the fact that biological consumption, particularly the remineralization of DET, was not taken into account by the NE model. The largest differences between the DO concentrations by the two experiments occurred below 200 m.

Although we usually expect the peak DO concentration to be located in the euphotic zone, the vertical DO distributions in the subpolar gyre (SPG) region and the Tsushima Warm Current (TWC) region differed in autumn. The peak DO concentration was still located at the subsurface in the SPG region (data not shown), whereas it was generally found below the depth of 100 m in the TWC region (Fig. 4.3E and F). We attribute the sudden increase in the depth of peak DO concentration in the TWC region to the influence of the advection of Tsushima Warm Water, which is characterized by its high salinity and low oxygen concentration. Takikawa et al. (2008) also found the low DO water with cold and saline properties at the depths between 75 and 100 m of the center part of TKS during autumn through hydrographic observation. On the other hand, the NE experiment did not represent the local reduction in DO concentration at all.

(2) Seasonal variation

Fig. 4.4 shows the seasonal variation in DO concentration at the surface and at 30 and 200 m depths averaged over the entire JS. At the surface layer, the DO concentrations of the JS were highest between winter and spring, then

FIG. 4.4 Seasonal variations in dissolved oxygen (DO) concentration (mL L^{-1}) at (A) the surface layer, (B) 30 m, and (C) 200 m of the Japan Sea. Dots represent observational data, and solid and dotted lines indicate values calculated using the control (CTL) and no-ecosystem (NE) experiments, respectively.

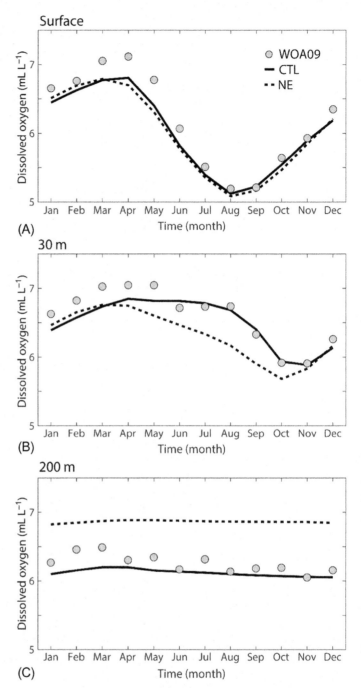

decreased between summer and autumn related to the oxygen solubility, as shown in Fig. 4.4A. The differences of DO concentrations between the two experiments were not significant throughout all seasons (Fig. 4.4A).

According to the observational data and the CTL experiment, the DO concentrations in the 30 m layer remained high from spring to summer, whereas the NE experiment showed that they started to decrease quickly from the middle of spring (Fig. 4.4B). This discrepancy was mainly because photosynthetic production exceeded biological consumption during this period. The most remarkable difference of DO concentration between the two experiments occurred in summer at 30 m, especially in August with 0.52 mL L^{-1}.

At 200 m, the DO concentration was weakly dependent on the season, as shown in Fig. 4.4C. The mean DO concentration of the CTL experiment (6.1 mL L^{-1}) was similar to the measurement result (6.3 mL L^{-1}), whereas the NE experiment produced much higher oxygen concentration (6.9 mL L^{-1}). This is because oxygen consumption processes, particularly the decomposition of DET, were not taken into account by the NE model.

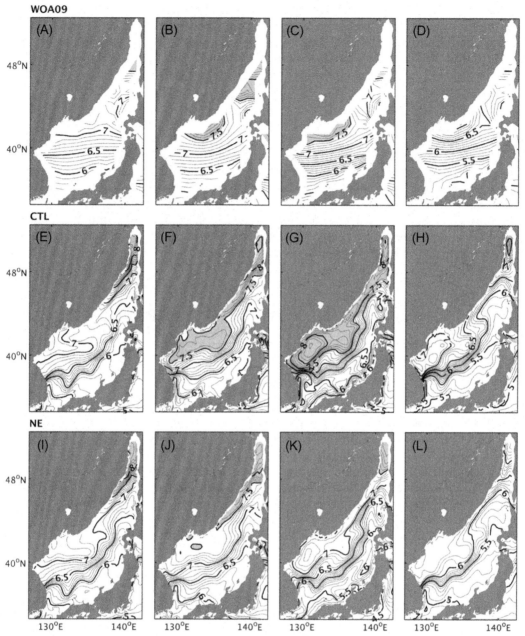

FIG. 4.5 Horizontal distributions of dissolved oxygen (DO) concentration (mL L^{-1}) at 30 m obtained from (A)–(D) the World Ocean Atlas 2009 (WOA09) data, (E)–(H) the control (CTL) experiment, and (I)–(L) the no-ecosystem (NE) experiment. From left to right, the plots show the distributions for January, April, July, and October.

(3) Spatial distribution

Fig. 4.5 indicates horizontal distributions of DO concentration at 30 m obtained from WOA09, the CTL experiment and the NE experiment. As the model had high horizontal resolutions, they yielded more detailed oxygen patterns than the WOA09 data.

In winter, the CTL experiment slightly underestimated the DO concentrations in parts of the northern JS (Fig. 4.5A, E, and I), this may be due to sparse observational data for this region. We considered the observed DO distribution in the northern part of the JS to be unreliable. The CTL experiment simulated the lower oxygen concentrations than the NE experiment in most of the JS, except around the northeastern part of the Korean Peninsula. This is because the water mass below the euphotic zone with overestimated oxygen concentration by the NE experiment appeared in the subsurface by winter mixing (Fig. 4.5A and E). In particular, deep convection in the SPG region caused significant negative differences of more than 0.2 mL L^{-1}.

From spring to autumn, the CTL experiment represented the high concentrations of DO in the SPG region (Fig. 4.5B–D and F–H), but the NE experiment did not keep up with the high DO feature of the SPG region (Fig. 4.5J–L). These positive differences between the two comparative experiments were weak and homogeneous in spring but strong from summer to autumn and marked throughout the SPG. Since the SPG region had abundant DIN owing to the deep winter convection, the supply of oxygen by photosynthesis exceeding the biological consumption was larger than elsewhere during summer season.

4.1.4 Summary and Future Works

We demonstrated that the representation of DO concentration in the JS can be reasonably simulated by using a coupled physical-biogeochemical model. The model performance is satisfactory as can be seen from the results of the model simulating seasonal features comparable with the observation for the subsurface layer of the JS. We will conduct long-term integration based on the climate change scenarios using this coupled model and predict how the DO concentrations and biogeochemical environments of the JS will change in the future. According to the recent researches, the DO concentration has been decreased in the deep layers of the JS (Gamo, 2001). We will investigate whether the DO concentration of the bottom water become anoxic or not in the future by taking the prediction results of the coupled model.

4.2 FUTURE PREDICTION OF NUTRIENT CONCENTRATION AND PRIMARY PRODUCTION IN THE JAPAN SEA

Akihiko Morimoto, Ryota Shibano, Masashi Ito, Tetsutaro Takikawa

4.2.1 Introduction

The Japan Sea (JS) is connected by four narrow and shallow straits to the East China Sea (ECS), the Pacific Ocean, and the Okhotsk Sea. The mean water depth of its interior is 1667 m, therefore the JS is relatively deep (Fig. 4.6). Water mass and materials flow into the JS mainly from the Tsushima Strait (TS) located in the southwestern part of the JS and flow out from the Tsugaru and Soya Straits that are located in the northern part. The Tsushima Warm Current (TWC), which flows on the continental shelf of the ECS and enters into the JS through the TS, plays an important role for material transport from the ECS to the JS and for material cycles in the JS. This fact is found from the following things: (1) decrease of surface salinity in the JS in summer and autumn is caused by enormous fresh water originating from Chinese rivers transport from the ECS to the JS, and (2) massive giant jellyfish which hatch at the Chinese coast is

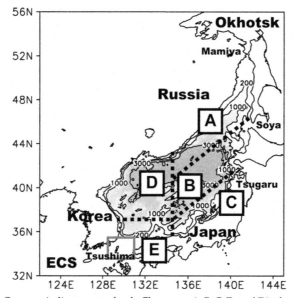

FIG. 4.6 Bathymetry in the Japan Sea. Contours indicate water depth. Characters A, B, C, D, and E in the figure show the areas where modeled and observed values are validated.

attributed to the damage of fixed fishing nets at fisheries on the Japanese coasts. Nutrients like nitrogen and phosphorus which are utilized by phytoplankton photosynthesis are also transported from the ECS to the JS by the TWC, and it is suggested that those nutrient transport might contribute to primary production in the southern part of the JS (Onitsuka et al., 2007). If nutrient flux from the ECS to the JS is varied due to marine environmental changes in the ECS, there would be a possibility of change in primary production and material cycles in the JS. Therefore, we have to know the variation in nutrient flux into the JS through the TS and elucidate lower trophic ecosystem response associated with the nutrient flux change through the TS.

It is known that nutrient concentration in the Eastern channel of the TS is low in winter and spring and high in summer and autumn. Interannual variation in nutrient concentration is large; the year with the higher concentration in summer is 5 µM and the year with the lower one is 3 µM. These kinds of variations in nutrient concentration are thought to be the change of nutrient source in the TS (Morimoto et al., 2012; Kodama et al., 2017). Although seasonal and interannual variations in nutrient concentration in the Western channel of the TS are unclear at the moment because of the limitation of observed nutrient data, mean nutrient concentration in the Western channel is 3 times higher than that in the Eastern channel which are estimated by the relationship between nutrient concentration and water temperature (Morimoto et al., 2013). It is also found that interannual variations in nutrient concentration both in the Eastern and Western channels are not synchronized with each other. In order to consider how nutrient concentration change in the TS contributes to the lower trophic ecosystem in the JS, nutrient flux should be considered not nutrient concentration. The nutrient flux through the Eastern channel of the TS has been estimated as 4 kmol/s (Morimoto et al., 2012). This value is 2 times larger than that from the Changjiang River in China. From this fact, the JS is regarded as a big river in the southwestern part which supplies enormous nutrients and also as a system like a coastal bay where primary production is occurred by nutrients supplied from land. However, there is a possibility of a different material cycle process because the horizontal scale of the domain is much larger and water depth is deeper than in a coastal area.

When we think about the marine environment of the JS in the future, we have to reveal quantitatively where and how the degree of marine environment change is associated with nutrient flux variations. In addition, we have to understand the cause of the nutrient flux change. In this section, seasonal and interannual variations in nutrient concentration and flux in the TS are shown based on observed data. We try to elucidate how the lower trophic ecosystem in the JS respond to nutrient flux change through the TS via a physical-ecosystem coupled model. At last, we show the origin of the nutrients in the TS via a model for ECS.

4.2.2 Seasonal and Interannual Variations in Nutrients in the Tsushima Straits

In order to know how much nutrient is transported by the TWC from the ECS to the JS, we have conducted hydrographic observations in the TS since 2005. Observed points in the Eastern and Western channel were six stations and two stations, respectively (Fig. 4.7). Since tidal currents can be predicted at any time along this observed line, it's

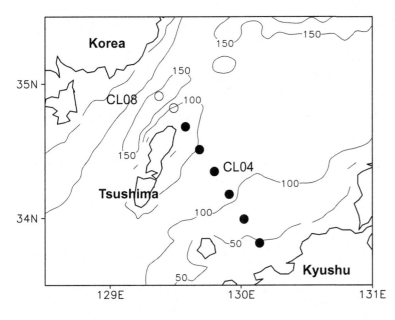

FIG. 4.7 Observed points in the Tsushima Strait. White circles denote observed points in the Western channel and black points denote those in the Eastern channel of the Tsushima Strait.

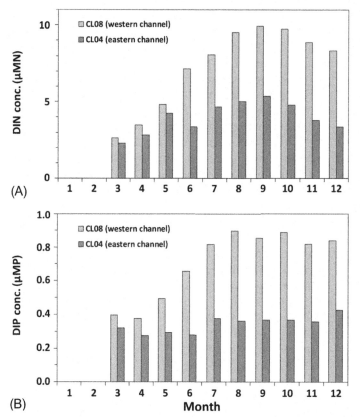

FIG. 4.8   Seasonal variations in (A) long term averaged (2005–2017) of DIN and (B) DIP concentration at CL04 and CL08.

possible to estimate the volume transport of the TWC at the observed period. Water samples for measuring DIN (Dissolved Inorganic Nitrogen) and DIP (Dissolved Inorganic Phosphorus) were taken at standard levels (0, 20, 20, 30, 50, 75, 100, 125, 150, 175, 200 m) and 5 m above the bottom. Current velocity was observed by using ship-mounted ADCP (Acoustic Doppler Current Profiler) and nutrient fluxes were calculated both by velocities normal to the observed transect and nutrient concentration at each observed point.

Fig. 4.8 shows seasonal variations in DIN and DIP concentration at CL04 in the Eastern channel and CL08 in the Western channel. DIN concentrations in both channels increase after March and reach the maximum in September, after that gradually decrease until December. Annual mean of DIN concentration in the Eastern and Western channels are 4.0 and 7.3 μM, respectively; DIN concentration in the Western channel is 1.8 times higher than in the Eastern channel. DIP concentration gradually decreases after March, and then increases in summer and keeps high concentration until December. Annual mean of DIP concentration in both channels are 0.34 and 0.70 μM, respectively. N/P ratio (Nitrogen Phosphorus ratio) in the Eastern and Western channels are 11.6 and 10.3, respectively. Those values are smaller than Redfield ratio of 16; the growth of phytoplankton in the TS might be limited by DIN.

Fig. 4.9 shows interannual variations in summer time (July to September) DIN concentration in both channels. Mean DIN concentrations in summer time at CL04 (Eastern channel) and CL08 (Western channel) are 5.2 and 10.4 μM, respectively, and those standard deviations are 0.76 and 0.84 μM, respectively. DIN concentration in summer at the Western channel is approximately 2 times higher than at the Eastern channel. This is because water depth in the Western channel is deeper than in the Eastern channel, and consequently subsurface DIN concentration in the Western channel is higher and bottom DIN concentration is much higher. DIN concentration in the Eastern channel in 2005 and 2013 were high, on the contrary in 2007 was low. DIN fluxes in 2005, 2013, and 2015 were large and those in 2007, 2009, 2011, and 2014 were small (Fig. 4.10). The range of DIN flux through the Eastern channel was 2–7 kmol/s with interannual variation. DIN flux in the Western channel cannot be calculated because there are only 2 stations there. However, the range of DIN flux through the Western channel might be 4–14 kmol/s that are estimated from the facts that DIN concentration in the Western channel is 2 times higher and assuming that volume transport in the Western channel is the same as in the Eastern channel. It is well-known that volume transport of the TWC in the Western channel is higher than in the Eastern channel. Therefore, it is expected that the interannual variation in DIN flux in the Western channel is much larger than the above estimation.

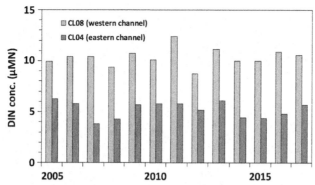

FIG. 4.9 Interannual variations in DIN concentration in summer time at CL04 in the Eastern channel and at CL08 in the Western channel.

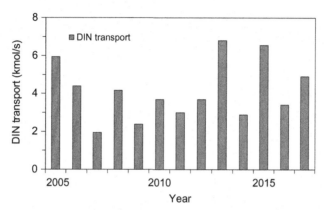

FIG. 4.10 Interannual variation in DIN flux in summer time through the Eastern channel.

4.2.3 Response of Lower Trophic Ecosystem in the Japan Sea Associated With DIN Flux Change

It was found that there is a year-to-year variability of DIN flux through the TS with a range of 2–7 kmol/s in the Eastern channel and 4–14 kmol/s in the Western channel from the repeated hydrographic observations in the TS. It will be important to know where and how much nutrient or phytoplankton in the JS would be changed while DIN flux varies in the TS. Since answering those questions is not so easy only from observed results, we tried to develop a numerical model to get the answers. Nutrients and planktons are advected and diffusive, and generated and removed by biological processes such as photosynthesis and grazing. In order to look into those physical and biological processes, we use a physical-ecosystem coupled model.

We want to know only lower trophic ecosystem response associated with variation in DIN flux in the TS. Climatological physical fields such as current velocity and water temperature are applied, namely, there are no interannual variations in the physical fields. We used the result of monthly mean of DREAMS_M as physical fields which is developed by Research Institute Applied Mechanics in Kyushu University (Hirose et al., 2013). As a biological process in the model, we considered nutrients (nitrate and ammonium), 2 types of phytoplankton, 3 types of zooplankton, dissolved organic nitrogen, and suspended organic nitrogen referred to in an ecosystem model NEMURO (Kishi et al., 2007) (Fig. 4.11). Zonal and meridional resolutions of the model are 1/12 and 1/15, respectively, and the vertical is divided by 38 layers, and the thickness of the layer shallower than 100 m is 4–20 m. At first, we calculated material cycle related to lower trophic ecosystem applying averaged seasonal variation of DIN flux in the TS. After validating the calculation results, we investigated response in lower trophic ecosystem in the JS associated with DIN flux change in the TS.

Calculated chlorophyll-*a* concentrations by the model are compared with observed ones derived from satellite ocean color data. Seasonal variations in modeled chlorophyll-*a* concentration are coincident with observation except area A (Fig. 4.12). Using the model, we investigated origin of nutrients for primary production in the JS. In this calculation, different labels attached to nutrients are supply from the TS, rivers in Japan, and the deeper layers of the JS, and we calculated how much nutrient is consumed at any points, and then estimated contribution ratio of DIN to primary production in each grid of the model. 40%–80% of primary production south of 38°N that is generated by DIN comes from the TS (Fig. 4.13A). The contribution ratio of Japanese rivers is much smaller, less than 5% north of the Noto Peninsula (Fig. 4.13B). On the contrary, more than 80% primary production north of 38°N that is contributed by DIN originated from the deeper layers of the JS (Fig. 4.13C). From the result, it is suggested that primary production in the northern part of the JS is strongly controlled by vertical DIN transport processes such as vertical mixing

FIG. 4.11 Lower trophic ecosystem in the model

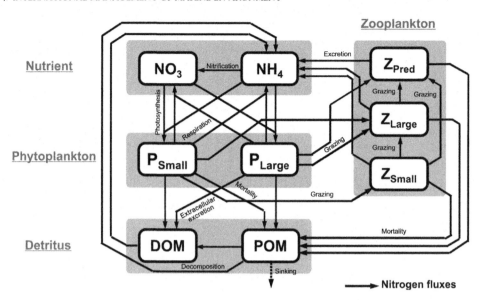

FIG. 4.12 Seasonal variations in modeled chlorophyll-*a* concentrations (black lines) and chlorophyll-*a* concentration of satellite data (+) at 5 areas of (A–E) whose area are depicted in Fig. 4.6.

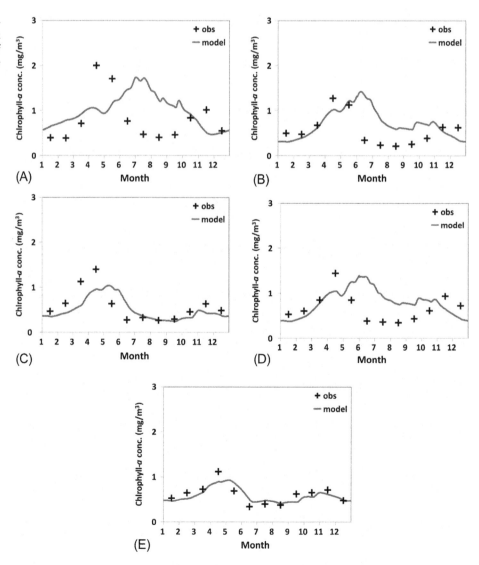

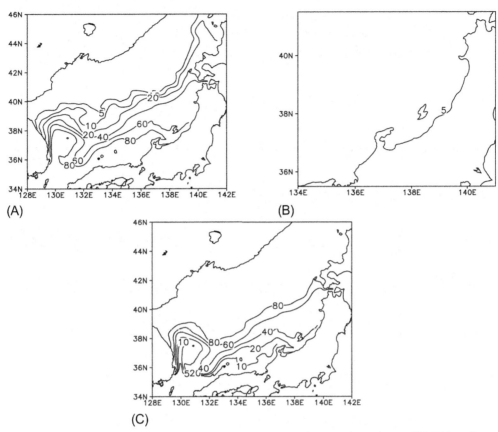

FIG. 4.13 Contribution ratio of each DIN source to primary production. (A) DIN from Tsushima Straits. (B) DIN from Japanese rivers, (C) DIN from deeper layer of the Japan Sea.

and vertical diffusion, very similar to the same processes in the open oceans, and primary production in the southern part of the JS is controlled by horizontal DIN transport. Therefore, it is expected that influences of DIN flux change through the TS appear in the southern part of the JS.

In order to know whether response of lower trophic ecosystem in the JS is associated with DIN flux change through the TS, sensitivity analysis was conducted. DIN flux through the TS increased by 50% to the average value. The condition corresponds to the DIN flux in 2013 shown in Fig. 4.10 when DIN flux through the TS was about 1.5 times larger than mean DIN flux. In the sensitive analysis, DIN flux from January to May and from November to December did not change, and DIN flux from June to October increased by 50%. It is clearly shown that DIN concentration in the southern part of the JS is increase related to DIN flux increase in summer (Fig. 4.14A). Especially, DIN along the Korean Peninsula shows large increase of DIN. Chlorophyll-*a* also increases along the Korean Peninsula with almost the same pattern as DIN. This result suggests that the high growth rate of phytoplankton is due to high DIN concentration (Fig. 4.14E). The area where DIN increases expands to the east in October but there are few increases of DIN along the Japanese coast (Fig. 4.14B). Chlorophyll-*a* concentration increases in large area except in TS and along the eastern coast of Korea (Fig. 4.14F). ΔDIN distribution in January is almost the same as in October and there is no increase of DIN along the Korea Peninsula because of no change in DIN flux after October (Fig. 4.14C). Δchlorophyll-*a* is low in the whole JS although DIN increases at any area (Fig. 4.14G). DIN concentration in April is increase with a zonal band at 38–40°N (Fig. 4.14D). Δchlorophyll-*a* is not in correspondence with ΔDIN; high Δchlorophyll-*a* appears relatively west of the Tsugaru Strait (Fig. 4.14H). As expected from Fig. 4.13, DIN flux change through the TS varies chlorophyll-*a* concentration in the southern part of the JS. Maximum Δchlorophyll-*a* was 0.1 mg/m^3 that is 25% of the chlorophyll-*a* concentration when averaged DIN flux is applied in the TS.

In the east coast of Korea, although both DIN and chlorophyll-*a* concentrations increase in July, only DIN increases in October but there is no change in chlorophyll-*a* concentration. We discuss the reason why there is no increase of chlorophyll-*a* in spite of DIN increases by about 5 µM. DIN concentration along the Korean Peninsula in summer is high (about 5 µM) because of costal upwelling caused by southern winds (Park and Kim, 2010). In addition, DIN concentration also increases due to large input of DIN through the Western channel of the TS. As a result, chlorophyll-*a* in summer increases due to high photosynthesis. DIN flux through the Western channel still increases in October. DIN concentration along the Korean Peninsula in October becomes higher (about 10 µM) because of vertical mixing related to surface

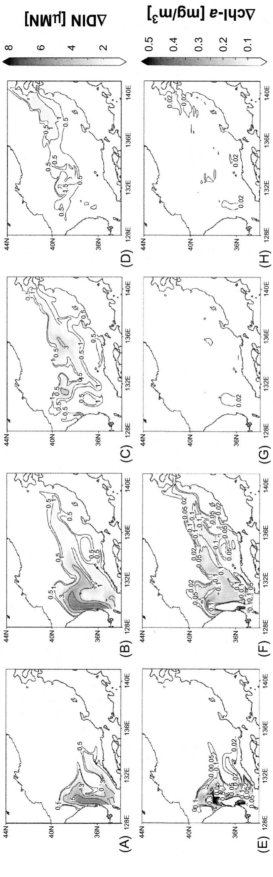

FIG. 4.14 Increase of DIN concentration averaged surface 100 m depth in (A) July, (B) October, (C) January, (D) April when DIN flux through the Tsushima Straits is increased by 50%. Increase of chlorophyll-*a* concentration averaged surface 20 m in (E) July, (F) October, (G) January, (H) April.

cooling. Because DIN concentration is high enough and the growth rate of phytoplankton decreases due to lower water temperature, it is assumed that chlorophyll-*a* concentration along the Korean Peninsula in October didn't change.

We also discuss the discrepancy of ΔDIN and Δchlorophyll-*a* along the Japanese coast. A number of DIN flow into the JS through deeper layers of the Eastern channel of the TS are transported to northeastward near Oki Island. Therefore, DIN transported along the Japanese coast decreases (Fig. 4.14A). Since DIN transported along the Japanese coast is consumed by phytoplankton (to be organic matter), ΔDIN decreases there (Fig. 4.14A). The organic matters decompose, and regenerated nutrients are consumed by phytoplankton quickly while transported along the Japanese coast. As a result, chlorophyll-*a* concentration increases along the Japanese coast in autumn but there is no change in DIN concentration (Fig. 4.14F). Model experiment shows that primary production increases south of 38°N are associated with DIN flux through the TS from summer to autumn although the increase processes are different along the coasts in Korea and Japan.

We assumed that DIN flux through the TS increase only from summer to autumn based on observed results. However, there is a possibility of change of DIN flux in other seasons. We conducted the case of DIN flux increase in winter. In this case, the increase area of DIN and chlorophyll-*a* concentration is limited in the southwestern part of the JS. This is because DIN flow into the JS is transported to the euphotic zone in short time due to winter vertical mixing, and the DIN is consumed by the phytoplankton near the TS. From the winter case result, it is found that change of DIN flux in summer contributes to change in primary production in large area.

4.2.4 Origin of DIN Flow Into the Japan Sea

In previous sections, we investigated how the response of lower trophic ecosystem in the JS is associated with DIN flux change through the TS, and it is revealed that the change of DIN flux through the TS contributes to change in primary production in large area. However, we cannot predict future marine environment in the JS if we do not know the reasons of DIN flux change through the TS. Fresh water flux through the TS estimated from our observed data in TS is comparable with the Changjiang River discharge; almost the same amount of fresh water from the Changjiang River flows into the JS. DIN concentration in the Changjiang River is larger than 100 μM (Wang, 2006). If DIN and fresh water have a same behavior, a number of DIN originated from the Changjiang River are supplied to the JS. There is a possibility of drastic increase of DIN concentration in the Changjiang River water and change of N/P ration because the economic growth rate in China is very high. Therefore, if DIN in the TS involved the DIN originated from the Changjiang River, DIN flux through the TS might be changed. We look into the origin of the DIN in the TS in this section.

We use almost the same physical-ecosystem coupled model for JS but the target area is the ECS. Since enormous DIN is supplied from the Changjiang River, N/P ratio near the river mouth might be high. Growth of phytoplankton is limited not by DIN but also by DIP there. Therefore, we consider both DIN and DIP as nutrients in the model for ECS. Contribution ratio of each DIN source to DIN in the TS is calculated by the same method in Section 4.2.3. We consider 6 DIN sources; (1) ECS and the Yellow Sea, (2) Taiwan Strait, (3) Kuroshio and the Pacific Ocean, (4) the Changjiang River, (5) the Yellow River, (6) Dust from the atmosphere. Physical field of current velocity and water temperature are used in 2006 (normal year of DIN flux in the TS), after getting a quasi-steady state, 6-labeled DIN are trajected.

Seasonal variation in contribution ratio of DIN source of Kuroshio and the Pacific Ocean and that of the Changjiang to DIN in the TS are shown in Table 4.2. Contribution of DIN of Kuroshio and the Pacific Ocean is high throughout the year and those are 77%–79%. On the other hand, contribution of DIN of the Changjiang is not so high; the maximum is approximately 10% and the average is 6.6%. Considerable freshwater originated from the Changjiang flows into the JS through the TS but less Changjiang-originated DIN transports to the JS. Therefore, it is predicted that DIN flux through the TS might not increase if DIN concentration in the Changjiang is drastically higher in the near future. From the above results, though we could not conclude the reason why DIN concentration in the TS varies year-to-year, observed variation of DIN concentration of 2 μM might be explained by variability in DIN contribution from Kuroshio and the Pacific Ocean. We have to do further research in terms of DIN flux variation in the TS.

TABLE 4.2 Seasonal Variation in Contribution Ratio of DIN in Kuroshio and the Pacific Ocean and DIN in the Changjiang to DIN in the Tsushima Straits

		Summer	Autumn	Winter	Spring	Yearly
Sources	Kuroshio+Pacific	79.1%	78.9%	78.9%	77.7%	78.6%
	Changjiang	9.9%	4.8%	6.5%	6.5%	6.6%

4.2.5 Summary

Primary production north of 38°N in the JS is sustained by vertical nutrient inputs from lower layers and south of 38°N in the JS is preserved by horizontal transport of nutrients from the TS. Nutrient supply from Japanese rivers is not so important for primary production in the JS except near river mouths. Nutrient concentration in the TS varies seasonally and interannually. Range of year-to-year variations in DIN flux through the Eastern and Western channel of the TS are 2–7 and 4–14 kmol/s, respectively. Such kinds of large variation in DIN flux attribute to large variability in chlorophyll-*a* concentration in the JS. If DIN flux through the TS is increased by 50% to the average DIN concentration in the JS averaged upper 100 m depth would be increased by about 5 μM and chlorophyll-*a* concentration averaged upper 20 m would be increased by 0.1 mg/m^3. Variations in DIN and chlorophyll-*a* concentrations associated with observed DIN flux change in the TS are not related to eutrophication. Since residence time of the DIN in the upper layer of the JS is approximately a year, marine environment in the JS might not be worse under the condition of current state.

Though a large amount of freshwater originated from the Changjiang flows into the JS through the TS, large parts of DIN in the TS is contributed by Kuroshio and the Pacific Ocean, and the contribution of DIN in the Changjiang is small. It is a concern that eutrophication in the ECS and the Yellow Sea due to drastic economy growth in China influences the marine environment in the JS. However, those influences on the JS is not so large under the current circumstance. However, if DIN concentration in the Changjiang further increases from now on, a changed lower trophic ecosystem in the JS is a possibility. N/P ratio in the Changjiang River would increase if DIN concentration further increases. Once N/P ratio increases in the ECS, DIP concentration in the ECS is depleted, and phytoplankton cannot consume DIN there. As a result, higher concentration of DIN water mass is transported to further east. Namely, DIN flux change through the TS is not a linear response to the change in DIN concentration in the Changjiang.

In order to understand the marine environment, especially the lower trophic ecosystem, in the JS, it is necessary to conduct continuous observation of nutrient concentration in the TS which is only one entrance of the JS, and to monitor nutrient concentration and N/P ratio from the Changjiang River mouth to the TS. We do not know the reason of interannual variations in DIN concentration in the TS at the moment. It is suggested from the results in our numerical model that variability of DIN originated from Kuroshio and the Pacific Ocean might contribute to the DIN change in the TS. In order to understand the future marine environment in the JS, we have to elucidate control factors to change DIN transport from Kuroshio and the Pacific Ocean to the TS.

4.3 DESIGN OF A MARINE PROTECTED AREA (MPA) BASED ON TRANSPORT AND SURVIVAL MODEL FOR LARVAE

Xinyu Guo, Yucheng Wang, Taishi Kubota, Naoki Yoshie

4.3.1 Introduction

In addition to the natural parks and natural environment conservation areas in Japan, marine protected areas (MPAs) are established to ensure preservation of fishery resources, coastal fishery resources development and common fisheries. Most MPAs are in coastal areas, and very few MPAs have been established in the open ocean as would be required from the viewpoint of biodiversity conservation and sustainable use of fishery resources. In the 2016 report by the Japanese Ministry of the Environment titled, "Ecologically or Biologically Significant Marine Areas Identified by Japan", an important offshore marine area was identified based on the life history of organisms. We expect additional MPAs of this type will be similarly identified in the future. In this study, we present results for MPAs established for Japanese common squid and snow crab, which are representative species in the Japan Sea, and we developed a method to establish a guide the development of new MPAs in a way that enhances larval resources.

4.3.2 Model for the Transport and Survival of Japanese Common Squid Larvae

Using results of the three-dimensional ocean circulation model named DREAMS_M (Data assimilation Research of the East Asian Marine System medium resolution model, Research Institute for Applied Mechanics, Kyushu University, Japan), we established a transport and survival model for simulating the feeding migration of Japanese common squid (*Todarodes pacificus*) larvae.

In the model, we treat the larvae of Japanese common squid as particles that are carried horizontally with the ocean current. Using Eqs. (4.10) and (4.11), the particle positions $(x(t+\Delta t), y(t+\Delta t))$ can be calculated based on the current position $(x(t),(y(t))$ and ocean current velocity $(u(x,y,z,t),v(x,y,z,t))$.

$$x(t + \Delta t) = x(t) + u\Delta t + \xi\sqrt{2A_H\Delta t} \qquad (4.10)$$

$$y(t + \Delta t) = y(t) + v\Delta t + \xi\sqrt{2A_H\Delta t} \qquad (4.11)$$

Here, Δt is the time step, u, v is flow velocity in the eastward and northward directions, A_H is horizontal viscosity coefficient, and ξ is a random number in the range of -1 to 1 with a standard deviation of 0.28. These variables were provided by DREAMS_M at 1-day intervals, and the calculation period was set from 1992 to 2013.

In the vertical direction, vertical movement of squid larvae was calculated using the following equations.

$$z(t + \Delta t) = z(t) + V_T\Delta t \qquad (4.12)$$

$$V_T(x, y, z, t) = w(x, y, z, t) + V_R + V_S \qquad (4.13)$$

Here, w is vertical flow velocity, V_R is velocity of the random walk movement, and V_S is velocity of Ontogenic Vertical Movement (OVM) based on life stage-specific directed swimming speed. w and V_R were calculated based on model results of DREAMS_M. V_S was set to 0.36 cm/s and its direction was toward the stage-specific preferred depth according to life stage (Kim et al., 2015). For particles already in the stage-specific preferred layer, Vs was set to 0.

Initial distribution of particles was determined by these three conditions.

(a) Potential spawning areas

According to Sakurai (2014), Japanese common squid usually spawns in a water depth of 100 to 500 m at the continental shelf and shelf slope from the East China Sea to the Japan Sea. Therefore, we set grid points in our model to depths of 100 to 500 m as potential positions for releasing particles.

(b) Sea surface temperature

Japanese common squid eggs usually hatch in areas with a sea surface temperature of 19.5°C to 23°C (Sakurai, 2014). Using this selection condition, we further selected grid points for hatching from among the grid points determined by the water depth condition in (a) and positioned nine particles at each grid point for hatching.

(c) Parent fish

Based on the assumption that the number of larvae increases with the number of parent fish, the number of particles determined by conditions (a) and (b) was further adjusted using the ratio of parent fish as a condition to consider the distribution density of larvae.

In this study, we performed a particle tracking calculation represented by Eqs. (4.10) to (4.12) for the period from October 1992 to March 2013. In the calculation for a specific year, the particles were released at 10-day intervals from October 1 to March 31 of the following year at grid points determined by conditions (a), (b) and (c) as described above. The tracking calculation for particles was carried out for 180 days from the release day. During the tracking process, survival conditions for each particle were based on temperature of the surrounding water. The lower temperature limit for survival was linearly changed from 15 to 10°C during the first month after release and fixed to 10°C for later stages; the upper temperature limit for survival was fixed to 23°C (Yamamoto et al., 2002; Yamamoto et al., 2007; Kim et al., 2015). The OVM condition for each particle was based on Kim et al. (2015) (Fig. 4.15).

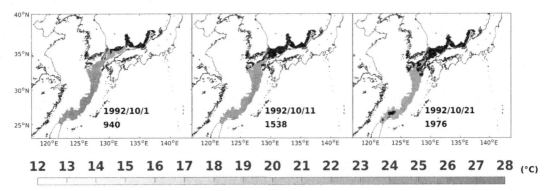

FIG. 4.15 Initial release position for particles on Oct. 1, 11, and 21, 1992. Black dots indicate locations for releasing particles, and numbers represent the total number of grid positions for releasing particles. Mean sea surface temperature is indicated by gray-scale color. From Oct. 1 to Oct. 21, 1992, the total number of particles increased and more particles were distributed in southward areas.

4.3.3 Model for the Transport and Survival of Snow Crab Larvae

The calculation methods for snow crab are similar to those for Japanese common squid. For the transport model of snow crab, current field and temperature were also necessary. Daily current and water temperature data outputs from DREAMS_M were used here.

For the initial distribution of snow crab larvae, grid points in the region from 35.3 to 38.3°N and from 131 to 137.7°E with a water depth from 200 to 500 m deep and sea surface temperature from 5°C to 16°C were chosen as locations for releasing particles. To consider the standing stock of parent fish, we divided the initial distribution area into seven sub-regions: Hamada, around Oki, north of Oki, Tajima, and the offshore areas of Wakasa, Kaga and Noto. The number of particles to release was based on the number of parent crabs (age 11) in each sub-region (Ueda et al., 2015).

For calculations of tracking particles, we considered current flow and random walk movement in the horizontal direction as well as vertical current flow, sinking velocity and diurnal vertical migration in the vertical movement of particles. Sinking speed (2 m/day) was referenced from sampling data of zoea and megalopa larvae in the Wakasa Bay off the coast (Kon et al., 2003). Diurnal vertical migration corresponding to the second zoea period (30 to 60 days after release) was assigned to particles in a time-dependent manner with velocity of 5 m/h from 11 p.m. to 9 a.m., 12.5 m/h from 9 a.m. to 1 p.m. and −10.0 m/h from 1 p.m. to 11 p.m. (positive values represent downward movement). The survival conditions were water temperature of 5°C to 16°C for particles up to 60 days after release and 5°C to 14°C after 60 days. If the particles are subjected to water temperatures beyond these survival ranges for more than 1 day, we treated it as a dead particle and stopped tracking it.

In reality, larvae of snow crab continue to grow after landing on the seabed in the molt between the megalopa and juvenile crab stages (Adachi, 1994). Therefore, only particles that settle onto the seabed can realistically become resource stocks, and the distribution of snow crab by water depth is from 200 to 500 m (Ueda et al., 2015). Based on these conditions, if a particle reaches the seabed where the depth is between 200 and 500 m after 90 days from release, we stopped tracking it. Finally, the particles that landed on the seabed in the area from 35.3 to 38.3°N and 131.0 to 137.3°E were counted and compared to snow crab numbers in the same area (Ueda et al., 2015).

We set the calculation period (1999–2013) based on two factors: a standing stock of age 11 female crabs were available for the period of 1999–2015 and model output for current and water temperature was available for 1993–2013. Particles were released on February 1 (Konishi et al., 2011) which is taken as the first day of the spawning season, and subsequently at 10-day intervals until the end of April. The total times of particle releases per year is 9.

The particle tracking period needs to cover the time required for larvae to hatch and settle. There are several reports on the period required for snow crab larvae to hatch and settle. Although not precisely determined, this is generally considered to take about 3 months (Adachi, 1994). In this study, the entire particle tracking period was set to be 120 days but at 3 months from release, we started to make judgments on whether the particles had settled on the seabed.

If a particle reached the water surface, the shore, or the ocean floor during the calculation period, we return it to its position on the previous day. However, at 90 days after the release of the particles, we stopped tracking the movement of the particles, provided the settlement conditions were satisfied.

4.3.4 Interannual Variation in Japanese Common Squid Larvae

In order to investigate the interannual variation of population size of Japanese common squid larvae transported into the Japan Sea, the five cases shown in Table 4.3 were considered. As an example, initial and destination positions of particles (after 180 days of tracking) released on November 20, 1992 are shown in Fig. 4.16 for all five cases. Larvae of

TABLE 4.3   Model Scenario Conditions

Case	Ocean current	Random walk	Survival condition	OVM	Number of parent fish
Case1	○	×	×	×	×
Case2	○	○	×	×	×
Case3	○	○	○	×	×
Case4	○	○	○	○	×
Case5	○	○	○	○	○

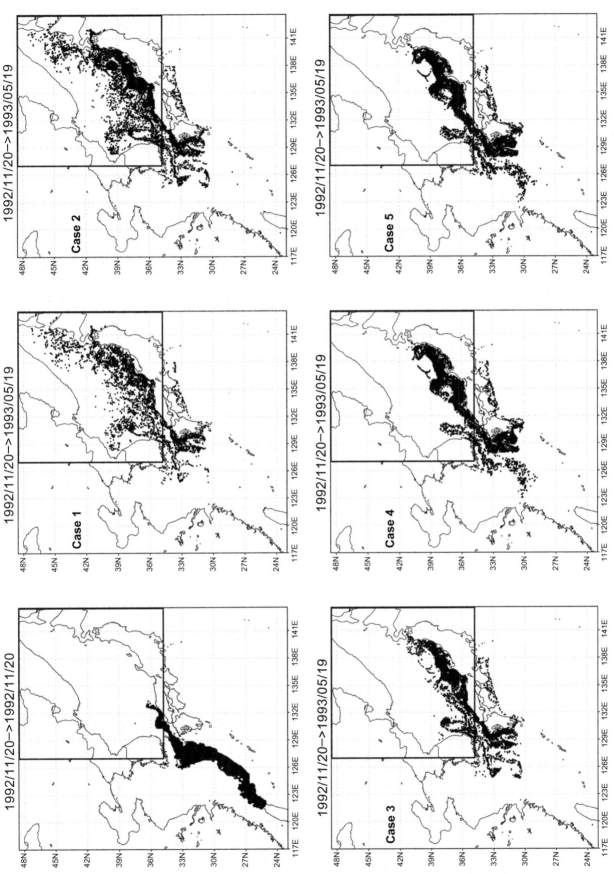

FIG. 4.16 Distribution of initial and destination (after 180 days) positions of larvae released on November 20, 1992 for Cases 1 to 5. Rectangles indicate the domain of the Japan Sea.

Japanese common squid released on November 20, 1992 were distributed in the regions from the East China Sea to the Tsushima Strait, and 180 days later, most of these larvae were transported into the Japan Sea while very few particles were transported to the south coast of Japan. There was little difference between Case 1 and Case 2, indicating the limited influence of random walk on the transport of larvae. For Case 3 in which the survival condition was introduced, the particles in the northern part of the Japan Sea died due to low temperature, and the particles that remained were in the area impacted by the Tsushima warm current. The result for Case 4 which introduced OVM was almost the same as for Case 3, indicating the limited effect of OVM on particle survival in the Japan Sea. On the other hand, many larvae born from January to March were transported to the area south of Japan (data not shown) and, consequently, the ratio of particles entering the Japan Sea to the total released was small. This seasonal variation coincides with results of a survey on the Japanese common squid (Kidokoro et al., 2015).

Interannual variation of larvae surviving in the Japan Sea is shown in Fig. 4.17. Although there are differences in particle number for Cases 1 to 4, the differences are slight and patterns are similar. Only Case 5, which considered variation in the number of parent fish, shows large interannual variation in the number of larvae surviving in the Japan Sea and has a much different pattern than for Cases 1 to 4. The surviving particles in the Japan Sea for Case 5 have a correlation coefficient of 0.84 with Japanese common squid resource information from surveys conducted in the Japan Sea (Kidokoro et al., 2015).

Based on these results, it is likely that transport and survival models for the Japanese common squid can accurately reproduce the interannual variations in resources in the Japan Sea. It also becomes clear that the number of parent fish is the most important factor affecting the Japanese common squid resource. Therefore, we can expect establishment of an MPA to positively impact the Japanese common squid resource if its establishment protects parent fish.

4.3.5 Interannual Variation in Snow Carb Larvae

In the initial distribution of snow crab larvae, a large number of particles were distributed between the 200 and 500 m isobaths of the target area of the sea, as indicated by the black shaded area (left panel of Fig. 4.18). There was some interannual variation in the number of initial particles due to differences in water temperature (data not shown). As an example, a portion of the particles released on February 1, 2004 shown in right panel of Fig. 4.18 were carried from the target area to the northeast by the Tsushima Warm Current and reached the vicinity of the Tsugaru Strait. Approximately 25% of particles remained in the target area.

The yearly distribution of the settled particles obtained from the particle tracking calculation is shown in Fig. 4.19. There is apparent interannual variation in the location and total number of settled particles. For example, the number of particles that settled on the seabed is large in 2004, and the particles are mainly distributed throughout the areas, except for the Hamada region. On the other hand, the number of settled particles was small in 2000, but settled particles

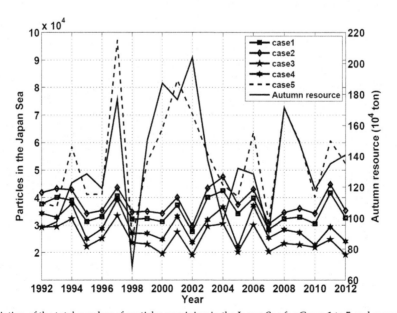

FIG. 4.17   Interannual variation of the total number of particles surviving in the Japan Sea for Cases 1 to 5 and resources of the Japanese common squid determined in autumn field surveys (Kidokoro et al., 2015).

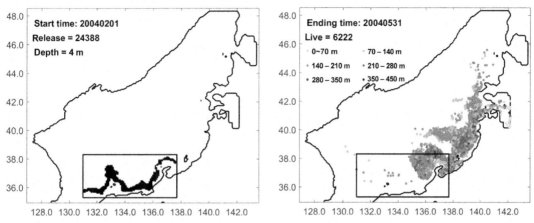

FIG. 4.18   Starting position *(left)* and destination position after 120 days *(right)* for larvae released on February 1, 2004. Rectangle indicates the target area, and particle numbers inside it are given at the left-top corner.

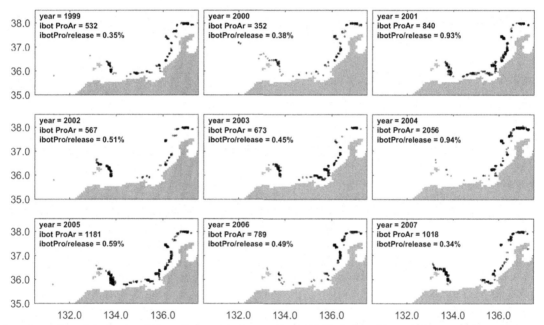

FIG. 4.19   Location and number of successfully settled particles from 1999 to 2007. Black dots indicate the locations of particles. Ratio of number of settled particles to the number of particles released is shows at upper part of each panel.

can be observed over the entire area. As a general trend, more particles settled in the area west of Wakasa and Kaga, more so than in the area west of Hamada and Oki.

Verification of the model was carried out by comparing the number of settled particles derived from model results and snow crab numbers. Snow crab molt several times per year until age 6; then, they molt only once per year (Ueda et al., 2015). Both male and female crabs take 6 or 7 years to reach age 10. By using this feature, the number of crabs at age 10 can be compared with model results.

In order to compare the number of settled particles with the snow crab numbers, it is also necessary to consider the time needed for the settled larvae to grow into crabs with an age of 10. A snow crab reaches the age of 10 in about 6 or 7 years from hatching, and a time lag of 6 years is considered to be needed between particle settling and recruitment into the resource class (Fig. 4.20). In other words, the number of settled particles in 1999 is comparable to the resource of snow crab of age 10 in 2005. Fig. 4.20 shows similar variation patterns between the number of settled particles in 2001 to 2008 and crab resources from 2007 to 2014. It is likely that transport and survival models for snow crab larvae are able to capture the interannual variation in the resource.

FIG. 4.20 Comparison of numbers of settled particles (bold line) and resource of snow crab at age 10. Data of snow crab resource from 1999 to 2014 is from Ueda et al. (2015).

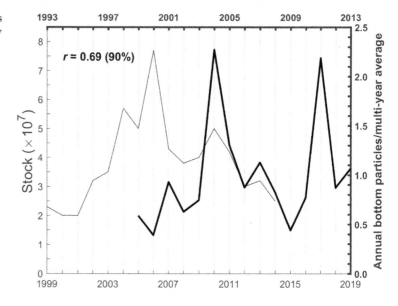

4.3.6 Concept and Method of MPA Design

Snow crab resources in the Japan Sea are protected by catch regulation which prevents overfishing of immature individuals and the existence of the MPA. The MPA for snow crab has an artificial fish reef made of concrete blocks to provide an area where fishing cannot be physically carried out. The MPA provides habitat not only for adults but also for immature individuals and areas for female crabs to reproduce. Since it is considered to be impossible to remove or rearrange the concrete blocks in the MPA, it is assumed that the trawl fishing will also not be possible in the future. Clearly, the cooperation of fishermen is indispensable for establishment of the MPA.

As the snow crab resource has been decreasing since 2007 when the MPA was established, current resource management in and around the MPA probably was not effective. Although catch regulations differ slightly among fisheries, the fishing period is basically the same every year. In this case, it is impossible to physically move the position of the snow crab sanctuary. Consequently, the current MPA offers protection that is always in the same place, but the distribution of snow crab changes every year. While it is ideal to set catch regulations in accordance with the distribution of snow crab, it is difficult to estimate the distribution of adult snow crab every year.

Although it is difficult to obtain the distribution of adult crab every year, it is possible to know the regions where released larvae readily settle onto the seabed through the use of transport and survival models of snow crab larvae obtained here. Using this information, it is possible to predict important areas for spawning. If we set the MPA to those areas important for reproduction, the protection will be highly effective and result in a greater number of snow crabs growing to adults.

Based on this idea, we propose a new resource management method. The model results suggest that the particles that settle later change depending on the release day because movement of particles is strongly influenced by ocean currents. For this reason, effective resource management for parent crabs should include changing the protected period and region every year in accordance of the estimated location of where the released larvae are likely to settle to the bottom. In this study, this concept is called a dynamic MPA. On the other hand, if catch regulation is set to a fixed area and enforced for a fixed period, it is called a static MPA.

In order to establish a dynamic MPA, we first need to find out the initial grid points of settled particles from the results of transport and survival models for snow crab. Next, the numbers of settled particles at each grid point in one year are counted, and the protection region and protection duration are determined so that the number of settled particles will be maximized. By carrying out such screening annually, we obtained a different protection areas and protection durations each year. In this screening, we set the size of the protection region to be $1° \times 4/5°$ and the protection duration to 1 month.

In order to evaluate the effect of the dynamic MPA, the protective effect of the static MPA was calculated, and the size and duration of the static MPA were set to be the same as for dynamic MPA. The static MPA (fixed protection area and fixed protection duration every year) had the highest average settled particles from 1999 to 2013.

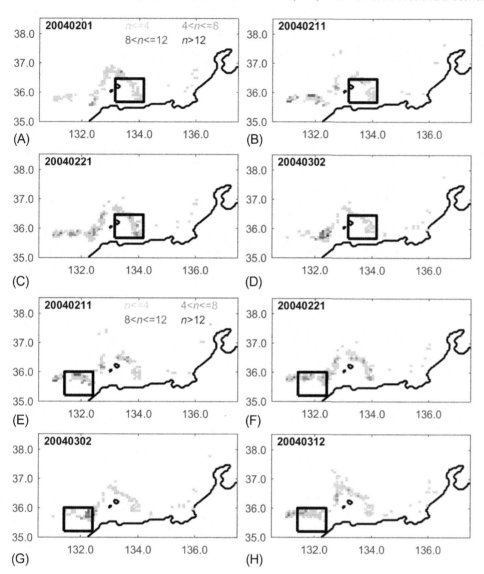

FIG. 4.21 Comparison of dynamic MPA (A–D) and static MPA (E–H) in 2004. The gray scale intensity denotes density of the settled particles. The rectangle encloses the position of the protection region.

Fig. 4.21 shows a comparison of the dynamic MPA and static MPA for snow crab. The protection period and region of the dynamic MPA and static MPA were different in 2004. The protection region of the dynamic MPA and static MPA are the same in some years, but the protection period is different (data not shown). In addition, there are some years when the protection period and region are the same for both the dynamic and the static MPAs.

In order to clarify the difference between dynamic and static MPAs, the percentage of particles protected by the dynamic and static MPAs each year were compared (Fig. 4.22). With an area of $1° \times 4/5°$ and a period of 1 month, the static MPA produced over 10% settled particles, while the dynamic MPA garnered about 20% settled particles. The difference between the dynamic and the static MPAs was most marked in 2011 when the protection ratio of the dynamic MPA was 13 times higher than that of the static MPA. The protection ratio of the dynamic MPA was generally high and stayed at over 20%. On the other hand, the protection ratio of the static MPA was less than 10% in 6 of the years, and it was even below 5% in 2011. On average, the protection ratio of dynamic MPA was 25.7%, and that of the static MPA was 17.7%. Therefore, using a dynamic MPA over a static MPA can improve the protection ratio by 50%.

We also examined the MPA for the Japanese common squid using a transport and survival model. With an area of 150×350 km and a period of 1 month, the dynamic MPA and the static MPA can protect 20% to 40% of the particles that ultimately survive in the Japan Sea (Fig. 4.23). Unlike the snow crab, there are few differences in efficacy between the dynamic MPA and the static MPA. Spawning and survival of squid larvae depend more heavily on water temperature than it does on current field. Since the interannual variation in the water temperature is less than in the flow field, the

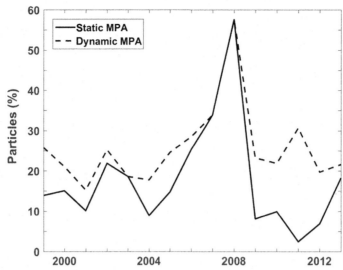

FIG. 4.22   Protection efficacy of particles in an MPA with an area of $1° \times 4/5°$ for snow crab. Vertical axis is the protection ratio of particles. Dashed line is for dynamic MPA and solid line is for static MPA.

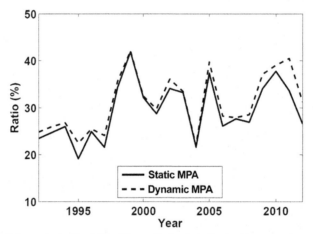

FIG. 4.23   The percentage of the particles protected by MPA. The vertical axis indicates the percentage of particles protected. The dashed line indicates dynamic MPA, and the solid line indicates static MPA. In the calculation, the area of the protection region was set to 150×350 km.

difference between the dynamic and the static MPA for squid was also small. Therefore, it is necessary to examine the efficacy of the dynamic and the static MPAs based on the type of the fishery resources being protected.

4.3.7 Summary

In this study, we developed a transport and survival model for the Japanese common squid and snow crab and demonstrated the predictive abilities by tracking the changes of these two fishery resources over the past 10 years. The numerical experiments examining factors affecting interannual variations of these resources suggested that the number of parent fish is important for predicting interannual variation of the Japanese common squid larvae while both parent fish and the distribution density are important for the snow crab. For both, the model suggested that protection of parent fish is important for increasing the larvae resource.

Establishing an MPA is an effective way to protect parent fish of species such as the Japanese common squid and snow crab. In order to improve the efficacy of the protection, we propose using a new dynamic MPA that can be moved to different areas and implemented for specific periods in response to different patterns in ocean currents on a year-to-year basis. Compared to conventional MPAs, which have a fixed place and time (static MPA), efficacy of the dynamic MPA was examined for snow crab. Compared to the static MPA, the protection ratio of the dynamic MPA was 1.45 times higher on average. In addition, for a smaller area of protection, efficacy of the dynamic MPA is increased more compared to that for the static MPA.

In order to achieve the target of protecting 10% of the sea area as proposed by the Aichi target, establishment of the MPA in Japan's coastal seas is necessary. MPAs must be established based on biodiversity conservation and scientific data (Mukai, 2009). In this study, the static MPA for Japanese common squid and dynamic MPA for the snow crab were introduced and evaluated using transport and survival models of hatching larvae. This approach makes use of field data to predict spawning, distribution and settlement patterns. However, the establishment of an MPA is not easy. It is necessary to solve technical problems, such as development of the numerical model and improving accuracy of future predictions while solving administrative problems such as obtaining the understanding of fishermen.

4.4 ENVIRONMENTAL MANAGEMENT OF TOYAMA BAY

Katazakai Saki, Jing Zhang, and Takafumi Yoshida

4.4.1 Introduction

The main sources of water and material balance to the ocean are contributions from the atmosphere, river runoff, and discharges of hot springs. However, from the viewpoint of water supply from the land to the ocean, submarine groundwater discharge linked with terrestrial groundwater, in addition to rivers, has also been regarded as a source. Fig. 4.24 illustrates previous research works on submarine groundwater discharge. As shown in the Figure, submarine groundwater discharge is a universal phenomenon in coastal zones around the world, and thus studies to clarify the supply of submarine groundwater discharge and its substances have been conducted all over the world.

Many studies on submarine groundwater discharge have been conducted in Japan. The sites of this research include, for example, Rishiri Island, Otsuchi Bay, Suruga Bay, Nanao Bay, Wakasa Bay, Obama Bay, Osaka Bay, Seto Inland Sea, Beppu Bay, and Toyama Bay. In particular, the amount of submarine groundwater discharge in Toyama Bay facing the Sea of Japan is one of the largest such volumes in the world, and it accounts for up to 25% of the freshwater supply flowing into this bay from the land. Furthermore, it has been revealed that the amount of nutrient supply to Toyama Bay is nearly equal to that of river water (Hatta et al., 2005). Therefore, Toyama Bay not only gets a large contribution from submarine groundwater discharge, but it is also assumed that this area has a strong connection with the land. However, climate change such as global warming has been remarkable in recent years, leading to changes in the ratio of rainfall and snowfall, including precipitation and the earlier snowmelt season. As a result, the water quality of freshwater and the recharge capacity of groundwater will change. In other words, the supply of freshwater and substances from land to Toyama Bay by river runoff and submarine groundwater discharge will likely change significantly due to global warming. Consequently, in order to conduct marine environmental management in the future, it is necessary to identify (1) long-term changes in water quality and (2) the resulting changes in the amount of water and its substance supply to Toyama Bay. In this section, we aim to understand the scientific features of river, groundwater,

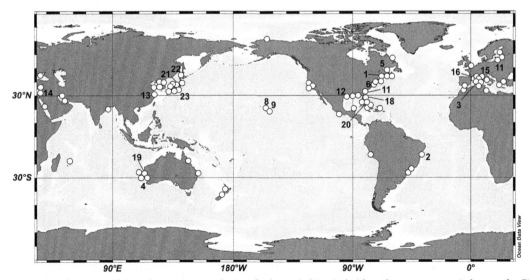

FIG. 4.24 Map showing the points of the submarine groundwater discharge (white circles) based on surveys carried out so far (Katazakai et al., 2018).

and submarine groundwater discharge in Toyama prefecture, the amount of freshwater and its substances transferred from land to Toyama Bay, and the long-term changes in these features.

4.4.2 Importance of Submarine Groundwater Discharge Linked to Groundwater System in the Land

Taniguchi et al. (2002) defined submarine groundwater discharge as SGD=SFGD+RSGD, where SFGD is the submarine fresh groundwater discharge and RSGD is the recirculated saline groundwater discharge. The amount of these two submarine groundwater discharges flowing into the ocean is estimated as 10%–31%. In particular, the submarine fresh groundwater discharge is estimated as 6%–10% of the total submarine groundwater discharge. Therefore, it can be said that the submarine groundwater discharge is an important route of water and substance transport from the land to the coastal zone.

Submarine groundwater discharge linked with the groundwater in land has attracted attention not only as a source of nutrients but also as a source of carbon in recent years. About 20% of the increase in carbon dioxide since the Industrial Revolution has dissolved to groundwater after the weathering process, which refers to its temporary storage in groundwater (Zhang and Mandal, 2012). However, if the water quality, the amount of water, and the residence time of the groundwater change dramatically, the supply of carbon dioxide originating in artificial activities toward the ocean also changes accordingly. As a result, the submarine groundwater discharge releases carbon dioxide to the ocean, that is, it may become one of the carbon sources and promote ocean acidification. Therefore, clarifying the long-term change in groundwater is important for understanding the various material transfers and its supply into the ocean.

Toyama Prefecture has a mountainous region of roughly 3,000 m altitude in the eastern part and extremely steep terrain, with the shortest distance from the mountains to the eastern coastline about 25 km. On the other hand, a slightly gradual alluvial fan spreads in the west of Toyama. In addition, Hatta et al. (2005) reported that submarine groundwater discharge is more prominent in the eastern part than in the western part. Furthermore, we found that the change in groundwater also shows different characteristics depending on the alluvial fan. From the next section, we describe the features of groundwater in the three alluvial fans located in the east-west direction of Toyama.

4.4.3 Features of Groundwater Linked With Submarine Groundwater Discharge and Its Long-Term Changes

In the Katakai River alluvial fan located in the eastern part of Toyama, many studies on groundwater and submarine groundwater discharge are underway. In the coastal zone, the presence of submarine fresh groundwater discharge is confirmed at water depths of 8 and 22 m (Fig. 4.25).

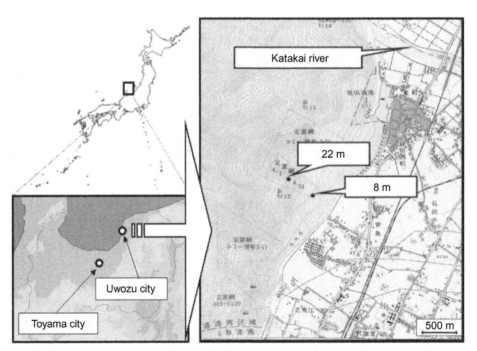

FIG. 4.25   Map showing the points of submarine groundwater discharge in Uozu.

In order to clarify the amount of submarine groundwater discharge, previous work used a measuring device (Tomidai-SGD Flux Chamber) adopting the chamber method as well as observations of the temperature over a wide area of the seafloor sediment. From the results, the amount of submarine groundwater discharge was estimated to be 0.5–1.3 L/min, and the difference in the amount of discharge between different points depended on the surrounding environment, such as the difference in the aquifer from the land to the sea level (Zhang et al., 2005; Koyama et al., 2005). Furthermore, Hatta et al. (2005) estimated from the box model that the amount of submarine groundwater discharge was about 25% of the river runoff. In addition, Hatta and Zhang (2013) also calculated the amount of submarine groundwater discharge using a 2-box model and the vertical distribution of salinity observed in Toyama Bay. From this calculation result, we can assume a high possibility that submarine groundwater discharge occurs even in locations deeper than depths of 0–40 m, where submarine groundwater discharge has been confirmed. Materials such as nutrients are also transported to Toyama Bay through river water and submarine groundwater discharge. The nutrient level supplied by submarine groundwater discharge linked with groundwater in the land is estimated to be equal to or higher than that in rivers (Table 4.4; Hatta et al., 2005). Therefore, the ecosystem of the coastal zone in Toyama Bay is largely supported by submarine groundwater discharge. Furthermore, from a wide-area survey of the Katakai River alluvial fan, it was found that groundwater penetrated the land from an altitude of about 850 to 1200 m passing through an old riverway, was retained for 10 to 20 years, and flowed as submarine groundwater discharge in the coastal zone (Zhang and Satake, 2003). That altitude is a forest area in Toyama, which suggests that the submarine groundwater discharge in the Katakai River alluvial fan is water discharged from the forest area and thus does not greatly influence the water that permeates the plain area.

It has also been reported that the Kurobe River alluvial fan in the eastern part of Toyama is influenced by the artificial use of groundwater. The monthly change in the groundwater level at the observation wells in the Kurobe River alluvial fan showed the lowest result in winter (January to March) (Fig. 4.26; Tebakari et al., 2013). This phenomenon can be seen not only in Kurobe but also in the entire prefecture of Toyama. In other words, it is understood that groundwater volume responds sensitively to anthropogenic influences such as groundwater use. While the groundwater level was the lowest in winter, the groundwater level tended to rise in all observation wells for the summer season. It seems that the groundwater level is restored by the infiltration of snowmelt water and precipitation occurring from spring to summer, and this is also supported by the isotopic composition of groundwater (stable hydrogen and oxygen isotope ratio) clarified by Mizutani and Satake (1997) and Mizutani et al. (2001). Since increases and decreases in the groundwater level can be seen in observation wells throughout Toyama (Toyama Groundwater Guideline, 2018), the discharge of groundwater, especially from spring to summer, is very important in Toyama.

On the other hand, in the Shonagawa alluvial fan in the western part of Toyama, Iwatake et al. (2013) evaluated the long-term change in groundwater using the stable oxygen isotopic ratio of groundwater, which was investigated

TABLE 4.4 Results of Nutrient Flux From Land to Toyama Bay

	Concentration (µmol/L)		Nutrient fluxes (kg/month)		
	River	SGD	River	SGD	SGD/River (%)
PO_4^{3-}	0.09	0.2	7.8	4.3	55%
$NO_2^- + NO_3^-$	13	70	639	851	133%

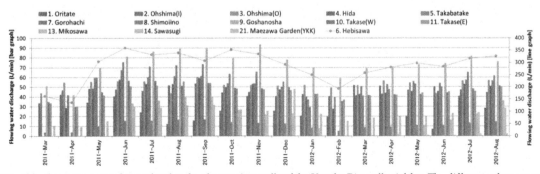

FIG. 4.26 Monthly change in groundwater level at the observation wells of the Kurobe River alluvial fan. The different colors represent different observation well.

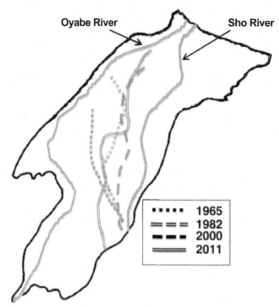

FIG. 4.27 Groundwater recharge status estimated from the $\delta^{18}O$ value. The line shows the boundary of recharge from the two rivers, and the difference in the type of line represents different years of investigation.

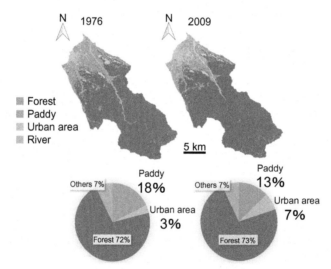

FIG. 4.28 Land use change in Uozu city from data of the National Land Numeral Information Download Service.

between 1965 and 2011 (Fig. 4.27). From this figure, it was found that the groundwater recharge situation changed greatly depending on the survey year. The reasons for changes in groundwater are considered to be (i) change in precipitation, (ii) change in land use, and (iii) change in groundwater usage. In fact, in Uozu City, located in the Katakai River alluvial fan, land use changed as shown in Fig. 4.28. It will be necessary to clarify the relationship between these change factors and the changes in groundwater and river water in the future.

4.4.4 Features of Submarine Groundwater Discharge and Its Long-Term Changes From Model Calculations

In recent years, it has been reported that the annual snowfall amount (average year; statistical period: 1962 to 2014, Japan Meteorological Agency) is decreasing in the Hokuriku region. This change may cause a change in the water quality of river and submarine groundwater discharge and the amount of substances supplied to Toyama Bay. Zhang et al. (2017) analyzed weather information and the groundwater level for about 30 years to examine the influence of climate change on the amount of submarine groundwater discharge. The point shown in Fig. 4.29 is the

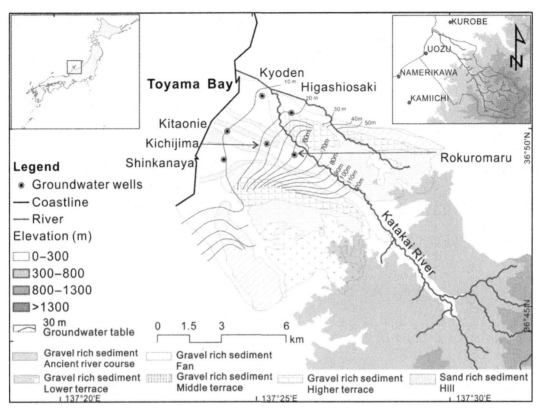

FIG. 4.29 Points of observation wells in the Katakai River alluvial fan.

observation well of Uozu City in Toyama. The weather observation values, such as precipitation and snowfall, are from the Meteorological Agency data.

Table 4.5 shows that the result of linear regression analysis of the groundwater level at observation wells from 1985 to 2015. Except for well No. 3, it became clear that the groundwater level around the fan has risen in the past 30 years. Furthermore, based on the analysis results in Table 4.5, Fig. 4.30 shows the results of estimating the amount of submarine groundwater discharge (water depth 8 m: U 8; water depth 22 m: U 22), and the long-term trend of each discharge amount is expressed by Eqs. (4.14) and (4.15). Y represents the discharge amount, and X represents the time.

$$\text{SGD flux at 8 m}: y = 0.45x + 25.28 \ \left(R^2 = 0.819\right) \tag{4.14}$$

TABLE 4.5 Results of Linear Regression Analysis of Groundwater Level at Each Observation Well

No.	Well	Depth (m)	Screen depth (m)	Data period (years)	Groundwater table (mean ± SD) (m)	Linear regression[a]
1	Shinkanaya	100	72–94	1985–2015	9.71 ± 0.75	$y = 0.01x - 4.75 \ (P = .12)$
2	Shinkanaya	33	17–28	1985–2015	9.69 ± 0.77	$y = 0.12x - 14.99 \ (P < .01)$
3	Kichijima	80	25–36	1985–2015	24.96 ± 2.44	$y = -0.07x + 171.88 \ (P < .01)$
4	Higashiosaki	42.5	9–20	1985–2015	19.77 ± 0.75	$y = 0.04x - 63.77 \ (P < .01)$
5	Kyoden	100	56–67, 78–89	1985–2015	9.86 ± 0.38	$y = 0.01x - 7.30 \ (P < 0.01)$
6	Rokuromaru	38	27–33	2004–2015	42.72 ± 4.08	$y = 0.41x - 780.20 \ (P < .01)$
7	Rokuromaru	80	64–75	2004–2015	40.96 ± 4.01	$y = 0.36x - 683.24 \ (P < .01)$
8	Kitaonie	70	59–71	2004–2015	6.83 ± 0.49	$y = 0.04x - 76.76 \ (P < .01)$

[a] *Linear regression between groundwater table (y) and month (x).*

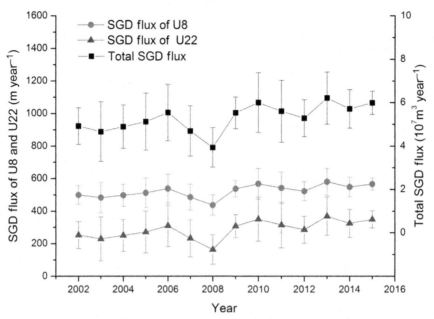

FIG. 4.30   Long-term change in the amount of submarine groundwater discharge released from the groundwater level. *Red* color is the depth at 8 m, *blue* color is the depth at 22 m, and *black* color shows change in total amount.

TABLE 4.6   Water Balance From 1976 to 2015

	Average (mean±SD) mm year^{-1}			Water budget^a 10^7 m^3 year^{-1}			
	Rainfall	**Snowfall**	**Precipitation**	**Evapotranspiration**	**River runoff**	**Groundwater usages**	**SGD**
1976–1996	2311±616	4492±2629	47±13	11±2.9	28±7.5	2.0±0.5	6.0±1.6
1997–2015	2652±300	3282±1346	54±6.1	13±1.5	32±3.6	2.0 ±0.2	6.0±0.68
1976–2015	2473±516	3850±2110	50±10	12±2.5	30±6.3	2.0±0.4	6.0±1.2
2010–2030	2949±150	2573±987	60±3.1	14±0.73	36±1.8	2.4±0.1	7.2±0.37
2030–2050	3147±695	970±387	64±14	15±3.4	38±8.5	2.6±0.6	7.7±1.7

Evaporation from the surface layer increased by 24%, river inflow amount by 60%, groundwater consumption by 4%, and submarine groundwater discharge by 12%, when precipitation is taken as 100%.

^{a} *Water budget is calculated by percentage of evapotranspiration (24%), river runoff (60%), groundwater usages (4%), and submarine groundwater discharge (SGD, 12%) to precipitation from 1976 to 2015 in Uozu.*

$$\text{SGD flux at 22 m} : y = 0.65x - 2.53 \ (R^2 = 0.819) \tag{4.15}$$

These results suggest that the amount of submarine groundwater discharge may increase as the groundwater level rises. In addition, the results of calculating the water balance from the land to Toyama Bay, from 1976 to 2015 based on the long-term weather data, are shown in Table 4.6. The amount of submarine groundwater discharge is estimated to have increased by 30% at maximum. This raises the possibility that the amount of discharge has increased due to the influence of climate change, such as increased rainfall. However, Hatta and Zhang (2013) also reported that the amount of submarine fresh groundwater discharge flowing at water depths of 40–100 m in Toyama Bay is 2–4 times that at water depths of 0–40 m. Thus, the result estimated by Zhang may actually be an underestimation. Therefore, it is necessary to reevaluate the water quality as well as the supply of water and substances flowing from the land to Toyama Bay by continuing surveys of river and submarine groundwater discharge.

4.4.5 Future Measures for Environmental Management of Toyama Bay

In Toyama Prefecture, projects targeting forest areas and groundwater are carried out to manage the environment. In particular, the Satoyama Rebuilding Development Project and the Midori Forest Rebuilding Project financed by the "water and green forest creation tax" are considered to lead to conservation of the submarine groundwater discharge originating from the altitude of the forest area. Regarding groundwater conservation projects, Toyama Groundwater

(A)   (B)

FIG. 4.31 (A) Groundwater recharge published by Toyama Prefecture, (B) example of groundwater recharge project in Toyama (Toyama Groundwater Guideline, 2018).

Guidelines (Toyama Groundwater Guideline, 2018) have been formulated to monitor the water level of observation wells, and groundwater recharge projects are underway. Among them, groundwater recharge projects are being actively carried out (Fig. 4.31A) to show areas suitable for groundwater recharge projects on a map. Furthermore, groundwater recharge projects using idle fields were implemented at four points in Toyama between November 2012 and February 25 using this map (Fig. 4.31B). As a result of the project, the recharge amount per day was found to be 200 m^3. As for penetration into the basement, it is thought that not only water but also substances, including nutrient salts, flow into the basement. Thus, we expect our project to also contribute to knowledge of the nutrient supply from the land area to Toyama Bay.

From the viewpoint of social science, it is also important that residents themselves frequently conduct activities to consider the environment by using the local people's love of their community. A symposium should be held as an event in Toyama, with the aim of generally informing the residents about changes in the water cycle due to the environmental load and the continuous use of the water circulation. It is thought that this symposium project would be an effective policy for creating a ripple effect by communicating the message of a lecture for children to adults.

4.4.6 Conclusion

Based on surveys made so far, it was found that groundwater in Toyama has been influenced by climate change, groundwater usage, and land use. It was also estimated that the amount of discharge due to climate change increased. By actually measuring the submarine groundwater discharge and analyzing its chemical components, we still need to quantify the current amount of submarine groundwater discharge output, the residence time until discharge and the amount of substance supply, and how the supply of water and substances changes. Furthermore, by complementing qualitative and quantitative assessments with the model calculations of water/substance supply in Toyama Bay obtained from these chemical analyses, it is possible to grasp a more accurate picture of the water circulation in the land-coastal zone. Finally, this work can contribute to the planning of adaptation measures for integrated land and ocean management.

4.5 THREE LAYERS MANAGEMENT OF THE JAPAN SEA

Takafumi Yoshida

4.5.1 Introduction

This study aimed to understand the impact of global warming and the environmental changes in the East China Sea on the environment of the Japan Sea as well as the mechanism of the impact. We have found that:

✓ Water temperature in the Japan Sea will continuously increase and could accelerate its speed in a specific scenario;
✓ Global warming will influence not only the rise of water temperature but also the physical and chemical environments of the Japan Sea. It may also change the ecosystem of the Japan Sea;
✓ The relationship between the East China Sea and the Japan Sea is stronger than we assumed. Especially the coastal environment of Japan is controlled by the environment of the East China Sea; and
✓ Land-sea integrated management is effective in several inner bays in the Japan Sea.

Based on these findings, a new coastal environmental management method for the Japan Sea was developed.

4.5.2 Impacts of Global Warming and the Environmental Changes in the East China Sea

In 2015, the Central Environment Council of the Ministry of the Environment, Japan developed the "Report on the assessment of impacts of climate change in Japan and challenges (recommendations)" (Central Environment Council of the Ministry of the Environment, Japan, 2015). This report showed the adaptive planning against the impacts of global warming in Japan, including the degree of influence of global warming on Japan and its severity, urgency of adaptation and information certainty based on scientific information. Referring to the items in the report, the impacts of global warming and the environmental changes in the East China Sea and their severity, urgency of adaptation and information certainty are summarized in Table 4.7 and 4.8.

As shown in Tables 4.7 and 4.8, the Japan Sea receives strong impacts from global warming and from the environmental changes in the East China Sea. Moreover, they may give multiple impacts to the Japan Sea, so the mechanism of the impacts is very complex. Most of the impacts cannot be controlled by current technology; therefore, the objective of this project is to propose several effective and efficient options for adaptation to potential changes.

Through our research, it is clear that an effective coastal management in the Japan Sea cannot be realized by existing local-level coastal managements but by a coastal management which has a wider view on global warming and various environmental changes in the East China Sea. So, we have proposed a new concept: "Japan Sea Three Layers Management".

4.5.3 Japan Sea Three Layers Management

Japan Sea Three Layers Management aims to conserve the coastal environment with three different scale managements. Three-scale managements mean the layers of the following three different geographical scales (Fig. 4.32):

1) Wide-scale management: focusing on the Japan Sea and the East China Sea
2) Middle-scale management: focusing on the flow of the Tsushima Warm Current
3) Local-scale management: focusing on each coastal area and bay on the Japanese coast

In each layer, predicted impacts and their mechanisms were studied and appropriate adaptations and management methods have been proposed.

4.5.3.1 *The First Layer: Wide-Scale Management*

The first layer is the wide-scale management focusing on the Japan Sea and the East China Sea. This area includes territorial seas and exclusive economic zones of littoral countries. Therefore, international management and cooperation among relevant countries are necessary and strongly recommended.

The impacts of global warming are shown in the entire area of the East China Sea and the Japan Sea. Numerical simulation predicted that water temperature in the East China Sea and the Japan Sea will increase due to global warming, and physical environment such as ocean circulation and chemical environment such as nutrient condition will also change. Distribution and migration of marine species in the Japan Sea will change due to the rise in water temperature. Therefore, in this area, it is suggested that the main adaptation should be to develop a surveillance network in order to detect changes and impacts as early as possible. To develop a surveillance network, international cooperation and information sharing among relevant countries are necessary. However, it is difficult for many countries to establish a new marine observation system for surveillance networks. Thus, it is reasonable to identify areas and items for specific monitoring attention and to utilize the existing regular monitoring by central and local governments and research institutes. Based on the findings of our project, we propose the following actions for monitoring the impacts of global warming and the environmental changes in the East China Sea.

Constructing a Surveillance Network on Global Warming

The first focused item is the rise in water temperature, the basic parameter to understand the impact of global warming. Numerical simulation predicted that the surface water temperature will increase by 2.75°C in the southern part of the Japan Sea and 3.13°C in the northern part by 2100, using global warming scenario RCP 8.5 (Fig. 4.33, Yoshida et al., 2018). The rise in water temperature will occur in the entire Japan Sea, so a surveillance network needs to cover the whole area of the sea. In this region, there are several international marine observation systems. One of them is the

TABLE 4.7 Assessment of the Impacts of Global Warming on the Japan Sea

Topics	Items	Current situation	Potential future impacts	Severity		Reason	Urgency	Credibility of information
				Point of View				
Environment	Japan Sea (Coastal area)	There is a clear trend of increasing water temperature (average: 0.039°C) in 207 monitoring sites from 1970 to 2010 (Central Environment Council of the Ministry of the Environment, Japan, 2015)	It is forecasted that water temperature in the Japan Sea will increase continuously The change of the Tsushima Warm Current due to global warming will cause change of physical, chemical and biological environments in the Japan Sea	◉	Environment	In the coastal areas in the Japan Sea, there are only a few enclosed inner bays and most of the coastal areas face the open ocean. It means the environment of the coastal areas are strongly influenced by the Tsushima Warm Current and the environment of the open ocean	○	○
	Japan Sea (Whole area)	JMA reported that sea surface temperature rose by 1.29°C in the southern part and 1.71°C in the northern part of the Japan Sea in the past 100 years	With two climate change scenarios (RCP 8.5 and RCP 2.6) of the IPCC sixth assessment report, water temperature will increase 2.75°C (by RCP 8.5) and 1.56°C (by RCP 2.6) in the southern part and increase 3.13°C (by RCP 8.5) and 1.04°C (by RCP 2.6) in the northern part of the Japan Sea. Global warming may impact not only water temperature but also the flow of the Tsushima Warm Current, and nutrient condition and ecosystem in the Japan Sea	◉	Environment	The speed of water temperature rise may be accelerated in a bad scenario The change of water temperature, ocean circulation, and primary production may change the distribution and migration of marine species in the Japan Sea	◉	○
Ecosystem	Japan Sea (Coastal area)	The number of research on the impact of global warming on primary production in the Japan Sea is small It is reported that low-temperature species are replaced by high-temperature species in many sea areas of the Japan Sea	Continuous water temperature rise is predicted and it causes change to high-temperature species and its expansion of distribution. The change of pathway of the Tsushima Warm Current could lead to change of migration and/or transportation routes	◉	Environment	Because of the possibility of change in primary production, the distribution and migration of species in the Japan Sea may change as well	◉	△

Continued

TABLE 4.7　Assessment of the Impacts of Global Warming on the Japan Sea—cont'd

Topics	Items	Current situation	Potential future impacts	Severity		Reason	Urgency	Credibility of information
					Point of View			
Fishery	Migration species (Japanese common squid)	Change of distribution and migration area due to water temperature rise was reported	Change of spawning ground and timing due to water temperature rise is predicted. It may impact stock biomass of species. Change of migration route (both adult and larvae fish) due to change of the Tsushima Warm Current pathway may cause change of fishery ground	◎	Society Economy	In recent years, decrease in fishery resource in Japanese common squid has been indicated, thus it is necessary to tackle this matter as early as possible	◎	○
	Benthic species (Snow crab)	There is little research on the impact of global warming on benthic species which live in deep waters. Change of water temperature in deep water is small compared with the surface, however, it is reported that the marine environment during the floating larval stage influences stock biomass	Water temperature rise in 200–500 m depth is not predicted, so the impact on adult crab may be small. On the other hand, water temperature rise and pathway change of the Tsushima Warm Current change the transportation and distribution of egg and larvae, and it is forecasted that the settlement area of larval crab will move northward	◎	Society Economy	Settlement rate of larval crab in the current main fishery ground in the western part of the Japan Sea may decrease, and in the eastern and northern parts of the Japan Sea may increase. If so, the current fishery environment may change	○	○
Water environment, water resource	Water supply (Groundwater)	Snowfall and timing may change due to global warming, which could influence the amount of groundwater	Due to the decrease of snowfall and early snow melting, the supply system of groundwater may change. In addition, groundwater is used for snow melting in city areas, and it causes decrease in groundwater levels	○	Society Economy	In recent research, it was reported that groundwater (Sub-marine groundwater discharge) contributes to coastal ecosystem and its production. Groundwater conservation may contribute not only to effective utilization of groundwater but also to the conservation of coastal ecosystems	△	△

【Severity】◎: High, ○: Middle, △: Lack of information, 【Urgency】◎: High, ○: Middle, △: Lack of information, 【Certainty】◎: High, ○: Middle, △: Lack of information.

TABLE 4.8 Assessment of the Impacts of the Environmental Changes in the East China Sea

Topics	Items	Current situation	Future impacts	Severity	View	Reasons	Urgency	Credibility of information
Environment	Japan Sea (Coastal area)	Low salinity water originated from river discharge in China flows into the Japan Sea. It is observed by monitoring at coastal set nets that low salinity water is transported by the Tsushima Warm Current to coastal areas of Japan	The environment of the coastal areas of the Japan Sea is controlled by the environment of the East China Sea. Therefore, environmental changes in the East China Sea may directly influence the environment of coastal areas of the Japan Sea. In addition, it is forecasted that the Tsushima Warm Current will be strengthened by global warming. It brings strong impacts from the East China Sea Water temperature rise and change in the Tsushima Warm Current influence distribution of marine species and their migration route in the Japan Sea Low salinity water is transported to the Tohoku region by the Tsushima Warm Current. In the river discharge water and coastal waters in the East China Sea, there are many Persistent Organic Pollutants (POPs); these POPs will be transported to the coastal areas of Japan with low salinity water. It is necessary to monitor the impact continuously	◎	Environment Ecosystem Society Economy	It was shown that the coastal areas of the Japan Sea are strongly impacted by the East China Sea 80% of coastal waters of the Japan Sea are originated from the East China Sea, and nutrient for primary production is also sourced from the East China Sea	◉	○
	Japan Sea (Whole area)	There is little research on the size of the impact from the East China Sea	It is predicted that change of inflow of nutrient through the Tsushima Strait causes change in primary production and its timing in a wide area of the Japan Sea The pathway of the Tsushima Warm Current will be moved northward by global warming. It means sea areas impacted by the East China Sea may change in the future	◎	Environment Ecosystem Society Economy	The extent of the impact of the East China Sea may change due to the fluctuation of the Tsushima Warm Current by global warming	◉	○

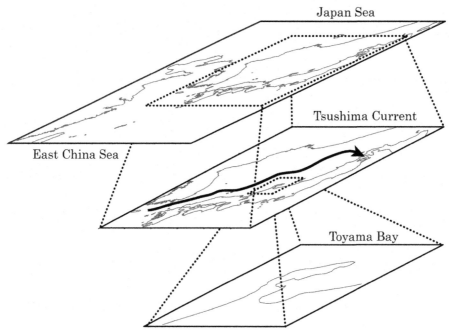

FIG. 4.32 Japan Sea three Layers Management. *Upper* shows the first layer (Wide scale), *middle* is the second layer (Middle scale), and *bottom* is the third layer (Local scale).

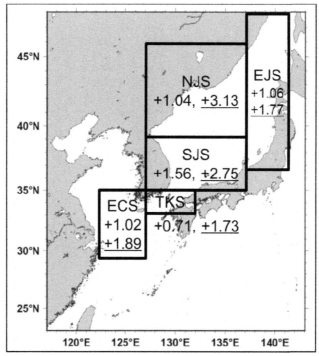

FIG. 4.33 Forecast of increase of sea surface water temperature by 2100. The numbers are forecasted average rise in the water temperature by scenario RCP 2.6, and the numbers with underlines are by scenario RCP 8.5. ECS means the East China Sea, TKS is the Tsushima Strait, SJS is the southern part of the Japan Sea, NJS is the northern part of the Japan Sea, and EJS is the eastern part of the Japan Sea.

North East Asian Regional Global Ocean Observing System (NEAR-GOOS) which China, Japan, Korea, and Russia participate in. NEAR-GOOS aims to develop the international ocean observing system, contribute to the understanding of the marine environment, and monitor changes in the environment. NEAR-GOOS developed a database for sharing and providing observed data, so it is efficient to use this database. In addition to NEAR-GOOS, a joint program by Japan, Korea, and Russia, Circulation Research of the East Asian Marginal Seas (CREAMS) conducts monitoring of the marine environment in the Japan Sea. Using existing international frameworks and databases is an efficient method to develop a surveillance network on global warming.

Regarding the sea surface temperature, the Ministry of the Environment, Japan and the Northwest Pacific Region Environmental Cooperation Center (NPEC) provide data of sea surface temperature monitored by satellite through the website "Marine Environmental Protection of Northwest Pacific Region". To provide wide-scale and long-term data on the sea surface temperature and to contribute to the development of a surveillance network on global warming is an important role for Japan.

The change in nutrient condition in the Japan Sea by the fluctuation of the Tsushima Warm Current is also an interesting phenomenon. Change in the Tsushima Warm Current will be observed mainly in the East Korean Warm Current which flows northward along the Korean Peninsula. The East Korean Warm Current changes its direction to eastward at 38°N. However, the point of the direction change will be shifted to 40°N with the strengthened flow of the Tsushima Warm Current by global warming. This shift will result in transport of warm and high salinity sea water to further northern areas. This high-density water is suggested as one of the reasons why nutrient concentration in the northern part of the Japan Sea will increase. Monitoring the change of the East Korean Warm Current is an effective method to understand the impacts of global warming on the Japan Sea. Korean ocean observation institutes conduct regular monitoring in the coastal areas of the Korean Peninsula, and it is important to share their monitored data.

In the past studies, global warming causes weakening of winter cooling and sinking of the surface water to the deep layer. In the deep layer of the Japan Sea, Japan Sea Proper Water (JSPW), which has cool and oxygen-rich water, is generated. Due to the weakening of sinking, decrease of oxygen concentration in JSPW is reported (Gamo, 2001). However, our prediction shows that vertical mixing will continue because high salinity water is transported by the East Korean Warm Current to the northern part of the Japan Sea and high-density surface water sinks to the deep layer. In addition, high water temperature enhances biological decomposition in the northern part of the Japan Sea. These changes cause increase of nutrient in this region. Such dramatic change of nutrient condition may influence primary production and ecosystem in the Japan Sea, and a surveillance network on nutrient would be necessary in the future. As increase of nutrient will occur mainly in the northern part of Japan Sea, a joint surveillance network between Japan and Russia is needed.

Impacts From the East China Sea

The impacts of the environmental changes in the East China Sea spread to the Japan Sea by the Tsushima Warm Current. We focused on the inflow volume of nutrients and low-salinity water through the Tsushima Strait. As shown in Section 4.2, when the inflow volume of nutrients changes, primary production in the southern part of the Japan Sea changes as well. It is observed by ship survey that the total inflow volume of nutrients through the Tsushima Strait drastically changes year to year. We have continued studying the mechanism of the volume change, and it was identified that the nutrients originate in the bottom water of the Kuroshio Current, namely the sub-surface of the Pacific Ocean. Thus, we propose establishing a surveillance network in the East China Sea and the Japan Sea. When the nutrient concentration is monitored, the nitrate-phosphate ratio (NP ratio) should be monitored as well. Phosphate is a rate-limiting factor for primary production in the Chinese coastal area, so, if the balance of nitrate and phosphate in river discharge is changed, excess nitrate may be transported into the Japan Sea. In addition, the change of NP ratio may cause a change in the composition of phytoplankton species, so it is required to monitor NP ratio carefully.

Regarding low-salinity water, it originates in the Taiwan Warm Current, flowing northward along the coast of China. The current is mixed with freshwater discharged through the Chang Jiang River and others to form low-salinity water. Then, this low-salinity water flows northeastward in the East China Sea and into the Japan Sea through the Tsushima Strait. The salinity of low-salinity water in the Japan Sea is inversely proportional to the volume of river discharge from the Chang Jiang River: when the amount of river discharge from the Chang Jiang River is high, the decrease of salinity intensifies. Our numerical simulation showed that low-salinity water flowed into the Japan Sea from summer to autumn and was transported to Tohoku coastal area with a speed of 350 km/month (Fig. 4.34). In addition, river discharge water and coastal water of China include high level of polycyclic aromatic hydrocarbon (PAH) and persistent organic pollutants (POPs), which means that one of the sources of pollutants in the Japan Sea is transported by the Tsushima Warm Current from the coastal area of China (Hayakawa et al., 2016). The Ministry of the Environment, Japan conducts marine environmental monitoring survey every year; however, the current monitoring takes eight years to be completed going around all the sea areas of Japan. In recent years, the concentration of PAH and POPs has a decreasing trend; however, the Ministry is expected to conduct the monitoring in summer and autumn annually in order to pay attention to the pollution of PAH and POPs.

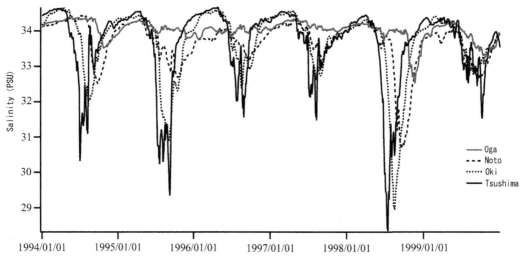

FIG. 4.34   Movement of low-salinity water along the coastal area of the Japan Sea. Change of salinity in four sites, Tsushima, Oki, Noto, and Oga.

FIG. 4.35   Image of surveillance networks. Gray part shows the surveillance network for monitoring of global warming, dotted part is for monitoring of nutrients and meshed part is for low-salinity monitoring.

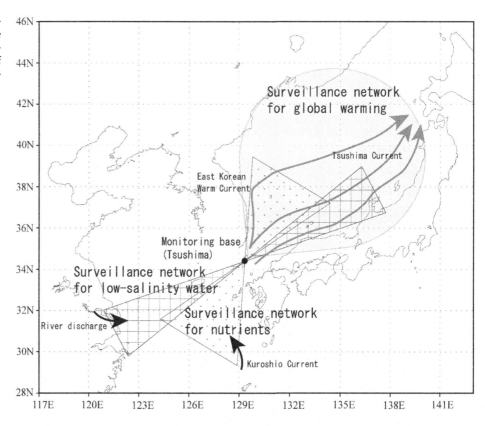

Fig. 4.35 shows the image of surveillance networks in the Japan Sea to monitor the impacts of global warming and the environmental changes of the East China Sea. For our proposed surveillance networks, the monitoring base is set on Tsushima Island. As the island is located in the entrance of the Japan Sea, it is the most appropriate site to monitor the impact from the East China Sea. In addition, the influence of the change in the Tsushima Warm Current to the Japan Sea caused by global warming can be predicted by monitoring at Tsushima Island. While the base of surveillance networks is established on the island, the area of networks can be changed according to the objectives of the observation. For example, when the impacts of global warming are monitored, the surveillance network covers the whole Japan Sea, and when nutrient is monitored, the western part of Kyusyu and the eastern coast of Korea are the center focus. In case of low-salinity water and POPs, the network extends from the estuarine area of the Chang Jiang River to the coastal area of Japan. In Table 4.9, observed items and observed methods are shown.

TABLE 4.9 Observed Items and Methods for Surveillance Networks

Surveillance networks	Observed items	Observed methods
Surveillance network for global warming	Water temperature Current direction and speed Dissolved oxygen Nutrients (Nitrate and phosphate)	Ship Satellite Argo float
Surveillance network for nutrients	Nutrients (Nitrate and Phosphate) Phytoplankton Chlorophyll-*a*	Ship Satellite
Surveillance network for low-salinity waters and POPs	Salinity PAH POPs	Ship

Future environmental changes in the Japan Sea and the relationship between the Japan Sea and the East China Sea were understood through this project, and we see most of the predicted changes causing severe environmental problems that have to be addressed immediately. Both the Japan Sea and the East China Sea are international sea areas, thus it is necessary to collaborate among relevant countries to tackle these issues. In this area, the Northwest Pacific Action Plan (NOWPAP), one of the Regional Seas Programmes of UNEP is adopted by China, Japan, Korea, and Russia in order to conserve the marine and coastal environment. Utilizing the NOWPAP framework is the most effective method to address the above-mentioned problems. However, NOWPAP hasn't initiated any global warming-related activities until now. It is expected that the Japanese government will report the findings of this project to NOWPAP while emphasizing global warming as one of the high-priority issues in this region. We hope this project contributes to the enhancement of international cooperation and the conservation of marine and coastal environment in this region.

Sustainable Use of Biodiversity and Fishery Resources in the Japan Sea

The East China Sea is a spawning ground for many marine species which live in the Japan Sea, such as Japanese common squid, yellowtail, tuna, sardine, and mackerel. Preferable bottom topography and seawater temperature are important factors to form the spawning grounds of these species. So, the increase of seawater temperature by global warming has a huge impact on marine species. In Section 4.3, the impacts of global warming on spawning of Japanese common squid and transportation and survival of its larvae are studied. The results show that the importance of spawning grounds in the East China Sea will be increased for Japanese common squid living in the Japan Sea (Fig. 4.36). Since the borders of EEZ of China, Japan and Korea are located in the East China Sea, international cooperation is necessary for the conservation of Japanese common squid. In recent years, rapid decreasing of the volume of landing has been reported. Japanese common squid is one of the key species which links low trophic species to high trophic species. Conserving Japanese common squid can contribute to the conservation of biodiversity in the Japan Sea. Our study determined the sea area and the timing to contribute reproduction of Japanese common squid living in the Japan Sea and suggested an effective management method. Conservation of a specific sea area and in a specific period of time is called Static Marine Protected Area (MPA). Based on the importance of Japanese common squid as a fishery resource, it is expected this Static MPA will be applied to the spawning and nursery grounds jointly by the three countries, China, Japan, and Korea. The setting of MPA for the conservation of specific marine species in high seas and

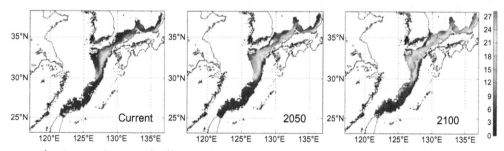

FIG. 4.36 Movement of main spawning grounds of Japanese common squid living in the Japan Sea. Forecast of the movement of main spawning grounds of Japanese common squid from now to 2100 due to global warming. White part indicates high egg density part which contributes reproduction.

offshore areas has been just started in several sea areas in the world. For the sustainable use of fishery resources and the achievement of Aichi Biodiversity Targets, it is desired that our effective conservation methods are actively applied in various areas.

4.5.3.2 *The Second Layer: Middle Scale*

Development of the Tsushima Warm Current Monitoring Network

The environment of the Japanese coastal area is strongly influenced by the impacts from the East China Sea through the Tsushima Warm Current. Thus, it is important to understand the behavior of the Tsushima Warm Current and implement effective management schemes based on the relationship between the current and the Japanese coastal area. Then, the pathway and volume of flow of the Tsushima Warm Current will change from global warming in the future. It means the relationship between the Tsushima Warm Current and the Japanese coastal area will change accordingly.

However, the observations and studies on the Tsushima Warm Current are limited compared with research on the Kuroshio Current and the Oyashio Current. We have to recognize the Japan Sea is a quite important sea area for Japan, and the Japan Sea is highly sensitive to the impacts of global warming and other external factors. So, it is necessary to enhance observations and research on the Japan Sea and the Tsushima Warm Current under government-academia cooperation. On May 2018, the third Basic Plan on Ocean Policy was approved by the Japanese cabinet. In this basic plan, conservation of biodiversity, measures against climate change and ocean acidification, protection against marine pollution and so on are recommended for conservation of marine environment. The Japan Sea is surely the sea area where such actions should be done.

Fortunately, local governments, universities and research institutes located on the Japan Sea side have conducted regular coastal monitoring. To develop an information sharing system among these stakeholders it is necessary to first develop a monitoring network of the Tsushima Warm Current and to enhance the collection of data and the under-standing of the current. The Japan Sea is a wide sea area and various stakeholders are involved. It is expected that the Japanese government will lead various stakeholders and commit to realize the development monitoring networks.

Sustainable Use of Biodiversity and Fishery Resources in the Japan Sea

Snow crab, one of the typical fishery resources in the Japan Sea, live in the depth of 200–500 m. In some areas, snow crab is a highly valuable commercial species. However, in recent years, the stock has a decreasing trend, and no-fishing zones and no-fishing periods are set for their protection mainly in the western part of the Japan Sea. The protection method is setting artificial fish reefs to avoid trawl fishery. In Section 4.3, we simulated transportation and survival of snow crab larvae and showed the sea area which can contribute to the reproduction of snow crab. It is also clarified that this spawning sea area changes annually based on the condition of the ocean environment. Protecting such specific sea areas will support conservation of this species more effectively. We called this protected area the Dynamic Marine Protected Area. However, the location of potential sea areas for Dynamic MPA should be simulated every year. In addition, snow crab is also influenced by global warming. The change in transportation of larvae causes northward shift to the distribution of the species. Now, the western part of the Japan Sea is a major fishery ground of this species; however, in the future, the main fishery ground may be shifted to the northern area. The thinking on the management of snow crab may be drastically changed in the future.

4.5.3.3 *The Third Layer: Local Scale*

On the Japan Sea side, there aren't so many bays and most of the coasts directly face offshore areas. Because the coastal area receives strong impacts from the East China Sea and offshore water, current local land-sea integrated man-agement policy is not working well. In order to understand the effects of the land-sea integrated management in inner bays on the Japan Sea side, we tried to quantify impacts from land and offshore in Toyama Bay (Fig. 4.37), and we studied the effectiveness of land area management. For this study, a new numerical ecosystem model for Toyama Bay was developed. Details of the numerical model is explained in Section 6.4.

The results show that the coastal area of Toyama Bay is influenced strongly by river discharge. Especially, in the euphotic zone (50–60 m depth), nutrient supply from rivers during summer to winter controls the coastal environment. Thus, management in land area can contribute to the conservation of the coastal environment.

In addition, we focused on the change in water circulation from global warming. In Toyama Prefecture, the volume of snowfall in winter is decreasing (Fig. 4.38 left). On the other hand, rainfall in winter has not changed (Fig. 4.38 right). This means that river discharge due to rainfall increase in winter. In past years, water was stocked as snow in the

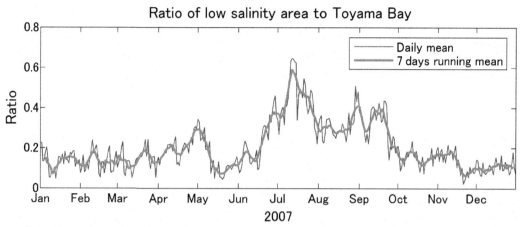

FIG. 4.37 Ratio of low-salinity area in Toyama Bay. The area under 32 PSU (salinity) is defined as low-salinity area influenced by river discharge. The line shows the change in the percentage of low-salinity area in Toyama Bay.

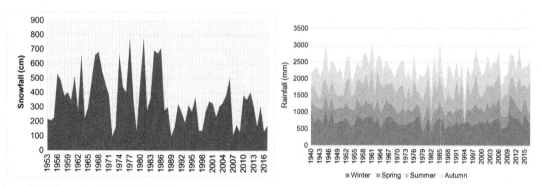

FIG. 4.38 Snowfall *(left)* and rainfall *(right)* in Toyama Prefecture.

mountain area and melted in spring and summer. However, global warming changes such water circulation in Toyama.

Huge snowfall in winter produces rich water and groundwater resources. In Toyama Bay, sub-marine groundwater discharge is identified. The volume of sub-marine groundwater discharge is 20% of the entire river discharge; however, the amount of contained nutrient is the same level as that in river discharge. This rich sub-marine groundwater discharge contributes to the high production in the coastal area (Hatta et al., 2005). However, the groundwater is used for melting snow in the city area in winter, causing a decrease in groundwater level. It is a seasonal phenomenon, yet decrease in snowfall and rice paddy area also led to less recharge of groundwater. In such climate and societial changes, how to manage groundwater is an important issue to be resolved. The Ministry of the Environment, Japan promotes the integrated environmental management from forest to ocean. Then, we propose adding a new component, groundwater management, into the ministry's plan for a new land-ocean integrated management method for sustainable use of coastal ecosystems.

4.5.4 Conclusion

Our group studied necessary/appropriate management for the Japan Sea, an international enclosed sea area, based on the impacts of global warming and environmental changes in the East China Sea. The study shows that the Japan Sea is more sensitive to external/internal changes than we imagined. The coastal environment of the Japan Sea is strongly influenced by these impacts, and the local coastal area management is not useful. Rather, coastal area management in the Japan Sea should consider the impacts of global and regional environmental changes. Our group proposed a new idea "Japan Sea Three Layers Management" based on the findings of our project. Unfortunately, the Japan Sea has not received any strong attention in the ocean policy of Japan until now. We have to see the Japan Sea in a different light and implement environmental conservation and marine management more actively in the future.

Japan Sea is an international semi-enclosed sea area and international cooperation among relevant countries is essential for its management. In this area, NOWPAP is established for international environmental cooperation to pursue marine environmental conservation. It is recommended that the Japanese government use this framework more effectively for the conservation of the Japanese coastal area. In addition, it is expected that Japan plays a leading role in enhancing international cooperation and conservation among the littoral countries of this area. We hope the outcomes of this project can contribute to the conservation of the Japan Sea.

References

Adachi, J., 1994. Biology and Stock of Snow Crab in the Western Part of Japan Sea. Report of Shimane Prefecture Fisheries Experimental Station, vol. 8, pp. 127–170 (in Japanese).

Boyer, T.P., Antonov, J.I., Baranova, O.K., Coleman, C., Garcia, H.E., Grodsky, A., Johnson, D.R., Locarnini, R.A., Mishonov, A.V., O'Brien, T.D., Paver, C.R., Reagan, J.R., Seidov, D., Smolyar, I.V., Zweng, M.M., 2013. Levitus, S. (Ed.), World Ocean Database 2013. In: NOAA Atlas NESDIS, vol. 72, p. 209. A. Mishonov, Technical Ed.

Central Environment Council of the Ministry of the Environment, Japan, 2015. Report on the Assessment of Impacts on Climate Change in Japan and Challenges (Recommendation) (in Japanese).

Gamo, T., 2001. Recent upward shift of the deep convection system in the Japan Sea, as inferred from the geochemical tracers tritium, oxygen, and nutrients. Geophys. Res. Lett. 28 (21), 4143–4146.

Hatta, M., Zhang, J., 2013. Temporal changes and impacts of submarine fresh groundwater discharge to the coastal environment: a decadal case study in Toyama Bay, Japan. J. Geophys Res: Oceans 118, 2610–2622.

Hatta, M., Zhang, J., Satake, H., Ishizaka, J., Nakaguchi, Y., 2005. Water mass structure and fresh water fluxes (riverine and SGDs) into Toyama Bay. Chikyuk-agaku (Geochemistry) 39 (3), 157–164 (in Japanese).

Hayakawa, K., Nagao, S., Aramaki, T., Takazawa, Y., Kameda, T., Sato, K., Inomata, Y., Ohizumi, T., 2016. Study on Potential Threat Caused by Organic Pollutants in the Japan Sea, Surrounding Sea and Atmosphere. Environment Research and Technology Development Fund of the Environmental Restoration and Conservation Agency of Japan. 5-1306 (in Japanese).

Hirose, N., 2011. Inverse estimation of empirical parameters used in a regional ocean circulation model. J. Oceanogr. 67, 323–336.

Hirose, N., Takayama, K., Moon, J.-H., Watanabe, T., Nishida, Y., 2013. Regional data assimilation system extended to East Asian marginal seas. Umi Sora 89, 43–51.

Iwatake, K., Mizoguchi, T., Tomiyama, S., Zhang, J., Satake, H., Ueda, A., 2013. Geochemical study of groundwater in the Sho river fan, Toyama Prefecture for heat usage by geothermal heat pump. Geochem. J. 47, 577–590.

Katazakai, S., Kambayashi, S., Zhang, J., 2018. Water discharge and material fluxes into the coasts via submarine groundwater discharge and its long-term change. Current studies and a case study of Toyama Bay. Kaiyo Monthly 61, 43–50 (in Japanese).

Kidokoro H., Goto, T., Takahara, H., Matsukura, R., 2015. Resource evaluation of Japanese common squid in Japan Sea in 2015, http://abchan.fra.go.jp/digests27/index.html (in Japanese)

Kim, J.-J., Stockhausen, W., Kim, S., Cho, Y.-K., Seo, G.-H., Lee, J.-S., 2015. Understanding interannual variability in the distribution of, and transport processes affecting, the early life stages of Todarodes pacificus using behavioral-hydrodynamic modeling approaches. Prog. Oceanogr. 38, 571–583.

Kishi, M.J., Kashiwai, M., Ware, D.M., Megrey, B.A., Eslinger, D.L., Werner, F.E., Noguchi-Aita, M., Azumaya, T., Fujii, M., Hashimoto, S., Huang, D., Iizumi, H., Ishida, Y., Kang, S., Kantakov, G.A., Kim, H.-C., Komatsu, K., Navrotsky, V.V., Smith, S.L., Tadokoro, K., Tsuda, A., Yamamura, O., Yamanaka, Y., Yokouchi, K., Yoshie, N., Zhang, J., Zuenko, Y.I., Zvalinsky, V.I., 2007. NEMURO—a lower trophic level model for the North Pacific marine ecosystem. Ecol. Model. 202, 12–25.

Kodama, T., Morimoto, A., Takikawa, T., Ito, M., Igeta, Y., Abe, S., Fukudome, K., Honda, N., Katoh, O., 2017. Presence of high nitrate to phosphate ratio subsurface water in the Tsushima Strait during summer. J. Oceanogr. 73, 759–769.

Kon, T., Adachi, T., Suzuki, Y., 2003. Distribution of snow crab, Chionoecetes spp., larvae off Wakasa Bay in the Sea of Japan. Fish. Sci. 69, 1109–1115.

Konishi, N., Seto, M., Yamamoto, T., Takahashi, Y., 2011. Development of numerical larval model based on swimming behavior for larval snow crab reared in the laboratory. Journal of Japan Society of Civil Engineers, Ser. B3 Ocean Engineering 67, 310–315 (in Japanese).

Koyama, Y., Zhang, J., Hagiwara, T., Satake, H., Asai, K., 2005. Flow measurement of submarine groundwater discharge at wide area in the Eastern Toyama Bay. Chikyuk-agaku (Geochemistry) 39 (3), 157–164 (in Japanese).

Lee, H.-J., Yoon, J.-H., Kawamura, H., Kang, H.-W., 2003. Comparison of RIAMOM and MOM in modeling the East Sea/Japan Sea circulation. Ocean Polar Res. 25, 287–302.

Mizutani, Y., Satake, H., 1997. Hydrogen and oxygen isotopes compositions of river water as an index of the source of groundwater. J. Groundwater 39 (4), 287–297 (in Japanese).

Mizutani, Y., Satake, H., Yamabe, A., Miyachi, H., Mase, N., Yamamura, K., 2001. Hydrogen and oxygen isotopes ratios of groundwaters in shallow aquifers beneath the alluvial fan. J. Groundwater 43 (1), 3–11 (in Japanese).

Morimoto, A., Goto, A., Takikawa, T., Senjyu, T., Ito, M., Onitsuka, G., Watanabe, A., Moku, M., 2013. Horizontal transport of dissolved inorganic nitrogen through the Tsushima Straits. Sea and Sky 89 (2), 69–77 (in Japanese).

Morimoto, A., Watanabe, A., Onitsuka, G., Takikawa, T., Moku, M., Yanagi, T., 2012. Interannual variation in material transport through the eastern channel of the Tsushima/Korea Straits. Prog. Oceanogr. 105, 38–46.

Mukai, H., 2009. Marine Protected Area: the effects and the present. J. Environ. Inform. Sci. 38 (2), 20–24 in Japanese).

MWR, 2011. Chinese River Sediment Bulletin 2010 (in Chinese). Ministry of Water Resources of the People's Republic of China, p. 80.

Onitsuka, G., Yanagi, T., Yoon, J.-H., 2007. A numerical study on nutrient sources in the surface layer of the Japan Sea using a coupled physical-ecosystem model. J. Geophys. Res. 112. C05042. https://doi.org/10.1029/2006JC003981.

Park, K.-A., Kim, K.-R., 2010. Unprecedented coastal upwelling in the East/Japan Sea and linkage to long-term large-scale variations. Geophys. Res. Lett. 37. L09603. https://doi.org/10.1029/2009GL042231.

Redfield, A.C., Ketchum, B.H., Richards, F.A., 1963. The influence of organisms on the composition of seawater. In: Hill, M.N. (Ed.), The Sea. 2. Wiley, New York, pp. 26–77.

Rho, T., Lee, T., Kim, G., Chang, K.-I., Na, T.H., Kim, K.-R., 2012. Prevailing subsurface chlorophyll maximum (SCM) layer in the East Sea and its relation to the physico–chemical properties of water masses. Ocean Polar Res. 34 (4), 413–430 (in Korean with English abstract).

Sakurai, Y., 2014. Breeding biology of Japanese common squid (*Todarodes pacificus*) and impacts of climate change on population dynamics. In: Fisheries Promotion. vol. 559. Tokyo Fisheries Promotion Foundation, p. 55, ISSN 1343-6074 (in Japanese).

Senjyu, T., Sudo, H., 1994. The upper portion of the Japan Sea Proper Water; its source and circulation as deduced from isopycnal Analysis. J. Oceanogr. 50, 663–690.

Takikawa, T., Morimoto, A., Onitsuka, G., Watanabe, A., Moku, M., 2008. Characteristics of water mass under the surface mixed layer. J. Oceanogr. 64, 585–594.

Taniguchi, M., Burnett, W.C., Cable, J.E., Turner, J.V., 2002. Investigation of submarine groundwater discharge. Hydrol. Process. 16, 2115–2129.

Tebakari, T., Mizoguchi, T., Motoyoshi, Y., Zhang, J., 2013. Discharge and water quality characteristics of flowing artesian wells in the Kurobe River alluvial fan, Japan. J. Jap. Soc. Civil Eng. Ser. B1 69 (4), 589–594.

Toyama Groundwater Guideline, 2018, Toyama Prefecture (in Japanese).

Ueda, Y., Youshou, I., Fujiwara, K., Matsukura, R., Yamada, T., Yamamoto, T., Honda, N., 2015. Resource Evaluation of Snow Crab in Japan Sea in 2014, http://abchan.fra.go.jp/digests26/index.html (in Japanese).

Wang, B., 2006. Cultural eutrophication in the Changjiang (Yangtze River) plume: history and perspective. Estuar. Coast. Shelf Sci. 69, 471–477.

Watanabe, Y.W., Watanabe, S., Tsunogai, S., 1991. Tritium in the Japan Sea and the renewal time of the Japan Sea deep water. Mar. Chem. 34, 97–108.

Weiss, R.F., 1970. The solubility of nitrogen, oxygen and argon in water and seawater. Deep-Sea Res. 17, 721–735.

Yamamoto, J., Masuda, S., Miyashita, K., Uji, R., Sakurai, Y., 2002. Investigation on the early stages of the Ommastrephid squid Todarodes pacificus near the Oki Islands (Sea of Japan). Bull. Mar. Sci. 71, 987–992.

Yamamoto, J., Shimura, T., Uji, R., Masuda, S., Watanabe, S., Sakurai, Y., 2007. Vertical distribution of Todarodes pacificus (Cephalopoda: Ommastrephidae) paralarvae near the Oki Island, southwestern Sea of Japan. Mar. Biol. 153, 7–13.

Yoshida, T., Zhang, J., Morimoto, A., Shibano, R., Hirose, N., Takayama, K., Guo, X., Wang, Y., Mano, T., Yoshie, N., 2018. Proposal of three-layer management in Japan Sea. Bull. Coast. Oceanogr. 56 (1), 31–38 (in Japanese).

Zhang, J., Hagiwara, T., Koyama, Y., Satake, H., Nakamura, T., Asai, K.,2005. A new flow rate measuring method—SGD (submarine groundwater discharge) flux chamber and its approach off Katakai Alluvial Fan, Toyama Bay, Central Japan, Chikyuk-agaku (Geochemistry), 39, 141–148, 2005. (in Japanese).

Zhang, J., Mandal, A.K., 2012. Linkages between submarine groundwater systems and the environment. Curr. Opin. Environ. Sustain. 4, 219–226.

Zhang, J., Satake, H., 2003. The chemical characteristics of submarine groundwater seepage in Toyama Bay, Central Japan. In: Land and Marine Hydrogeology, pp. 45–60.

Zhang, J., Zhang, Z.F., Liu, S.M., Wu, Y., Xiong, H., Chen, H.T., 1999. Human impacts on the large world rivers: would the Changjiang (Yangtze River) be an illustration? Glob. Biogeochem. Cycles 13, 1099–1105.

Zhang, B., Zhang, J., Yoshida, T., 2017. Temporal variations of groundwater table and implications for submarine groundwater discharge: a three-decade case study in Central Japan. Hydrol. Earth Syst. 21, 3417–3425.

5

Environmental Economics, Culture, and Negotiation in the Coastal Sea

Takahiro Ota*, Katsuki Takao[†], Takuro Uehara[‡], Keito Mineo[§], N. Obata[¶], Kenichi Nakagami[¶], T. Yoshioka[‖], Ryo Sakurai[‡], Takeshi Hidaka[#], and Satoquo Seino**

*Graduate School of Fisheries and Environmental Sciences, Nagasaki University, Japan
[†]Specially Appointed Professor, Ritsumeikan University, Osaka, Japan
[‡]College of Policy Science, Ritsumeikan University, Osaka, Japan
[§]Graduate School of Agriculture, Kyoto University, Kyoto, Japan
[¶]Professor Emeritus, Policy Science, Ritsumeikan University, Ibaraki, Osaka, Japan
[‖]OIC Senior Researcher, Ritsumeikan University, Osaka, Japan
[#]School of Humanity-oriented Science and Engineering, Kindai University, Iizuka, Japan
**Ecological Engineering Laboratory, Graduate School of Engineering, Kyushu University, Fukuoka, Japan

© 2019 Elsevier Inc. All rights reserved.

In fishing villages in Japan, a maritime nation, depopulation and the super aging phenomenon are proceeding rapidly, which is an unprecedentedly serious situation. To consider conservation and maintenance of coastal zones as not only an issue for fishermen, who account for 0.2% of the Japanese population, but an issue for all Japanese people, it might be effective to socially realize the concept of Satoumi, by aiming to cooperate with local residents. The Japanese term, Satoumi, refers to coastal zones that have sound bioproductivity and biodiversity through human activities, which are composed of five elements. Three factors support the conservation and revitalization of coastal zones: material circulation, ecosystems, and communication. The other two facilitate the realization of Satoumi: the field of activities and the executors of this activity.

Twenty years have passed since the concept of Satoumi was defined by Dr. Tetsuo Yanagi in 1998 as, "coastal zones where biodiversity and productivity improve through human activities," and the concept is now recognized in and out of Japan and is accompanying many practices (Yanagi, 2008).

The concept is being expanded globally, such that the ideal future of the Earth, with its limited resources, should be realized by aiming for a sustainable world both naturally and socially, by decreasing the transportation costs of products and carbon emissions through Chisan-chisho, which means the wise use of natural and human resources through Satoyama and Satoumi (Matsuda, 2013). Many discussions have been held about how to use coastal zones, but we also need to realize the sustainable management of coastal zones. From this perspective, evaluation of the ecosystem services of coastal zones and the pursuit of approaches for sound integrated Coastal Zone Management (ICZM) is urgently required.

5.1 THE ECONOMIC VALUE OF A CLEAN AND PRODUCTIVE COASTAL AREA: THE CASE OF BEACH AND OYSTER AQUACULTURE IN NORTHEASTERN HIROSHIMA BAY, JAPAN

Takahiro Ota

5.1.1 Introduction

Since ancient times, the Seto Inland Sea has had many places with a good Satoumi environment (Shirahata, 1999). During the period of rapid economic growth, however, marine pollution increased remarkably, leading to the enactment in 1973 of the Law for Special Measures for the Conservation of the Environment of the Seto Inland Sea, which is a special section of the law of the Water Pollution Control Law. This law and its water quality regulations imposed limits on total emissions of nitrogen and phosphorus to the ocean, which have had some positive effects, such as fewer red tides (Setonaikai Research Conference, 2007). Although some places in the sea still suffer from eutrophication, there are also places where marine products are being affected by less nutrients (Yamamoto, 2015). Other factors that affect Satoumi, such as climate change, biodiversity loss, and mainstreaming of the ecosystem service concept, also have been changing.

The Ministry of the Environment consulted the Central Environment Council about the "future direction for Setonaikai, and the environment, conservation, and restoration" on July 20, 2011, and received a reply on October 30, 2012 (Central Environment Council, 2012). This reply said "a desirable Seto Inland Sea" includes "a clean sea," "a sea in which diverse organisms can live," and "a healthy sea." These goals require, respectively: water quality control in accordance with the ocean area and season; conservation of biodiversity; and collaboration with city residents amid the declining population in fishing villages and continuing sustainable fishery.

In order to achieve "a desirable Seto Inland Sea," it is necessary to quantify the value of Satoumi and assess the current condition. Because the various values of not only Satoumi, but also the natural environment have not been widely recognized, the origin of decision-making has been ignored. The lack of understanding of such value has led to the loss of Satoumi and poor management. Although there have been studies about the motivation, such as financial, for local residents to conserve the sea (Sakurai et al., 2016a), no studies have evaluated the economic value of clean and productive Satoumi. The current research examined the inner part of Hiroshima Bay, which is highly enclosed and suffers water quality and fishery production problems.

5.1.2 Methods

5.1.2.1 Environmental Valuation Method

This research used the stated preference method, because clean and productive Satoumi also has nonuse value, and it is difficult to observe the economic actions of consumers. To make the mailed questionnaires easier for respondents, this research used the contingent valuation method (CVM), which can be robust with a small sample size.

5.1.2.2 Outline of the CVM Investigation

Goods and Services for Valuation, WTP Questions, and WTP Scenarios

This research focused on the inner part of Hiroshima Bay of the Seto Inland Sea, where the enclosed level is high. This research used the transparency of the sea water at a beach as an index of environmental quality and coastal cleanliness. Although there are various indices for marine cleanliness, such as water quality, this research assumed water-bathing, which is when most people come in contact with sea water directly and notice the water quality. Regarding marine productivity, focusing on aquafarming of oysters, which are a popular and abundant marine product in this area, this research used the existing stock and stripped weight of oysters. The value of an oyster has two aspects, however, its utility value as an edible marine product in the market and its nonuse value as a symbol of marine productivity. This research did not examine the individual utility of consuming goods, which can be evaluated using market prices, but examined the change of quantity of production and consumption by the entire community. The transparency of sea water as an index of coastal water quality was calculated at Bayside Beach Saka, which is located inside Hiroshima Bay and has relatively poor water quality compared with other bathing beaches in the bay (the shaded area indicated by arrow 1 in Fig. 5.1).

The oyster-farming quantity, which represents marine productivity, was calculated at the oceanfront in Saka-cho, where Bayside Beach Saka is located, and in eastern Kanawajima as farming places (zone of the vertical lines marked by the Farming arrow in Fig. 5.1). The numerical results of a simulation using a model built by professionals and researchers on the material circulation of the ocean were used before and after the improvement of the environmental

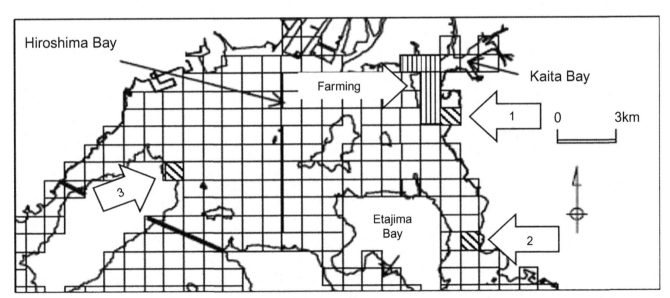

FIG. 5.1 Target coastal area. Diagonal line areas show bathing beach used in questionnaire. Vertical line areas show oyster farming sites used in questionnaire. Arrow 1 is Bayside Beach Saka, Arrow 2 is Karugaham, Arrow 3 is Miyajima Tsutsumigaura. *Source: revised Fukken Co., Ltd. 2015. Hiroshima wan syuhen no saiteki kaki yousyokuryo tou no suiteikekka gyomu houkokusyo, Fig. 2.1.*

quality shown to respondents in the CVM scenario (Fukken Co., Ltd, 2015, p. 1-1). This research by Fukken (2015, p. 1-1) aimed to simulate the production structure from minerals to plankton-eating fish to shells (oyster farming), and the optimum amount of oyster farming. This provides a useful model for analyzing and managing the optimum nutrient input and water quality depending on usage pattern, and for identifying the nutrient burden and production structure depending on the ocean area.

The scenario uses numerical values of surface (0–3 m) chlorophyll a computed by this model, and the values of "amount of oyster existing" and "weight of rippled oyster body." The conversion formula, which estimates transparency from the value of chlorophyll a, was used to determine the transparency of the sea water representing marine cleanliness. This formula was derived from water quality and flow simulation carried out in Osaka Bay, which also is located in the Seto Inland Sea (IDEA Consultants, Inc., 2015). Formula 2.1 expresses the relationship between transparency and chlorophyll a using the results of the water survey "Kouiki sogo suishitsu chosa kekka" ($R_2 = 0.6606$). Formula 2.2 expresses the same relationship using the results of the water survey "Kokyo yousui iki suishitsu sokutei kekka" ($R_2 = 0.4445$).

$$y = 7.9801x(-0.\hat{3}96) \text{ (IDEA Consultants, Inc.2015 : Fig.2.54)} \tag{2.1}$$

$$y = 6.5329x(-0.\hat{4}14) \text{ (IDEA Consultants, Inc.2015 : Fig.2.55)} \tag{2.2}$$

where, y denotes transparency (m) and x denotes chlorophyll a (μg/L). The average transparency obtained from these two formulas is shown as transparency. The value of the 1×1 km mesh shown with the diagonal line and arrow 1 in Fig. 5.1, which is Bayside Beach Saka, among the simulated values throughout the Hiroshima Bay area, was used as the transparency of the bathing beach. To compare the numerical values of other beaches of Hiroshima Bay, a reference value is shown (Fig. 5.2). These bathing beaches are Karugahama (arrow 2) and Miyajima Tsutsumigaura (arrow 3). The quantity of oysters and rippled oyster body weight were used to define the quantity of oyster-farming, representing marine productivity. Oyster farming locations, the total or average value of five meshes of vertical lines marked by the Farming arrow was used.

To denote improvements of environmental conditions, the scenario assumed a 50% decrease of nitrogen and phosphorus release in the summer (June to September) and a 50% increase in the winter oyster harvesting (October to May)

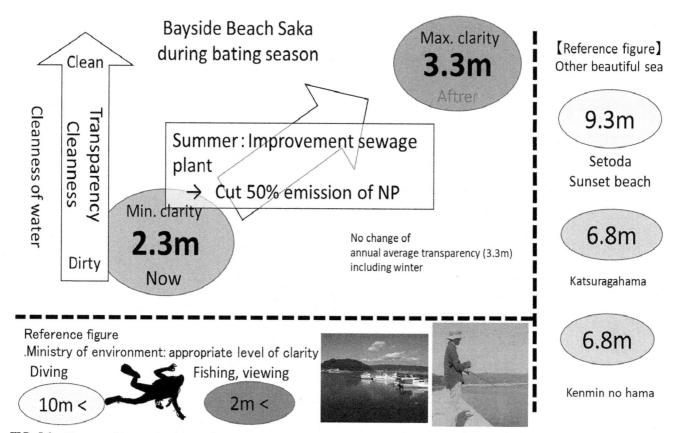

FIG. 5.2   Figure used in questionnaire to show transparency change.

season. It is assumed the total discharge throughout the year does not change. The researchers and professionals who made the model used in this research selected this setup after much discussion. This scenario considered the optimal amount of oyster farming without influence on the productivity of Hiroshima Bay and surrounding sea area when the nutrient loading from upstream changes (Fukken Co., Ltd, 2015, p. 4-1). The discharge of nutrients between summer and winter is adjusted, and the total amount in a year remains constant because it is considered undesirable to greatly reduce the nutrient discharge based on the recent history of the water quality control plan (Fukken Co., Ltd, 2015, p. 4-1). In order to attain these environmental changes, the survey considered improving the equipment of the existing sewage plant for advanced treatment. This method is the most feasible and cost-effective to reduce the discharge of nitrogen and phosphorous in summer by 50%. Although controlling discharge in agricultural fields that are the sources of nitrogen and phosphorous and further regulating direct drainage from factories are more suitable than advanced treatment, they are not realistic because the cost would be prohibitive because of the large area and diverse entities involved. Integrated operation of sewage drainage to increase the discharge of nitrogen in winter is carried out in a plant in Kagawa Prefecture (Miyagawa et al., 2015); increasing the discharge in winter is estimated to impose no additional cost (Tanda and Harada, 2011). Furthermore, because the transparency of the sea water of this coastal area decreases in summer, reducing the nutrient discharge in summer improves the environmental quality for water use, including sea-bathing.

In the simulation, the values for the 10 years from 2004 to 2013 were used for variables such as the water discharge from land areas and weather data (Fukken Co., Ltd, 2015). As the value (Q_0) before improving ocean cleanliness, the average transparency value of formulas 2.1 and 2.2 using the 10-year average of maximum chlorophyll a value each August was used. As the value (Q_1) after improvement, the average transparency value of formulas 2.1 and 2.2 using the 10-year average of minimum chlorophyll a value each July was used. These two transparency values are the maximum and minimum during the bathing season when the nutrient discharge is reduced by 50% in summer. This research aims to facilitate understanding and evaluation for respondents by using scenarios showing a relatively large difference in transparency before and after environmental improvement. The current sea water transparency of 2.3 m in August is close to the environmental standard of Hiroshima Bay 6, which is nearest to Bayside Beach Saka in "Kokyo yousui iki suishitsu sokutei" (example: on August 7, 2012, the transparency at depth 0.5 m was 2.3 m; Hiroshima Prefecture, 2014a), confirming its validity.

Using ocean productivity as one indicator of existing stock in the target coastal area, the value (Q_0) before improvement is the summed value of the five meshes (1×1 km, Fig. 5.1 diagonal area) of the 10-year average in each, with business as usual, i.e. the scenario of 0% decrease of nutrient discharge. The value (Q_1) after improvement is the equivalent of a 50% reduction. Using the weight of the oyster's bare body as the second oyster-farming indicator, Q_0 is the average value of five meshes of the 10-year average in each from October to March, optimal season for harvesting farmed oysters, with the scenario of 0% decrease of nutrient discharge. Q_1 is the equivalent of a 50% reduction.

Before asking the questions for WTP to indicate the current cleanliness of Bayside Beach Saka and the current productivity of the target sea, the survey showed the numerical value before improving environmental quality and showed background information on a photograph. The questionnaire also included an explanation and figure to indicate the difference between the states before and after the improvement (Fig. 5.2).

In these figures, because quantitative values for a beach or ocean space alone do not give a concrete image about the target values, the questionnaire included reference numerical values for comparison. For marine cleanliness, it showed the water transparency of a clean beach (Setoda Sunset Beach, Katsuragahama, and Ken min no Hama) in Hiroshima Prefecture. This transparency was calculated based on the formula for transparency and COD in Osaka Bay (IDEA Consultants, Inc., 2015: Fig. 2.54 and Fig. 2.55) based on COD (mg/L) obtained by measuring the water quality of the bathing area in Hiroshima Prefecture in July 2015. The questionnaire also showed the target value of transparency for recreational use of water given by the Ministry of the Environment: >10 m for diving and >2 m for fishing, hiking, and viewing (Enclosed Sea Bodies, Water Environment Division, Water and Air Environment Department, Ministry of the Environment, 2010, p. 50) The minimum and maximum transparencies from July to August of Miyajima, Tsutsumigaura, and Karugahama can be simulated by the model. The annual quantity of oysters harvested in Hiroshima City (Hiroshima Prefecture, 2014b) and bare body weight per oyster in Hiroshima Prefecture (average weight from 2003 to 2014) (Hiroshima Prefecture, 2014b, p. 7) were given as a reference for marine productivity. The questionnaire also showed the change of total existing stock of farmed oysters in the whole of Hiroshima Bay and the bare body weight per oyster in Hiroshima Prefecture.

The WTP questions in the questionnaire were as follows.

"In order to improve the water quality in summer, one method that has a significant effect at low cost is to improve the treatment level of a sewage plant. However, in order to remove much nitrogen and phosphorous from sewage, it is necessary to install more efficient equipment, which is expensive. Another method, which incurs no additional cost, is to drain water at a lower treatment level with a 50% increase in nitrogen and phosphorous in winter. It is assumed that the local government charges an additional xx yen per month as a sewage fee for 10 years, which is the time required to pay for installing the advanced treatment equipment.

Would you be willing or not willing to pay an additional xx yen per month (xx yen per year) as a basic sewage fee to improve the sewage plant to make the coastal area clean and productive, by reducing pollution by 50% in summer and increasing pollution by 50% in winter? (If you use drinking water, you must pay the basic sewage fee.)"

Here, the payment method was "additional payment for basic sewage fee." This was chosen in order to make the result of this CVM the most realistic payment method possible for future coastal management policy. In Hyogo Prefecture and Kagawa Prefecture, they began conducting experiments on controlling nutrients by sewage plant control and the same management approach could be done in coastal areas in the future. To improve a sewage plant, annual revenue from a local bond is needed in practice, but it is also acceptable to increase the beneficiary charge simultaneously. Because this is not what is usually imagined by the basic sewage fee in daily life, however, some explanation was added to the WTP question as follows.

- The improvement effect is immediately acquired from the year when this measure is implemented, and the target sea becomes clean and productive.
- The additional revenue is used only for improving the equipment of the sewage plant.
- All sewered houses must pay the basic sewage fee.
- Basic sewage fee for one month (excl. tax): Hiroshima City: 695 yen, Saka-cho: 783 yen, Kaita-cho: 745 yen.
- When using 20 m3 of tap water in one month in Hiroshima City for example, the sewage fee per household increases from 2219 yen to 2419 yen per month (excl. tax). (The usage fee per discharge does not change.)
- It is required by national government that the cost of adding equipment for a sewage disposal plant is obtained from both fiscal expenditure (local bond etc.) of the national and local government and fee revenue. Expenses must be charged to users of sewage plants.

The survey assumed a payment term of 10 years (the period for which the basic sewage fee increased). The cost of improving sewage treatment can be repaid by the higher usage fees for a fixed period. Although the life of a pumping station is about 30 years, because respondents might not understand the situation if the increase lasted far into the future, the survey limited repayment to 10 years. This research did not evaluate the market value of oysters but asked questions supposing an increase in the quantity and quality of oysters, which many people living around Hiroshima Bay consume. Notes added to the questions were as follows.

Note 1: Please assume that you yourself do not purchase and eat the oysters which become larger as a result of the improvement. Please consider only the improvement effect by which many people living around Hiroshima Bay can eat many larger oysters than currently and judge the pros and cons.
Note 2: When agreeing, it will be necessary to suppress other expenditures.
Note 3: When enforcing a policy, not only you but all people living around Hiroshima Bay will pay an additional fee.

This research used a double bounded dichotomous choice. This involves showing a higher fee when the respondent agrees to pay for the first offer, and a lower fee when the respondent does not agree to pay for the first offer. Based on existing studies at 2000 to 3000 yen per year and the tendency for compulsory collection by payment form like this time to reduce WTP, the survey set four monetary values.

The first offer was an increase of 30 yen, 50 yen, 100 yen, and 200 yen every month. When the first offer was 30 yen, the minimum fee was 10 yen. When the first offer was 200 yen, the highest fee was 400 yen, and others used fees on the upper and lower sides of the first offer.

Samples of the Investigation

The beneficiaries of the goods and services are residents in local administrative areas facing the coastal area. These areas around Bayside Beach Saka can easily be accessed by car. These local jurisdictions are Minami-ku, Hiroshima-shi, Aki-ku, Hiroshima-shi, Kaita-cho, Fuchu-cho, and Saka-cho.

5.1.2.3 Survey Outline

The CVM survey was conducted by post. The questionnaire was distributed for about one week from September 18, 2015, and was collected by October 11, 2015, using the postal service that could deliver to all residents in a selected postal code zone. The sampled postal code zone was selected from the zones located within 6 km from the target area, except for Aki-ku and Fuchu-cho (northern part) which are >10 km from the target area. One or more postal code zones from Minami-ku, Aki-ku, Kaita-cho, Fuchu-cho, and Saka-cho were selected, such that the zones were arranged geographically equally. The distance from the target area was within 4 to 6 km except for Saka-cho. The number of recipients was 1120 in total, and the number of respondents was 188 (16.8%).

5.1.2.4 Analysis Method

In this research, the WTP function model proposed by Cameron and Quiggin (Cameron and Quiggin, 1994; Kuriyama, 1998, pp. 122–126) was used. The WTP that might be paid to implement an environmental preservation measure assumes that it is divided into an observable portion and a nonobservable portion. This was computed using Terawaki's (2000) program by using Limdep 10/Nlogit5.

5.1.3 Results and Discussion

The valid response rate was 13.9% (156 of a total of 188 responses, excluding invalid responses). Reasons for not replying were selected by respondents who disagreed with both the pros and cons: "Although I agree with pollution control, I don't want to pay through a higher basic sewage fee," "There is a better way of controlling the cleanliness of northeastern Hiroshima Bay than improving plant treatment ability in summer," and "Users of beaches and suppliers and consumers of oysters should pay for the cost." Responses that gave no opinion were excluded, such as: "I need more information to determine the pros or cons about changing the basic sewage fee in order to improve pollution control."

In this research, the valid response rate was as low as 13.9%. The questionnaire contained 10 pages, which is long for a postal survey, however, this type of survey can generate many valid replies that can be used for the analysis, and many respondents answered sincerely. Because the sampled population might not have been fully representative, the results should be interpreted with caution. In the future, authoritativeness should be improved by obtaining authorization from the prefecture. Measures to request a reply, improve the return rate, and reflect the answers in actual polices should be taken in the future.

The estimation results are shown in Table 5.1. In the full model, including individual attributes, only income showed a significant t value ($P < .05$). The tendency for WTP to increase with income has been pointed out by many researchers (e.g., Yoshida et al., 1997).

Total WTP is defined as the value obtained through WTP per household of the whole beneficiary area multiplied by the total number of households. This total WTP signifies the total benefit value of a clean and productive bay in northeastern Hiroshima Bay. The median of this value was 30,099,392.2 yen per month, and the average was 31,955,226.2 yen. Assuming a payment period of 10 years, the median price was 3,611,927,064.0 yen, and the average was 3,834,627,139.2 yen.

5.1.4 Conclusions

In this paper, by double-bounded dichotomous-choice CVM, we assessed the economic benefit of a clean and productive coastal area for residents living around northeastern Hiroshima Bay. The estimated median value was about 30 million yen per month for all households of the beneficiary area, and about 3.6 billion yen for 10 years. The suggested measure for the CVM scenario was feasible and the payment vehicle was an imposed fee to reduce contingency. The CVM scenario of controlling nutrient input by sewage plant control already is practiced in other parts of the Seto

TABLE 5.1 WTP Estimation Result

Variables	Coefficient	P value
Constant	1.5824	0.1848
Male (1) or Female (2)	−0.1375	0.5227
Age (20s=2、 30s=3、 …、 70s=7、 more than 80s=8)	0.1194	0.1625
Education (Junior high=1、 High school=2、 …、 Graduate school=6)	0.1002	0.1990
Number of family living with	0.0485	0.4779
Income (log) (10,000 yen)	0.4073[a]	0.0036
N	156	
Log likelihood	−178.71	
AIC	371.40	
WTP median (yen/household/month)	220.90	
WTP average (yen/household/month)	234.52	

[a] $P < .05$.

Inland Sea. Reliable simulation modeling was used to determine cleanliness and productivity after improving the environmental condition by controlling water quality. Considering these points, the benefits are clear and the scenario contingency is low. Moreover, we evaluated the benefits to the community as a whole in terms of coastal productivity. Therefore, if the market price of oyster-farming is added, the economic value will be even larger. In order to create clean and productive Satoumi in the Seto Inland Sea, the current economic assessment indicates the need for discussion and appropriate management measures through sewage control and higher sewage fees.

5.2 VALUING SATOUMI: A RENT APPROACH TO MIXED NATURAL-MANMADE ECOSYSTEM SERVICES

Katsuki Takao

5.2.1 Introduction

In rural and suburban areas, we find forests, grasslands, lakes, ponds, and seacoasts from which we receive benefits such as food, firewood, and even cultural and aesthetic satisfaction. Through mutual interactions, intimate ecosystems make up a unique mixture of natural and manmade elements. We can call them mixed natural-manmade ecosystems (mixed ecosystems), sometimes called Satoyama in Japan. As human influence keeps spreading all over the Earth, pristine flora and fauna become harder to find, and many ecosystems seem to fall, more or less, under the mixed category.

In recent years, a considerable amount of research has been directed to quantifying the economic value of services provided by natural ecosystems. Although these studies tend to focus on free services from ecosystems such as clean air and water, relatively little attention has been paid to the cost elements included in receiving these services. In reality, many services from mixed ecosystems are, to be fair, joint products of natural and human efforts with substantial cost paid by human individuals and/or communities to foster, maintain, and harvest them.

This chapter tries to examine the economic nature of the services from mixed ecosystems and to fill the gap between the theories of classical economics and environmental economics. We then try to explore ways to measure the economic value of the services from mixed ecosystems, taking coastal ecosystems as an example.

5.2.2 Economics of Mixed Ecosystems

In view of the degree of intervention from human activities, we can classify ecosystems into three broad categories, as shown in Fig. 5.3.

- Untouched or pristine natural ecosystems, in which human intervention is either minimal or nonexistent. Examples include national park reserves, polar regions, and high seas.
- Mixed natural-manmade ecosystems, in which human intervention is visible but not yet dominant. Examples include ecosystems of forests, pastures, lakes, and coastal seas located close to human settlements.
- Manmade systems, in which human activities are dominant. Vegetation and animals are largely designed and controlled by human beings. Examples include urban areas, industrial areas, and farmland.

Services from mixed ecosystems have some salient characteristics from an economic viewpoint.

First, the supply of services from these ecosystems depends on the level of human care and on natural fluctuation. If an ecosystem is maintained healthily to ensure reproduction, with adequate density of young trees and fish, the

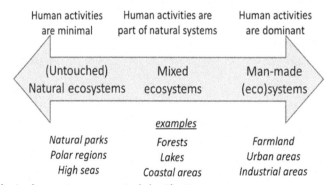

FIG. 5.3   Natural, manmade, and mixed ecosystems, conceptual classification.

ecosystems enable abundant harvesting for us. If the condition deteriorates, however, typically by overgrazing, over-fishing, or environmental pollution, the supply of services can dwindle or even disappear altogether. This relationship is delicate and nonlinear. The relationship between input and output can be subject to the common law of diminishing returns up to some point, just like in farmland and other manmade activities. Beyond this point, however, an additional input can result not only in progressively smaller returns, but also in zero or even negative ones, leading to a depletion of entire ecosystem resources.

These unique characteristics of mixed ecosystems were modeled by Gordon (1953) in his diagram about fishery economics, as shown in Fig. 5.4.

The upper panel of the figure shows the relationship between the fish population (stock size) and the net annual growth rate, $G(x)$, in the long run. The midway peak represents the maximum rate of growth of the fish population, or from the fishery viewpoint, the maximum sustainable yield (MSY). The lower panel shows the profile of fishery revenue, $R(E)$, and the cost, $C(E)$, as the functions of fishing effort, E. The vertical distance between $R(E)$ and $C(E)$ indicates excess profit or rent[1]. The rent reaches the maximum when the fish catch is optimally kept at the maximum economic yield (MEY).

This sea rent is different from land rent in a number of ways. Although land rent is common in terms of market rentals, sea rent is rarely traded in markets. Because it is difficult to draw maritime borders, property rights and rent are seldom developed on seas or lakes. The sea rent began to be realized in the market only recently, because limited property rights were created by such policies as individual tradable quota (ITQ) schemes in a few countries.

When sea rent is realized, its nature tends to be manmade. Although sea rent primarily depends on a natural capacity to yield output, it is created only by a governmental policy to establish a planned monopoly in accessing resources and selling their products.

Although sea rent varies depending on policies and other institutional frameworks, we mainly focus on the maximum, representing the potential capacity of the specific sea ecosystem to make profits for human society, hereafter called an "inherent" rent.

In the following sections, we focus on Japan's coastal seas as an example of mixed ecosystems and try to measure their inherent rent.

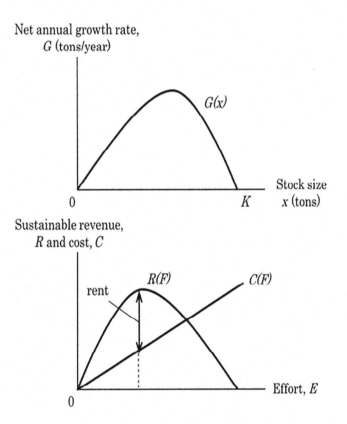

FIG. 5.4   The Gordon model.

---

[1] The term of rent is defined in this article as those payments to a factor of production that are in excess of the minimum payment necessary to have that factor supplied (after Varian, 2014).

5.2.3 Field and Data

We take up an entire off-coastal area around the Japanese archipelago, including territorial seas and exclusive economic zones (EEZ) up to 200 nautical miles from the coast. This area of the sea is one of the most productive marine ecosystems in the world, and Japan has a long history of catching and eating fish, with many coastal villages traditionally depending on fishing as their livelihood. Yet the coastal sea areas are not much cultivated, except in limited near-shore areas allocated to aquaculture of fish, oysters, and seaweeds.

We focused on the entire off-coastal area for two technical reasons. First, because of the migratory nature of fish, broader sea areas are easier to relate fishing efforts and catches than disaggregated ones. Second, statistical data about fishing are traditionally collected and published more systematically at the national level than at local levels in Japan.

We used two statistical data sources: "Sea Surface Fishery Production Statistical Survey"[2] and "Fishing Boats Statistical Tables Based on Fishing Boat Registration—Comprehensive Report." The former, available for the period after 1956, is used for fishery output, in both weight (metric tons) and value (yen), and the latter, available after 1952, is used for fishing effort.

After excluding aquaculture and pelagic fisheries outside the EEZ, we compiled these statistical data by fishing methods: namely, fixed nets, angling, long lines, gill nets, surrounding nets, lift nets, trawls, boat seine, shell and seaweed collection, and miscellaneous fishery.

Among other statistics, we focused on "gross boat horsepower" as our main indicator of fishing effort. We opted for this indicator rather than the number of fishermen, work-hours at sea, numbers of boats, or boat tonnage because engine power seems to reflect the consistently growing pressure on ocean ecosystems. Although human labor input indicators show continued decline since the 1960s, technical innovation, such as sonars, powered winches, and automated angling systems, have made it possible to catch more fish than ever before. The gross engine horsepower, which inherently reflects such technical innovation, indicates a trend more in line with these change in fishery input than any other indicators.

5.2.4 The Result

Fig. 5.5 shows the relationship between the fish catch and the gross horsepower of fishing boats in the Japanese off-coastal seas, plotted for the period between 1956 and 2012. The result shows a drastic rise and fall: Until the mid-1980s, a parallel growth of both variables is seen, reflecting a rapid postwar growth of the fishery sector. After a peak at 9.44 million tons in 1984, the fish catch started to plunge toward the bottom at 3.30 million tons in 2012.

It appears that the pattern of change generally appears to conform to the inverse U-shape of the Gordon model. Compared to an estimated trend curve, there are two periods of divergence, denoted as areas A in the upside and B in the downside. Area A is characterized by short-term overfishing, which started to depart from the estimated long-term equilibrium from the late 1970s. Apparently, high levels of landing had resulted from a reckless catch to

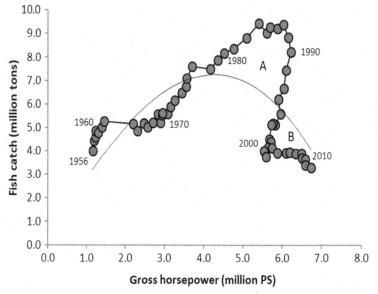

FIG. 5.5 Fish catch and gross horsepower in coastal and off-coastal fishery Japan, 1956–2012.

[2] http://www.maff.go.jp/j/tokei/kouhyou/kaimen_gyosei/index.html.

eat into the resource base of the fish population. This drastic rise and fall of fish catch might have been magnified by a global boom of the sardine fishery between 1973 and 1996 (The Fishery Agency of Japan, 2016).

Area B represents a backlash of the overfishing in the preceding period. The fish catch fell well below the estimated long-term equilibrium resulting in a shock of poor catch, supposedly because of hampered reproductive capacity of the fish stock. The more recent situation after 2008 seems to indicate a return to the equilibrium in spite of a much smaller catch than MSY or MEY, and the gross horsepower indicates a creeping increase.

We also plotted the disaggregated diagrams for the 10 fishing methods, and the results are shown in Fig. 5.6. We found seven out of 10 methods indicate similar inverse U-shaped patterns, with the estimated trend curve added in each figure.

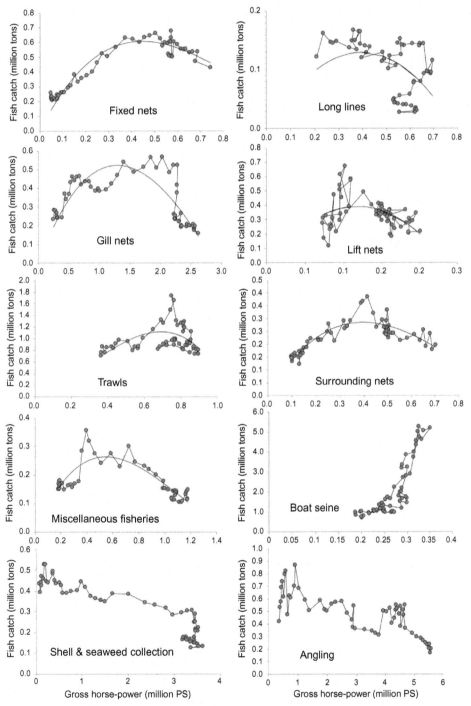

FIG. 5.6   Fish catch and gross horsepower by fishing method, 1956–2012.

Among the exceptional three, surrounding net fishery appears to have gone through a boom-and-bust phase with synchronized rise and fall of engine power and catch, which is typically observed in pelagic fishery. Shell and seaweed collection and angling indicate a slow but monotonic fall in catch as compared to a monotonic increase in fishing effort. The fall of fish catch, however, seems to have accelerated recently for both methods.

5.2.5 Rent Based on the Gordon Model

In this section, we estimate the inherent rent for the entire Japanese coastal sea based on the Gordon model. To proceed with the estimation, we make the following assumptions.

(1) There is no current rent in the fishery sector, meaning that the cost of fishing operation is equal to the sales proceeds of landed fish.
(2) The cost function of the fishery sector, $C(E)$, is proportional to the gross boat horsepower.
(3) Both production and cost functions are subject to the boundary condition that $R(E)=0$ when $E=0$.
(4) The average fish price is fixed to the average price per weight (metric tons) in the base period between 2002 and 2006, i.e., the total value of landed fish divided by the weight.

The production function, $R(E)$, is modeled by a number of different equation forms. Each model is fitted to statistical data by the ordinary least squares (OLS) method, and the results are shown in Table 5.2. The fitness of those models to actual data are compared in Fig. 5.7.

We selected a simple polynomial function of degree three, as follows, as our standard model because it is simple and its fitness is among the best.

$$R(E) = aE^3 + bE^2 + cE + \varepsilon$$

in which a, b, and c are constants, and ε is an error term.

The inherent rent for the Japanese coastal seas is estimated by the following formula.

$$\text{inherent rent} = \max \{R(E) - C(E)\}$$

The estimation result of the annual inherent rent for the Japanese coastal seas is 1.173 trillion yen/year (approximately US $10.2 billion with an exchange rate of 115 yen/US$) as shown in Table 5.3. Assuming a discount rate at 1%, this annual rent roughly translates to an asset value of 117.3 trillion yen. The maximum economic yield, MEY, at which the inherent rent is realized, is estimated at 7.08 million tons per year, roughly twice as much as today's annual catch.

This inherent rent can be classified into two components by sources: through the catch increase (volume effect) and through the improved efficiency by eliminating excess capacity (efficiency effect). The volume effect is estimated by multiplying the increase of fish catch and its average price, and is estimated at 807 billion yen, or 68.8% of the inherent

TABLE 5.2   Estimated Models

Model	Equation
Model 1 (Base model)	$R(E)^a = -0.0187E^3 - 0.2734E^2 + 3.051E$
	$(t = -0.606)\ (0.827)\ (4.821)$
	$R^2 = 0.9309\ F = 242.5\ n = 57$
Model 2	$R(E) = -0.4095E^2 + 3.408E$
	$(t = 9.98)\ (14.69)$
	$R^2 = 0.9304\ F = 367.8\ n = 57$
Model 3	$R(E) = 3.101E - 0.3191E^2 - 0.00054E\ e^E$
	$(t = 9.544)\ (-0.4.033)\ (-1.334)$
	$R^2 = 0.9327\ F = 249.3\ n = 57$
Model 4	$R(E) = 4.997E\ e^{-0.2871\ E}$
	$(t = 16.279)\ (-13.896)$
	$R^2 = 0.7782\ F = 193.09\ n = 57$

[a] *R(E) is indicated as the weight of landed fish, in million tons per year, and F in million PS.*

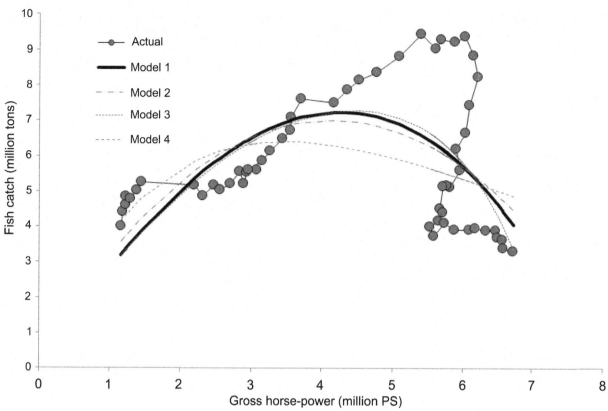

FIG. 5.7   Goodness of fit for $R(E)$ by different models.

TABLE 5.3   Inherent Rent for Japanese Coastal Seas (in Billion Japanese Yen Per Year)

Rent and yield	Annual rent	(High, Low)[a]
Base scenario	1173	(1929, 416)
MEY (million tons per year)	7.080	(10.35, 3.81)
MSY (million tons per year)	7.243	(10.52, 3.97)
Cost reduction scenario	1621	(2377, 864)
Price dwindling scenario, 50% lower	769	(1147, 391)
80% lower	527	(678, 376)

[a] 95% confidence range.

rent. The remainder is deemed as the efficiency effect, estimated at 366 billion yen per year, or 31.2%. The volume effect is found to be greater than that of the efficiency effect for the base scenario, as shown in the upper belt of Fig. 5.8.

For comparison, other estimates of fishery rent found in various studies are listed in Table 5.4. Most of these estimates are derived from bio-economic models rather than empirical statistics, but the underlying theory, the Gordon model, is common to all sources. A recent estimate by the World Bank indicated US $51 billion per year for the world fishery, and our rent estimate for the sea areas around Japan accounts for approximately 20% of this world total (World Bank and FAO, 2009)).

5.2.5.1 Sensitivity analysis

In order to test the reliability of our estimate, a few sensitivity analyses are done. First, our assumption on the cost function might lead to underestimation of rent. The assumption that there is no current rent, meaning that the cost is equal to the sales proceed of fish, might be too pessimistic from the fishermen's point of view, and the cost might be smaller and the profit positive. In addition, there is a possibility to further cut the cost if optimal fishery management is introduced to remove unnecessary capacity.

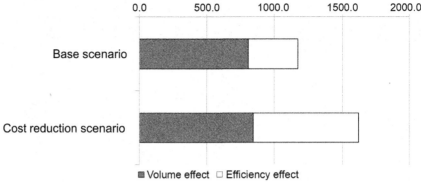

FIG. 5.8 Sources of implied rent for Japanese coastal seas (billion yen/year).

TABLE 5.4   Estimated Fishery Rent

Target area	Estimate	year	Source
World fishery	$51 billion	2009	World Bank/FAO
World fishery	$80 billion	2005	Wilen
World fishery[a]	$90 billion	2002	Sanchirico and Wilen
World fishery	$46 billion	1997	Garcia and Newton
World fishery	$50 billion	1993	FAO
Gulf of Tonkin, Vietnam	$52 million	2006	Nguyen and Nguyen (2008)
Iceland cod	$426 million	2005	Arnason
Namibia hake demersal trawl	$94 million	2002	Sumaila and Marsden (2007)
Peru anchoveta purse seine	$163 million	2006	Paredes (2008)
Bangladesh hilsa artisanal	$115 million	2005	(not cited at source)

[a] *Includes projection.*

Source: World Bank and FAO. 2009. The Sunken Billions: The Economic Justification for Fisheries Reform, Agricultural and Rural Development, World Bank/FAO, Washington, DC/Rome.

It has been argued that current practice in Japanese fisheries involves plenty of inefficiencies because of excessive capacity in the fishery sector under current fishery management. A study by Managi (2009) concluded that <1% of vessels, fishing gear, and fuels are sufficient to sustain the current level of Japanese fishery. We examine the possibility of a cost reduction of 90% per landed fish, allowing a little more capacity with a view to the weather-dependent nature of fishery operation (the cost-reduction scenario).

With this cost-reduction scenario, the rent is found to rise 38.2% to 1.620 trillion yen as shown in Table 5.3. In this scenario, the efficiency effect accounts for almost half of the estimated rent as the volume effect did, as shown in the bottom belt of Fig. 5.8.

Second, fish prices could fall in response to a rise in supply. An average fish price could fall even further if the share of low-graded fish increases. The latter factor stems from the fact that the catch of low-graded fish varieties, typically sardines, had fallen more than higher-graded ones. If the catch of low-graded fish recovers and the average fish price falls, then the rent might fall, possibly making our rent estimate too optimistic.

We thus considered following price-dwindling scenarios: falls of 50% (moderate fall) and 80% (sharp fall) in average price for additional portion of fish catch compared to the base period (2002–06). It was assumed in both scenarios that the prices remain constant up to the current supply, and only additional supply will face lower prices.

With the price-dwindling scenarios, the inherent rent is estimated at 769 billion yen per year with a moderate fall (50%) and 527 billion yen with sharp fall (80%), respectively. These rent estimates are smaller, as expected, by 34% and 55%, respectively, than the base scenario, yet our inherent rent remains substantially positive.

In summary, although there are both upside and downside variations, we estimate the inherent rent for Japanese coastal seas at 1.173 trillion yen per year as a middle estimate, with overall range of variation from 149 billion at the low end to 2.000 trillion yen at the high end, based on alternative scenarios examined.

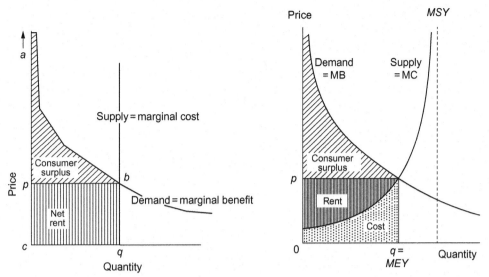

FIG. 5.9 Consumer surplus, rent, cost, and willingness to pay. *Left panel by Costanza, R., R. d'Arge, R. de Groot, S. Farberk, M. Grasso, B. Hannon, K. Limburg, S. Naeem, R. V. O'Neill, J. Paruelo, R. G. Raskin, P. Sutton, M. van den Belt, 1997. The value of the world's ecosystem services and natural capital. Nature 387, 253–260.*

5.2.6 Discussion

In this section, we consider the inherent rent from a viewpoint of economic evaluation. The left panel of Fig. 5.9 shows a common explanation of rent in the market of ecosystem services. Assuming the supply of ecosystem services is fixed and the production cost is zero, the market equilibrium between their supply and demand will be given at point b.

The area below the demand curve to the left of the vertical supply line represents the gross benefit of the ecosystem services, in terms of consumers' aggregated willingness to pay (WTP); the area is divided further into the ones representing rent (below) and consumer surplus (above).

In cases where the cost is not zero, as we have discussed for services from mixed ecosystems, the figure should be redrawn with a nonvertical supply curve like in the left panel of Fig. 5.9. In a case of fishery-based services, however, the situation might need an additional explanation.

The left panel of Fig. 5.10 shows the Gordon model, as copied from Fig. 5.4 (Costanza et al., 1997). The central panel swaps the horizontal and vertical axes of this diagram, indicating a total cost (TC) curve as a function of the output = catch quantity.

This TC curve has an unusual characteristic in that it indicates a multivalued function. This means that a long-run fishery output can be made possible under two long-term equilibria, that is, one under an optimal management with fishing effort smaller than E_{MSY}, and the other under an overfishing situation with effort greater than E_{MSY}.

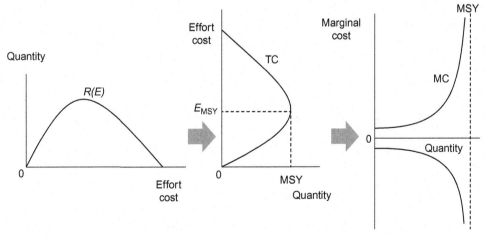

FIG. 5.10 Fishing effort, cost, and marginal cost.

Based on this multivalued TC curve, implied marginal cost (MC) can be drawn in two curves, one above the horizontal axis and the other below. The latter curve might point to a paradoxical case of negative MC, meaning that an additional input will result in a decrease of output. This is exactly what Graham (1935) described as "fishermen spend time and money to reduce their catch."

Only with the MC curves above the horizontal axis, the MC = supply curve meets the demand curve with a long-run equilibrium at *MEY*. This equilibrium is possible only by an optimal fishery management and over the long run.

The right panel of Fig. 5.9 shows only the positive part of the MC curve, defined in the horizontal range between zero and *MSY*. The gross benefit of ecosystem services is shown by the area under the demand curve to the left to the market equilibrium, *MEY*, and this area has three components as:

$$\text{(gross benefit of ecosystem services)} = \Sigma\text{WTP}$$
$$= \text{(consumer surplus)} + \text{(rent)} + \text{(cost)}$$

Or, in net terms as:

$$\text{(net benefit of ecosystem services)} =$$
$$= \text{(gross benefit of ecosystem services)} - \text{(cost)}$$
$$= \text{(consumer surplus)} + \text{(rent)}$$

We find that the net benefit of ecosystem services has two components: consumer surplus and rent. The former benefit falls on consumers as extra satisfaction from consuming services exceeding their prices, and the latter falls on the resource owners in terms of excess profit, just like the cases of oil or mineral resources.

When fishery reform is deliberated to realize this potential rent, the question of to whom the rent belongs is particularly important. This is because the rent is created only by a governmental policy to authorize monopoly in accessing resources and in supplying fish to the market in favor of a limited group of quota holders. In this sense, the rent is artificial, such as in the case of taxi licensing and professional qualifications of lawyers, certified accountants, etc. Inadequate approach to quota allocation might trigger intensive rent-seeking activities among potential stakeholders.

This study has focused so far on fishery services from mixed coastal ecosystems, the rent approach might be broadened to include other types of mixed ecosystems, among which pastures and communal forests are possible examples, but these are left to further study.

5.2.7 Conclusion

This section tries to explore a way to value services from mixed natural-manmade ecosystems. We focused on rent, specifically on the inherent rent, as potential financial gains from the fishery in coastal sea ecosystems.

Mixed natural-manmade ecosystems provide us with various goods and services, which have unique characteristics to be distinguished from those of untouched natural ecosystems or manmade systems. Mixed ecosystems require some sort of effort for human individuals and communities to obtain the most services from them.

Rent might provide a useful scale of resource value of the services from mixed ecosystems. The Gordon model of fishery economics offers an analytical basis of the rent in fishery services and the inherent rent, its maximum.

Based on the Gordon model and long-term statistical data, the inherent rent of the Japanese coastal sea areas is estimated at 1.173 trillion yen (US $10.2 billion) a year. This translates to 117.2 trillion yen (US $1.02 trillion) at a discount rate at 1%.

Fishing ground ecosystems, such as over-exploited one in Japanese Satoumi, could be a special case that clearly indicates the existence of rent, and thus whether rent can be found in other mixed ecosystems should be investigated further.

5.3 REGIONAL SUSTAINABILITY ASSESSMENT FRAMEWORK FOR INTEGRATED COASTAL ZONE MANAGEMENT: SATOUMI, ECOSYSTEM SERVICES APPROACH, AND INCLUSIVE WEALTH

Uehara Takuro and K. Mineo

5.3.1 Introduction

The goal of this study is to advance methodology for development of a regional sustainability assessment framework for integrated coastal zone management (ICZM or ICM). ICZM has been globally adopted. The European Commission adopted a new initiative on ICM on March 12, 2013 (European Commission, n.d.). The Japanese Basic Act on

Ocean Policy (Act No. 33; April 27, 2007) stipulates that a state shall adopt ICZM. The aim of ICZM is to coordinate the application of different policies and activities affecting coastal zones, which covers coastal sea and land areas, in order to develop the coastal zones in a sustainable manner. It respects the limits of natural resources and ecosystems and involves transdisciplinary knowledge and all stakeholders across the different sectors (European Commission, n.d.; Japanese Basic Act on Ocean Policy: Act No. 33, April 27, 2007; Maccarrone et al., 2014).

These ambitious aims provide a wide scope for improving ICZM. The European Commission, for example, conducted a formal evaluation of ICZM efforts across Europe and suggested the need to make ICZM more operational (Ballinger et al., 2010). Syme et al. (2012) point out that there is a gap between science and coastal management, and it is unclear how to integrate research outcomes into coastal management schemes. We need to develop methodologies for addressing a variety of difficulties including a lack of engagement among scientists, practitioners, and policymakers, a lack of data, uncertainty, contrasting terminologies, and the need to integrate results from a broad spectrum of methodologies and disciplines (Reis et al., 2014).

This study presents a novel sustainability assessment framework that uses indicators to bridge this gap. Indicators for ICZM have been in increasing demand, and most of the existing best practices for ICZM outline their use (Maccarrone et al., 2014). Various social and ecological indicators have been developed (Duraiappah et al., 2012; Hattam et al., 2015; Turner et al., 2014); however, shortcomings remain. For example, Maccarrone et al. (2014) raise four principal shortcomings of ICZM indicators: lack of reflection of socioecological interactions, lack of feedback from the responses to previously adopted management actions, lack of consistency with the ICZM objectives, and lack of the availability and homogeneity of the indicator set. Current indicators often lack practical ICZM insights but provide a description of a state or the past; their contributions to sustainability often are not explicitly explained, but implicit. Our proposed framework attempts to overcome some of these shortcomings.

Our framework is systematic, not ad hoc, and transdisciplinary. It integrates three separately developed approaches that explicitly reflect stakeholders' values: Satoumi (the Japanese concept of socioecological production landscapes (SEPLs)), the ecosystem services approach (ESA), and inclusive wealth (IW). Each approach makes a different contribution to sustainability assessment, and their integration is complementary and makes the assessment framework credible and operational. Ad hoc approaches (Harris and Pearson, 2004) or a lack of theoretical foundation (Kulig et al., 2010) has hindered the development of a sustainable assessment framework. Without a theoretical or systematic foundation, it is difficult to validate and further develop the framework. IW is grounded in neoclassical growth theory but can be extended beyond the neoclassical framework. IW is realistic because it allows for an imperfect economy in which government maximizes intergenerational social welfare (Arrow et al., 2003). It integrates environmental, economic, and social aspects through capital assets, which contrasts with most attempts to measure sustainability (Singh et al., 2012). The integration is transdisciplinary, in that it is intended to construct an issue-oriented, interdisciplinary, and participatory assessment framework[3]. The transdisciplinary design of the framework makes it operational; it provides practical inputs to ICZM to attain or sustain a desired coastal zone. The actual practice of this type of research "has long lagged behind theory" (Fischer et al., 2015, p. 146).

The framework was tested in Japan's Seto Inland Sea (SIS) and analyzed in detail. We discuss the outcomes' insights for ICZM of the SIS and future research needed to further develop the framework. This case study makes an important contribution to the scientific literature, which is composed of predominantly European and North American cases and comparatively few Japanese examples (see Liquete et al., 2013 for a review). The case study aims to provide sufficient guidance for readers to apply the framework to their areas of interest.

5.3.2 Regional Sustainability Assessment Framework

Three approaches with seven steps comprise our framework (Fig. 5.11). It extends IW to compute a sustainable development path index by adding sustainability conditions, and adopts Satoumi and the ESA to make IW credible and operational as a regional sustainability assessment tool. Satoumi, ESA, and IW play different but partially overlapping or related roles. Their integration is complementary and increases the effectiveness of all three approaches when applied to coastal zone management and decision making.

IW is calculated as the sum of the net current value of social welfare, which is provided by the product of capital assets (i.e., productive base) and the assets' shadow prices (i.e., social welfare weights; Pearson et al., 2013) instead of the uniform weights that are commonly used in ecology (Banzhaf and Boyd, 2012). The original IW, however, does not guide us to choose a desired state of coastal zones or rank capital assets for the evaluation targets as an input to ICZM.

---

[3] See Common and Stagl (2005) for the connotations of transdisciplinarity.

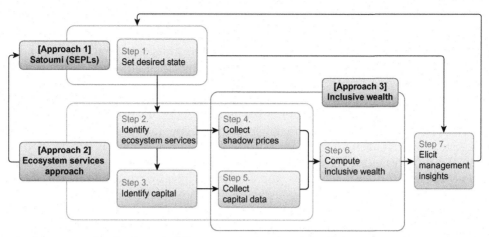

FIG. 5.11   Schematic of the proposed regional sustainability assessment framework for ICZM.

Satoumi pictures the desired state, which helps rank the evaluation and management targets. The ESA can bridge the gap between IW and Satoumi by translating the desired state of coastal zones into a list of capital assets and their shadow prices. Although some researchers claim that the ESA can provide normative knowledge, there is "not a single accepted normative framework" (Abson et al., 2014, p. 35) and it cannot be substituted for Satoumi. Although Satoumi is a Japanese concept, the framework could be applied to other countries where similar SEPLs exist (Gu and Subramanian, 2014).

The following sections define the three approaches, describe their relative strengths and weaknesses for the sustainable use of coastal zones, and demonstrate their application to the Seto Inland Sea, Japan (Steps 1–7).

5.3.2.1 Approach 1: Satoumi

Satoumi[4] can be defined as "a coastal area where biological productivity and biodiversity has increased through human interaction" (Ministry of the Environment Government of Japan, n.d.-a). This has been generalized into SEPLs, referring to "dynamic mosaics of habitats and land uses that have been shaped over the years by interactions between people and nature in ways that maintain biodiversity and provide humans with goods and services needed for their well-being" (Gu and Subramanian, 2014). Satoumi is a multifaceted concept, and its various aspects have been elucidated since it was proposed by Yanagi in 1998 (Yanagi, 2012). Among them are three major aspects: normative, managerial, and context-dependent. First, Satoumi is about the normative state of the coastal area, such as "the state of the coastal sea that the local community desires" (Berque and Matsuda, 2013, p. 192), "preferable coastal area environment" (Ministry of the Environment Government of Japan, n.d.-a), or "the representation of an ideal relationship which should be created between human beings and the sea" (Duraiappah et al., 2012, p. 25). Second, it is about coastal zone management regimes (Duraiappah et al., 2012) that enable the desired coastal state to be sustained. Satoumi is "managed with a mix of traditional knowledge and modern science" (Duraiappah et al., 2012, p. 26), and involves local communities (Abson et al., 2014). Although Satoumi uses traditional knowledge, it adopts a form of co-management that is adaptive to changes in socioecological contexts (Henocque, 2013) (arrow from Step 7 to Satoumi in Fig. 5.11). In Satoumi, management involves various stakeholders such as the community, and municipal, prefectural, and national administrations, each of which makes different contributions (Hidaka, 2016a). Third, Satoumi reflects socioecological contexts (Duraiappah et al., 2012; Hidaka, 2016a), varying according to place and time. Although the term is relatively new, Komatsu and Yanagi (2015) found examples of Satoumi since at least the Edo period (1620–1850 CE). Because Satoumi, in a sense, is based on local memory, the desired state and management are not fashioned anew but can give them a certain transparency (i.e., familiarity). The transparent presentation of the values ensures the credibility and robustness of the sustainability assessment methods (Sala et al., 2015).

The main contribution of Satoumi to the framework is to picture a stakeholder-engaged and socioecological view of a desired coastal area that helps rank capital assets for IW to assess the sustainability of the coastal area and elicit insights for ICZM. This normative aspect is of crucial importance, since without it there is no explicit guidance to

---

[4] "Sato" means a local community or village where people live their life and "umi" is the most common word meaning the sea (Shimizu et al., 2014, p. 209).

identify elements of the coastal area that should be targeted. As discussed in later sections, neither IW nor the ESA provide guidance for ranking capital assets. Satoumi also provides management insights (arrow from Step 1 to Step 7 in Fig. 5.11) that reflect previous experiences in socioecological contexts.

Satoumi, however, is still in need of quantitative outcome measures or metrics (Henocque, 2013; Millennium Ecosystem Assessment, 2003). Satoumi has been shaped over the years without providing quantitative analysis or explanation. Therefore, the desired state envisioned by Satoumi requires quantitative translation (assisted by the ESA) into a list of targeted capital assets.

5.3.2.2 Approach 2: Ecosystem Services Approach

The Millennium Ecosystem Assessment (2003) defines ecosystem services as the benefits people obtain from ecosystems, and emphasizes the linkages between ecosystem services and human well-being.[5] The ESA reveals such linkages. Various cascade models address the linkages (e.g., Haines-Young and Potschin, 2010; van Oudenhoven et al., 2012; Bennett et al., 2015); a cascade model could describe the linkages as a stream: starting from a biophysical structure or process, to functions, to ecosystem services, to benefits, and ultimately to economic values (van Oudenhoven et al., 2012). Hattam et al. (2015) developed generic marine ecosystem service indicators that can be tailored to multiple contexts. Although ecosystems are central to the ESA, social systems comprising manmade capital and human capital also are emphasized because ecosystem services are often a joint production of these capital assets (Costanza et al., 2014), and the ESA should be discussed within a social-ecological context (Reyers et al., 2013).

The ESA makes two contributions to the framework: translation and facilitation. First, the ESA translates a desired state envisioned by Satoumi into a list of capital assets with their shadow prices, which can be used directly for IW computation. The ESA is highly compatible with Satoumi (a Japanese SEPL), because it adopts a systems perspective and views landscapes as a system (Naveh, 2000). The ESA sheds light on ecosystems rather than single species (Daniel et al., 2012). Using developments in landscape ecology,[6] Andersson et al. (2015) proposed service-providing units (SPUs) to clarify the landscape-ecosystems services (ES) connection. SPUs are defined as "the smallest distinct physical unit that generates a particular ES and is addressable by planning and management, to explore the dimensions of ES generation within landscapes" (Andersson et al., 2015, p. 158). Barbier (2012) claims that a landscape could serve as the basic unit (a quantifiable land unit (Zonneveld, 1989)) for measuring changes in natural capital because the landscape influences ecological processes and services provided by such (see Fu and Jones, 2013) for recent developments in coupling landscape patterns and ecological processes). The SPUs or land units could be used, in part, for capital asset selection (e.g., particularly for natural capital including seagrass beds and mudflats) within the assessment framework, and it is relatively straightforward and practical to adopt them as a target for ICZM. ICZM works not on goods and services but on the physical components on which they are made or based upon. Second, the ESA helps people understand what they want and receive from SEPLs because people are often not aware of what they want or are receiving (Costanza, 2015), as indicated by the arrow from Step 1 to Step 7 in Fig. 5.10.

The linkage between the ESA and sustainability, however, is "implicit and vague at times" (Abson et al., 2014, p. 35). Based on a survey of 1388 peer-reviewed publications on ecosystem services from 1997 to 2011, Abson et al. (2014, p. 35) conclude that there is "not a single accepted normative framework" that would "both broaden and deepen our understanding of the role of the ecosystem services concept in relation to the broader societal goal of sustainability." Sustainability involves a normative argument (Derissen et al., 2011). An increase in ecosystem services often is deemed good implicitly, and is not challenged in relation to the broader societal goal of sustainability in the face of tradeoffs such as economic development and conservation. Without a normative framework, however, the ESA might fall into "a technocratic discourse" (Abson et al., 2014, p. 35).

5.3.2.3 Approach 3: Inclusive Wealth

IW is theoretically rigorous, realistic, comprehensive, and practical; it is a suitable base for developing a regional sustainability assessment framework for ICZM. IW is not ad hoc as is often the case for sustainability indicators (Harris and Pearson, 2004; Sala et al., 2015) but is based on the theory of welfare indices (Arrow et al., 2003) "which provides a clear rationale for using changes in aggregated capital values as a measure of sustainable development" (Harris and Pearson, 2004, p. 29). Therefore, IW can be extended further in a logical and consistent manner to fit into a context. IW is

---

[5] ESA has been growing and promising as a field. See Bennett et al. (2015) for the state-of-the-art and current issues of the ESA, and Gómez-Baggethun et al. (2010) for ESA history.

[6] Landscape ecology is somewhat similar to the ESA and Satoumi. Landscape ecology evolved in Central Europe as a result of a holistic approach, and deals with the interrelationships between humans and their open and built landscapes (Naveh and Lieberman (2013). For recent developments in landscape ecology, including its relationship to the ESA (see Fu et al., 2013, for example).

realistic because it assumes imperfect economies in which government is not successful in maximizing intergenerational welfare. In contrast to most economics arguments, IW theory can be applied to nonconvex ecological-economic systems (Dasgupta and Mäler, 2004; Uehara, 2013). IW is comprehensive because its theory covers various practical issues, including shadow-price estimation, population dynamics, exogenous influences, and uncertainties, and it is practical because it is composed of three indices (capital assets, their shadow prices, and exogenous components) whose considerations lead directly to management insights.

In IW,[7] social welfare W at time t is expressed by a value function V as

$$W_t = V(\boldsymbol{K_t}, \alpha, t) \tag{5.1}$$

where $\boldsymbol{K_t}$ is a comprehensive list of capital assets at time t, α is a resource allocation mechanism, and V is a linear combination of the assets and their shadow prices awarded as the weights. The shadow price of the ith asset at time t is defined as

$$p_{it} = \frac{\partial V(\boldsymbol{K_t}, \alpha, t)}{\partial K_{it}} \equiv \frac{\partial V_t}{\partial K_{it}}. \tag{5.2}$$

The shadow price represents the current discounted value that would arise from a marginal change in the quantity of the asset.

A sustainable development path at time t is defined as

$$\frac{dV_t}{dt} = \sum_i p_{it} \frac{dK_{it}}{dt} + \frac{\partial V_t}{\partial t} \geq 0. \tag{5.3}$$

That is, a region is on a sustainability development path if the sum of changes in aggregate capital asset values and external influences (called a drift term) is greater than or equal to zero. Because the drift term reflects external influences, the shadow prices and/or the assets to satisfy the condition are the main direct ICZM targets. For example, improving the quality of an asset (e.g., seawater purity) could result in a higher shadow price (P_i) (Banzhaf and Boyd, 2012). Restoring seagrass beds results in a larger asset (K_i). The reflection of the shadow prices, which capture the contribution of each asset to social welfare, is a strength of IW.

IW has been adopted as a well-established sustainability assessment framework at the national scale (e.g., Arrow et al., 2012; UNU-IHDP and UNEP, 2012, 2014); however, it raises various theoretical and practical complications when applied to a regional open system (Harris and Pearson, 2004; Pearson et al., 2013) and needs to elicit practical insights for management. To make IW better fit into a regional open system, we propose that a coastal zone is on a sustainable development path only if it satisfies the following two conditions:

Sustainable development path condition 5.1: desired state

$$\begin{cases} \dfrac{dV_t}{dt} > 0 \text{ if } V_t < \underline{V} \\ \dfrac{dV_t}{dt} \leq 0 \text{ otherwise} \end{cases}, \tag{5.4}$$

where $\underline{V}$ is a desired state of the coastal zone. Sustainable development path condition 5.2: strong sustainability

$$p_{jt} K_{jt} \geq \underline{v}_j \, j \in (1, ..., m), \tag{5.5}$$

where $\underline{v}_j$ is the threshold of an asset, valued with its shadow price.

The first condition concerns desired social welfare. The original condition $\frac{dV_t}{dt} \geq 0$ can be reasonable for a national scale because V_t covers all the social welfare that people benefit from the assets available in the country. A coastal zone, which is a subsystem of a country (or even a region), however, supplies only part of the social welfare people gain. Therefore, there are tradeoffs among benefits, between the coastal zone and other subsystems. For example, maximizing the benefits from ecosystem services within the coastal zone may limit opportunities for industrialization or urbanization in adjacent and/or overlapping areas. Economic programs that maximize benefits from the coastal zone without considering other subsystems may not necessarily result in the maximization of the total social welfare. It may be misleading to assume that subsystem level (e.g., a coastal zone) sustainability will aggregate up to generate system-level (e.g., national level) sustainability (Anderies et al., 2013). Therefore, we need $\underline{V}$, a desired state of the coastal zone or level of benefit people want from the coastal zone with its tradeoffs considered.[8] This point often is

---

[7] For a comprehensive description of IW, see Arrow et al. (2003).

[8] Ideally, $\underline{V}$ must be derived as part of the maximization of total social welfare such as at the national level. Such multiscale and multilevel considerations might seem formidable, but Anderies et al. (2013) proposes a framework to deal with that complexity.

missing in ecosystem management literature, where it is implicitly agreed that the decline in natural capital is always undesirable. The second condition is strong sustainability, whereas the original IW assumes weak sustainability (Arrow et al., 2003). A coastal zone is not on a sustainable development path if some of the asset in monetary value (i.e., shadow price multiplied by the amount of the asset) is below its threshold. Whereas strong sustainability often indicates a physical threshold (e.g., Barbier and Markandya, 2013), we propose a monetary value of an asset that allows management to choose either improving quality (reflected in the price) or quantity (in physical units) of the asset to satisfy the condition. With this condition, we intend to capture two aspects of threshold. One aspect is to have a threshold value over which people want to benefit. For example, people might prefer a small beautiful beach to a large dirty beach. The other aspect is to have a threshold that is key to the resilience of a system (Uehara, 2013). Resilience is key to sustainable SEPLs (Gu and Subramanian, 2014). Pearson et al. (2013) applied IW to the Goulburn-Broken Catchment in Australia and incorporated a resilience of the catchment by computing the threshold value of groundwater tables, which are key to salinization of agricultural land and land under native vegetation. Because the $\underline{V}$ and $\underline{v_j}$ criteria are value laden, a consensus must be reached among stakeholders through a participatory process catalyzed by Satoumi.

IW provides a computational framework and criteria for sustainable development, but it is not sufficient alone, as it does not provide guidance to select capital assets of importance and set $\underline{V}$ and $\underline{v_j}$. Satoumi and the ESA can complement IW.

5.3.3 Example: Seto Inland Sea

5.3.3.1 Seto Inland Sea as Satoumi (Step 1)

The SIS (Seto Naikai) is an enclosed coastal sea in western Japan (Fig. 5.12). The sea has an area of 12,203 km^2; an average depth of 38 m (The Association for the Environmental Conservation of the Seto Inland Sea, 2014); and is economically, culturally, and ecologically important to Japan. It is surrounded by 11 prefectures, whose population is ~35 million (National Institute of Population and Social Security Research, 2013). The regional gross domestic product is about $940 billion US (Cabinet Office, Government of Japan, 2015).

There were Satoumis during the Edo Period from 1620 to 1850, though it was not coined as such at that time (Komatsu and Yanagi, 2015). However, during the rapid economic progress that began in the mid-20th century, ecosystems in the SIS were destroyed or degraded, resulting in, for example, declining fish catches, destruction of coastal zones for landfills and other anthropocentric uses, and water pollution (The Association for the Environmental Conservation of the Seto Inland Sea, 2015). In response to these severe situations, laws (e.g., Act on Special Measures concerning the SIS Conservation effective in 1973) and organizations (e.g., Governors and Mayors' Conference on the Environmental Conservation of the SIS in 1971) have been established. Today, Satoumi has been accepted widely as a desired state of the SIS among its stakeholders. The Governors and Mayors' Conference on the Environmental Conservation of the Seto Inland Sea (2007) proposed Satoumi. The recent amendment of the Act on October 2, 2015 shifted focus to Satoumi for water-pollution prevention.

Scientific studies about the SIS have been conducted, including a thorough assessment of the socioecological changes by the Japan Satoyama Satoumi Assessment (Duraiappah et al., 2012). The findings are important, but do not explicitly provide ICZM insights leading to a sustainable development path toward a desired Satoumi.

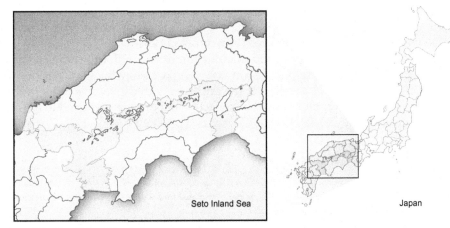

FIG. 5.12   Location of the Seto Inland Sea in Japan.

TABLE 5.5   Example Capital Assets as ES Sources for the SIS Satoumi

ES classifications	Marine ESs (proposed by Hattam et al., 2015)	SIS-specific ESs, (provided and/or envisioned by the Satoumi)	Capital assets	Type of capital
Provisioning services	Food provision: Farmed sea food	Oysters	Oyster farmers	Human
	Food provision: Farmed sea food	Oysters	Oyster rafts	Manufactured
	Food provision: Wild capture sea food	SIS Sawara	SIS Sawara	Natural
Regulating services	Waste treatment and assimilation	Water purification	Mudflats and seagrass beds	Natural
	Climate regulation	Carbon sequestration	Mudflats and seagrass beds	Natural
	Coastal erosion prevention	Prevention of beach erosion	Mudflats and seagrass beds	Natural
Habitat services	Migratory and nursery habitat; Gene protection	Biodiversity maintenance	Mudflats and seagrass beds	Natural
Cultural services	Leisure, recreation, and tourism	Recreational and cultural services	Mudflats and seagrass beds	Natural
	Information for cognitive development	Education	Seagrass beds	Natural
	Cultural diversity; Spiritual experience	Cultural and religious services	SIS Sawara	Natural

5.3.3.2 Identify Capital Assets as a Source of ESs (Steps 2 and 3)

We selected oyster farms, oyster rafts, SIS Sawara, mudflats, and seagrass beds as key capital assets for the SIS Satoumi, whose contributions to ESs are listed in Table 5.5. The list is indicative and by no means comprehensive.

Applying IW at the regional scale raises additional and nonnegligible complications with capital asset selection (Harris and Pearson, 2004; Pearson et al., 2013). Pearson et al. (2013) chose capital assets for their regional application of IW, saying that it is "not practically feasible to include all stocks" (p. 18). We believe that striving to include all capital assets is not necessarily favorable for a regional assessment. A Satoumi comprises only part of the social welfare and corresponding capital assets in a region. If we include all the assets in the region surrounding the SIS (including Osaka, the second-largest economic power in Japan), IW would include a huge set of capital assets that do not contribute to the SIS Satoumi, and such an IW would highlight aspects beyond its scope.

We propose two criteria for capital asset selection in response to these complications: being relatively immobile and being critical for the attainment and sustainability of a Satoumi (or in general, a desired state of a SEPL).[9]

Selected assets should be less substitutable, or mobile, between a coastal zone and other systems. If an asset can be replenished or substituted easily with one from outside the system, the IW means little because it is beyond ICZM control.

What "being critical" means needs further clarification. Because Satoumi is a dynamic mosaic of habitats and land uses (Duraiappah et al., 2012), it should include various components. Each stakeholder might have a different preference for goods and services provided by Satoumi, and the assessment should reflect such diversity and not just the majority's preferences. "Being critical" also include the passage of time and having thresholds whose crossing leads to significantly different consequences for goods and services (Pearson et al., 2013).

We address three forms of capital (human, natural, and manufactured) critical for the SIS Satoumi, including the reasons they were chosen. For social capital, an important form of capital, we adopt the same position as Pearson et al. (2013, p. 17) in that "it is an 'integrating' capital: the greater the positive social capital, the more productive and sustainable the other capital stocks. Thus, social capital should be reflected in the values assigned to the capitals included in the IW measurement."

---

[9] Double counting might not be an issue because the purpose of IW here is not to be consistent with the System of National Accounts such as the System of Environmental-Economic Accounting (Hein et al., 2015). Additionally, the absolute size of IW, which is influenced directly by double counting, is not the focus. Rather, we are interested in the changes in IW.

The importance of capital assets is determined based on expert consultation, laws, and regulations for the SIS, and from the literature. This limits the policy insights obtained in this study: to make the result fully useful for ICZM, the selection is targeted at the desired state of Satoumi (i.e., v_j and $\underline{V}$), which is agreed upon among stakeholders. This is a large undertaking beyond our scope. We believe that $\overline{our}$ result still is useful as a methodological contribution to regional sustainability assessment and is insightful for the SIS.

Although Satoumi includes various components, we emphasize the assets related to the SIS's prosperous fishery. The SIS Satoumi cannot be realized and maintained without a prosperous fishery and active fishermen (personal communication with an expert, Takehiro Tanaka, 2015).

Human Capital

Generally, it is difficult to assess "regional human capital" (Harris and Pearson, 2004, p. 8) because it is highly mobile. Our test case focused on oyster farmers because of their importance and relative immobility, because of their specialized skills. Oyster production is economically important because it is one of the most important fisheries in the SIS. It is also important for the nation, because it accounts for ~80% of oyster production in Japan on a monetary basis (Setonaikai Fisheries Coordination Office, 2012). Yanagi (2012) claimed "fishermen are the guardians of the coastal sea" (p. 18). Without them, the coastal zones "will be used in a disorderly manner by people on land who do not understand the characteristics of the coastal sea" (Yanagi, 2012, p. 18). Berque and Matsuda (2013, p. 191) also emphasized "the voluntary contribution of significant labor by ecosystem users, mostly fishers" as the key to Satoumi.

Natural Capital

We selected mudflats, seagrass beds, and Sawara (Japanese Spanish mackerel, *Scomberomorus niphonius*) for our study.

Mudflats and seagrass beds are two major ecological landscapes and management targets for the realization of Satoumi in the SIS (e.g., Central Environmental Council, 2012 and Governors and Mayors' Conference on the Environmental Conservation of the Seto Inland Sea, 2007) and all over Japan (Panel for the Visions of Seagrass Beds and Mudflats, 2015). They have been destroyed because of coastal development during the high-growth period of the Japanese economy. The mudflats declined by 20% from 1969 to 2006, and the seagrass beds declined by 72% from 1960 to 1990 (Ministry of the Environment Government of Japan, n.d.-b).

Mudflats and seagrass beds provide structure and function for water purification, biodiversity maintenance, carbon sequestration, prevention of beach erosion, and recreational and cultural services, among others. Educational benefits categorized in cultural services are a crucial component for realizing Satoumi. In Hinase, junior high school students participated in a seagrass bed restoration project, in which students who had not had contact with the sea nurtured their connection to the area (Hinase Junior High School and Network for Coexistence with Nature, 2016). Nurturing a connection with an area is crucial in the long term where decline in the number of the residents is severe. Seagrass beds are also potentially crucial in terms of thresholds. A seagrass bed expert, long involved in the ecological systems in the region, said a threshold area exists under which seagrass beds cannot sustain and recover on their own (personal communication with an expert, Takehiro Tanaka, 2015). Although there is no solid scientific evidence, it is an important notion that requires investigation.

SIS Sawara are important culturally and religiously, as well as economically (about 7% of the oysters in terms of market value) for the SIS Satoumi. Because Sawara is cooked for both *hare* (special days for religious and cultural events) and *ke* (average days), it is both sacred and soul food (Innami, 2015). Although it is a challenge to capture their monetary values such as shadow prices, cultural ecosystem services are of crucial importance (Chan et al., 2012, 2016). The SIS Sawara has a regional population with habitat inside and near the SIS (Uehara and Mineo, 2016), making them relatively immobile and a good ICZM target. Sawara is a management target because its population has declined rapidly. From 1987 to 1998, the population declined by 95%, but recovered to 30% of the 1987 level as of 2013, following the implementation of a resource management plan (Fig. 5.13). The plan limits a population threshold (Blimit) below which a restoration plan is implemented (Fisheries Agency of Japan and Fisheries Research Agency, 2014).

Manufactured Capital

Although manufactured forms of capital are largely immobile, their inclusion in IW should be selective because most manufactured capital can be replenished easily by entities outside the region, such as the Japanese Government and large companies. Additionally, most of this capital (e.g., roads and bridges) is not used exclusively for Satoumi. The manufactured capital in our case study consisted of oyster raft areas, which are specific to Satoumi and largely immobile.

FIG. 5.13 Estimated population and catch of Seto Inland Sea Sawara (Setonaikai Fisheries Coordination Office, 2012).

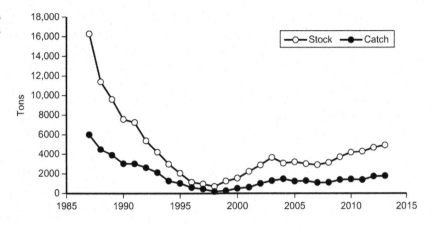

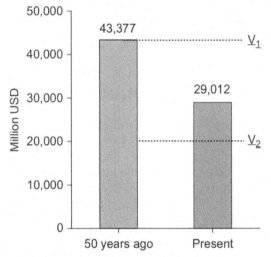

FIG. 5.14 Changes in IW over the past 50 years in the Seto Inland Sea. V_1 is the desired state (the same level as the one 50 years ago), and V_2 is a desired state below the current IW.

5.3.3.3 IW Computation (Steps 4, 5, and 6)

The study's primary objective is not to pursue the accuracy of IW in the SIS. Rather, it is to demonstrate the applicability and insights for ICZM. For replicability, the computational details are provided in a supplemental document.

5.3.4 Results and Their Insights for ICZM (Step 7)

Fig. 5.14 shows the changes in IW in the SIS and Fig. 5.15 shows the IW components. Over the past 50 years, the value of the SIS Satoumi measured by IW has decreased by 33.1% (from $43,377 million US to $29,012 million US) as shown in Fig. 5.14. The changes in its composition vary: Although seagrass beds have decreased by 68.9%, SIS Sawara has recovered by 16.2% (Fig. 5.15). To reiterate, a change in a capital's total value is because of the changes in a product's quantity and/or value.

We focused on four topics directly related to the insights for ICZM: sustainable development paths, shadow prices as a direct ICZM target, demographic changes, and drift term.

5.3.4.1 Is the Seto Inland Sea on a Sustainable Development Path?

We cannot answer the question with certainty because we lack sufficient information to check the sustainability conditions (Eq. 5.4 and Eq. 5.5). We could not set them here, because this should involve a consensus of v_j and $\underline{V}$ among beneficiaries of the SIS Satoumi, which is beyond the scope of this study. The currently available documents about the desired states proposed (e.g., Governors and Mayors' Conference on the Environmental Conservation of the Seto Inland Sea, 2007) are not sufficient to derive these specific criteria.

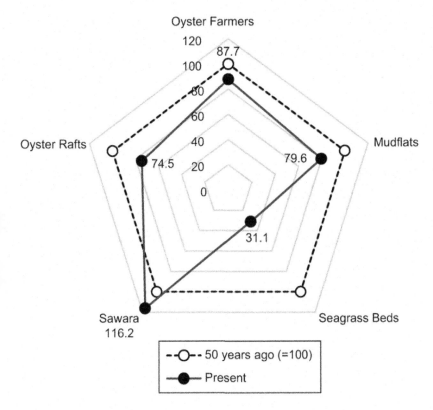

FIG. 5.15 Changes in the total values of capital assets comprising the IW (pjtKjt). Baseline data for 50 years ago are normalized as 100.

The decline in the IW during the past 50 years does not necessarily mean that it is not on a sustainable development path. For example, people might not desire to restore the Satoumi from 50 years ago (i.e., V_1 for example) if it means giving up the economic development that has occurred during that time. It is on a sustainable development path if people are willing to accept additional loss (i.e., V_2 for example) for a further economic development.

5.3.4.2 Shadow Prices as a Direct ICZM Target[10]

In addition to capital assets (K_i), which are a straightforward management target, some shadow prices (P_i), particularly of natural capital, also can be ICZM targets. An asset captures a quantity of natural capital, but a shadow price captures the quality in terms of its social welfare contribution. Both can contribute to IW. Because of the tradeoffs between economic development and ecosystem conservation, there is a limit to the quantitative expansion but not a significant limit to the quality.

A Panel for the Visions of Seagrass Beds and Mudflats organized by the Ministry of the Environment Japan points out the importance of the quality aspects (Panel for the Visions of Seagrass Beds and Mudflats, 2015). Their preliminary report estimates that 78% of mudflats and seagrass beds in Japan are unhealthy and need maintenance and conservation (Panel for the Visions of Seagrass Beds and Mudflats, 2015). Another estimate shows that, although mudflats did not change much in area, their water purification services drastically declined because of the decline of shellfish (Fig. 5.16).

Although it is unavoidable to adopt the benefit transfer method when a shadow price for a target area is not available, it should be done with great care. Uehara and Mineo (2016) collected previous studies' valuation of Japanese coastal zone ecosystem services and found huge variations in the valuation of mudflats (Fig. 5.17).

A benefit transfer using metaregression analysis could be more accurate than a simple unit value transfer (Rosenberger and Loomis, 2003), but the sample size (17 previous studies about Japanese mudflats, Uehara and Mineo, 2016) might not be sufficient for a robust estimate.

---

[10] See Arrow et al. (2012), UNU-IHDP and UNEP (2012, 2014), and Pearson et al. (2013) for theoretical and practical discussions about the methodological and theoretical issues of shadow prices.

FIG. 5.16 Mudflat areas and water purification services (Fisheries Agency of Japan, 1998).

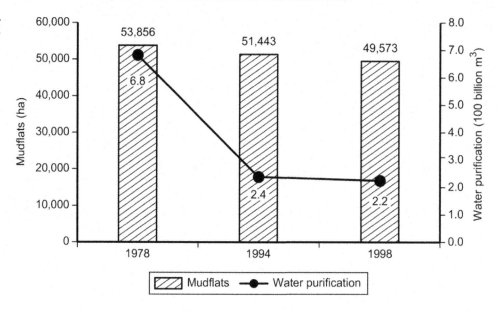

FIG. 5.17 Comparison of mudflat valuation in various studies. The values are obtained from Uehara and Mineo (2016), Costanza et al. (1997), and de Groot et al. (2012). The databases by Costanza et al. (1997) and de Groot et al. (2012) include valuation studies all over the world.

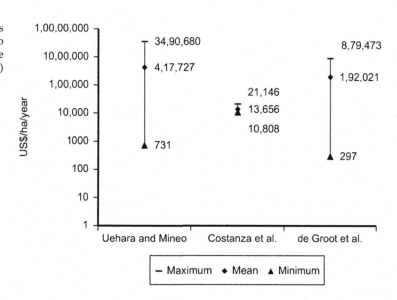

Because of the great importance of shadow prices for this framework, original estimates are highly recommended to capture the linkage between ICZM practices, ecosystem changes, and shadow prices. Previous studies about social and ecological indicators (e.g., Duraiappah et al., 2012; Hattam et al., 2015; Turner et al., 2014) played an important role in building that linkage (Table 5.6).

5.3.4.3 Impacts of Demographic Changes on the Sustainable Development Path

Demographic changes are meaningful and should be incorporated in the evaluation. The theoretical discussion about the use and interpretation of IW per capita, however, "still remains unsettled" (Arrow et al., 2003, p. 676). Changes in IW (Eq. 5.3) should include a term that captures the current value of demographic changes (Arrow et al., 2003). Nevertheless, with certain assumptions, the comparisons of wealth per capita adjusted for demographic changes could be valid (Arrow et al., 2003), and we adopted this position. The population in the SIS area is projected to decline (Table 5.7). Assuming the total IW remains the same 25 years from now, IW per capita should increase from $0.83 million US to $1 million US.

TABLE 5.6   Estimation of the Value of Seto Island Sea's Tidal Flats Using the Benefit Transfer Method

Area	Area of tidal flat (ha)	Unit (JYen/ha/Year)		Value of flow (100 M JYen/Year)	Current value of stock (100 M JYen, discount rate 5%)
Japan	49,380	**Current Study**	**50,603,440**	**24,988**	**499,760**
		Costanza et al. (1997)	1,654,344	817	16,338
		de Groot et al. (2012)	23,261,400	11,486	229,730
		Ministry of the Environment Japan (2014)	108,205,478	53,432	1,068,637
Seto Inland Sea	11,943	**Current Study**	**50,603,440**	**6044**	**120,871**
		Costanza (1997)	1,654,344	198	3952
		de Groot et al. (2012)	23,261,400	2778	55,562
		Ministry of the Environment Japan (2014)	108,205,478	12,923	258,460

Ref. Hiroshima Prefecture, 2014a, Table 4.

TABLE 5.7   Total IW, IW Per Capita, and Population Changes in the Seto Inland Sea Area (National Institute of Population and Social Security Research, 2013 for population)

	50 years ago	Current	25 years from now
Total IW (million USD)	43,377	29,012	29,012
Population (thousands)	28,707	34,748	28,992
IW per capita (million USD)	1.51	0.83	1.00

TABLE 5.8   Fisheries Cooperative Associations' Opinions About Conservation Activities and the Membership (Ministry of Agriculture, Forestry and Fisheries of the Government of Japan, 2006)

Relationship between conservation activities and the membership	Seagrass beds	Mudflats
Conservation activities will be difficult to maintain in future, because of aging and dwindling membership	54.4%	52.1%
Conservation activities already are difficult to maintain, because of aging and dwindling membership	31.0%	38.5%
There is no problem because there are enough conservation activity leaders.	9.2%	7.3%
Other	5.4%	2.1%

5.3.4.4 Drift Term

The drift term ($\frac{\partial V_t}{\partial t}$) in Eq. 5.3 vanishes, assuming a time-autonomous resource allocation mechanism (Arrow et al., 2003). It might not vanish, however, for a small and open region because of exogenous influences, such as growth in factor productivity, export prices, exogenously led inflows and outflows of capital assets, and population changes.

Because it is exogenously given, the drift term generally is not targeted in ICZM. It is possible, however, to internalize an exogenous influence into ICZM by association with other management levels. Because coastal zones are a complex socioecological landscape, a multiscale approach (Ostrom, 2010) is required (e.g., Cummins and McKenna, 2010). Hidaka (2014) proposed the combination of dual management systems (e.g., prefectures and municipalities with communities) and network governance in the context of Satoumi.

One of the most important exogenous factors in this context is volunteers living outside the Satoumi. Berque and Matsuda (2013) emphasized the voluntary contribution of significant labor by ecosystem users, mostly fishers as a key to success. With the declining fisherman population, however, more volunteers who can support their conservation activities, such as those from nonprofit and nongovernmental organizations, are required (Henocque, 2013). A recent survey supports this point (Table 5.8). The current system, in which conservation activity leaders are mostly the association members (i.e., fishers), is not or will not be sustainable. With a multiscale approach that covers a larger system, it is possible to internalize such volunteers as part of ICZM.

5.3.5 Conclusion

This section proposed a novel regional sustainability assessment framework that integrates three separately developed approaches (IW, Satoumi, and the ESA) to make ICZM more operational. Although ICZM has been adopted globally, opportunities remain to improve ICZM practices for achieving its ambitious goals. This section focused on the methodological advancement of ICZM indicators, which is one of the most pressing research realms, to attain or sustain desired coastal zones through ICZM.

The three approaches complement one another to provide practical insights for ICZM. IW is a sustainability path indicator grounded in economic theory. It computes the net current value of social welfare based on the products of capital (human, natural, and manufactured) and their shadow prices. Shadow prices, which differentiate IW from pure ecological indicators, are important for ICZM, because they capture the quality of natural capital assets especially. To make it work at regional scale, we proposed two sustainability criteria: desired state and strong sustainability. Because neither IW nor the ESA provides guidance for ranking capital assets, Satoumi (a traditional Japanese SEPL that envisions a desired state of coastal zones) was introduced. The ESA facilitates envisioning Satoumi because people might not be fully aware of its benefits and translates the desired state envisioned by Satoumi into the list of capital assets and their shadow prices, which can be used in IW. Implementation of the framework involves stakeholder involvement, especially in Satoumi, and adopts an iterative process for adaptive comanagement.

We discussed the framework and its implementation along with Fig. 5.11, and tested it in the SIS. Although the results cannot be applied readily to ICZM because of inaccuracies and lack of stakeholder agreement through a participatory process, this section makes a meaningful proposal for the full implementation of the framework and demonstrates the insights for ICZM. IW helps ICZM rank capital assets to target those of highest priority and to plan appropriate policy measures (e.g., whether to change the quantity or improve the quality). Demographic changes are of great concern. The drift term might be a management target, depending on how ICZM is defined.

Crucial future research topics relate to scenario analysis, which were not discussed in detail. First, more research about the dynamic mechanism of shadow prices is needed for scenario analysis, because they are a direct ICZM target rather than a mere monetization of capital assets. Shadow prices are dynamic (or instable) (Banzhaf and Boyd, 2012; Bastian et al., 2013). Although the investigation of such complex dynamics of ecosystem services and their human welfare that can be reflected in shadow prices is of huge importance, such studies remain scant (Bennett et al., 2015). Second, because of the complexity and uncertainty of the issue and the paucity of factual data, further development of ecological (e.g., Fu and Jones, 2013), ecological-economic (e.g., Eppink and van den Bergh, 2007), and socioecological (e.g., Schlueter et al., 2012) modeling and simulation that explain the mechanism of ICZM impacts is required.

5.4 DYNAMIC SUSTAINABILITY ASSESSMENT OF SATOUMI (NAKAGAMI ET AL., 2018)

Kenichi Nakagami

5.4.1 Framework of Sustainability Assessment of Satoumi

(1) Meanings and limits of sustainability assessments

The concept of Satoumi is that the value of the sea is not a simple correlation between coastal zones and people, and that it is important for many people, not only stakeholders such as fishermen, to expand the area of activities, and to argue how to establish activity entities. By doing so, the diversity (material circulation, ecosystems, and communication) and sustainability (activity areas and activity entities) of Satoumi are assured. The increase in diversity leads to the development of the fishing industry, which is on the wane. To aim for community development during the rapid acceleration of depopulation and aging, it is fundamental that diversity and sustainability strengthen regional characteristics, and to do that, it is necessary to establish an implementation process. If we define Satoumi as "clean, productive, and prosperous," it is necessary to organize the current elements in consideration of each area's situation. Clean reflects material circulation and ecosystems in diversity; productive reflects activity areas and activity entities in sustainability; and prosperous reflects communication in diversity. We organize the elements of sustainability assessments of Satoumi as follows based on the environment, society, and economy.

Clean

Water quality standards: maintain healthy material circulation and show the environmental situation of high- and low-quality water.

Landscape standards: show the landscape situation of coastal zones, and the natural and historic landscape of the coastal zones, including seaweed beds and tidal flats.

Productive

Population standards: show the situation of the population, which is fundamental to urban activities.
Fish catches standards: show fishermen's activities in Satoumi.

Prosperous

Cultural standards: show symbiosis with nature, history and culture.
Communication standards: show regional cooperation and communication between fishing villages and urban cities.

(2) Sustainability of Satoumi

The sustainability assessment is done to measure various aspects of the environment, society, and economy. In the planning process to achieve sustainability, we assume potential sustainability by measuring the environmental carrying capacity using general indices such as the ecological footprint in the pre-assessment of planned target areas. We conduct sustainability assessments using special indices for the environment, society, and economy, based on a general understanding. Then we decide a sustainability plan based on the assessment of the situation and abilities. These three steps of sustainability assessment are needed to understand the state of the target areas in sustainability planning, to measure the ability to realize sustainability, and to assess the will as a decision to conduct a sustainability plan.

In Fig. 5.18, we show the sustainability elements of the state, the first step related to a sustainability assessment of Satoumi. Although the elements of the environment, society, and culture vary, depending on each Satoumi, the issues are environmental policy planning for the environment and society, the development of environment friendly fisheries for the environment and economy, and the development of the community for society and the economy, when we focus on Satoumi.

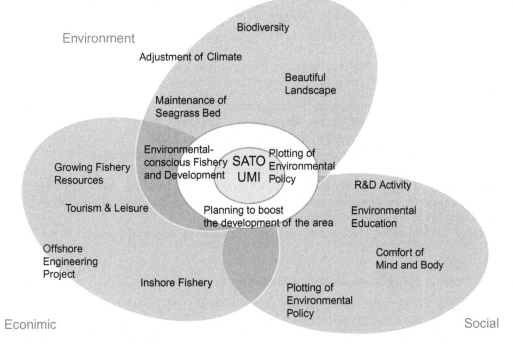

FIG. 5.18 Sustainability assessment factors of Satoumi.

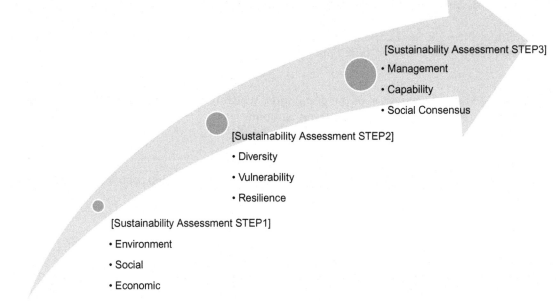

FIG. 5.19   Dynamic sustainability assessment process.

(3) The process of dynamic sustainability assessments

Various aspects of the environment, society, and economy of sustainability assessments of Satoumi can to be listed according to the region and period. To make these elements ensure the diversity and sustainability of Satoumi, however, it is necessary to have the ability to realize sustainability, and a will to use such ability based on the recognition of the condition. The process of dynamic sustainability assessments is shown in Fig. 5.19.

First, we list environment, society, and economy as the elements that show the state of sustainability. Even if we understand the current state of sustainability, however, the ability to maintain that condition is required. The elements to maintain the condition of sustainability are diversity, vulnerability, and resilience. After achieving such abilities comprehensively, collaboration with urban cities as well as regional communities, and movement toward the future, is necessary to realize the sustainability of Satoumi.

The elements of the will to realize sustainability are management, capability, and social consensus.

Based on the state of sustainability, the dynamic assessment to realize the sustainability of Satoumi is as follows.

First step: setting target areas (Satoumi) and problem structures.
Second step: organizing past basic data of the target areas (1950, 2015, 2050).
Third step: economic assessment of the ecosystem services of target areas.
Fourth step: first phase of sustainability assessment of target areas (five-grade assessment).
Fifth step: A (ideal state) − B (current state) = C (gap assessment).
Sixth step: second phase of sustainability assessment of target areas/deciding management level.
Seventh step: third phase of sustainability assessment of target areas/deciding attainment level.
Eighth step: deciding the level for clean, productive and prosperous.
Ninth step: offering guidelines for coastal zone management and reflecting the results to environmental policy.

(4) Elements and criteria of dynamic sustainability assessments

The elements and criteria of dynamic sustainability assessments are shown in Tables 5.9, 5.10, and 5.11. Regarding the criteria, there are subjective elements because we created it based on the current condition of Satoumi.

(5) Calculation procedure of sustainability assessments

The calculation procedure of sustainability assessments is shown in Fig. 5.20.

TABLE 5.9 Sustainability Assessment Indicators Step 1 State

	Criterion	Component of evaluation	Item of evaluation	Index	Evaluation of criterion		
					A	B	C
Clean/ Beautiful	Water quality (WQ)	Economic (EC)	Economic activity (BWQEC)	DID (Relative to Nationwide Average)	High	Medium	Low
		Environment (EN)	Water quality standard (BWQEN)	Achievement rate of TN environmental standard	Achievement rate (high)	Achievement rate (medium)	Achievement rate (low)
		Social (SO)	Construction and improvement of water treatment (BWQSO)	Ratio of construction and improvement of sewerage	$\geqq 90\%$	70%~90%	$\leqq 70\%$
	Landscape (LS)	Economic (EC)	Tourism industry (BLSEC)	Number of visitors/ increase-decrease rate of Visitors	Superior	Medium	Inferior
		Environment (EN)	Landscape conservation (BLSEN)	Level of Landscape Conservation/ Achievement of Ordinance	Superior	Medium	Inferior
		Social (SO)	Institution of landscape conservation (BLSSO)	Number of Ordinance	Superior	Medium	Inferior or nothing
Productive	Population (PO)	Economic (EC)	Population of coastal area (PdPOEC)	Increase-decrease rate of Population	$\geqq 0\%$	$0\% \sim -10\%$	$\geqq -10\%$
		Environment (EN)	Inhabitant consciousness (Environment) (PdPOEN)	Satisfaction rating related amenity	High	Medium	Low
		Social (SO)	Inhabitant consciousness (Social) (PdPOSO)	Level of the will of continuous habitation	High	Medium	Low
	Fishery (FP)	Economic (EC)	Activity of fishing ground (PdFPEC)	Transition of fishery yield	Plus	Same Level	Minus
		Environment (EN)	Fisheries environmental conservation (PdFPEN)	Transition of the area of tideland and seagrass bed	$\geqq 0\%$	$0\% \sim -10\%$	$\geqq$-10%
		Social (SO)	Society of fishery area (PdFPSO)	Number of persons engaged & young successor	$\geqq 0\%$	$0\% \sim -10\%$	$\geqq$-10%
Prosperous	Culture (CU)	Economic (EC)	Culture and economy (PsCUEC)	Number of cultural facilities, transition of number of users	$\geqq 10\%$	$\pm 10\%$	$\geqq$-10%
		Environment (EN)	Rural cultural program (PsCUEN)	Historical cultural assets	Superior	Medium	inferior
		Social (SO)	Rural cultural education (PsCUSO)	Cultural education program in primary and junior high school	Multi-existence	Existence	Non-existence
	Exchange (EX)	Economic (EC)	Rural exchange program (PsEXEC)	Economic activity of tourism industry	Active	Medium	Static
		Environment (EN)	Rural exchange promotion Program (PsEXEN)	Know-how of cultural exchange	Existence	Unclear	Non-existence
		Social (SO)	Exchange rural and cities (PsEXSO)	Number of alignments of municipal government	$\geqq 5$	3–5	$\leqq 3$

TABLE 5.10   Sustainability Assessment Indicators Step 2 Ability

	Criterion	*Component of evaluation*	Item of evaluation	Index	Evaluation of criterion		
					A	**B**	**C**
Clean/ Beautiful	Water Quality (WQ)	*Diversity (D)*	Balance of industry (BWQD)	Production value by industry, ratio of primary sector of industry	High	Medium	Low
		Vulnerability (V)	Climate change strategy (BWQV)	Variation of fishery species	Inferior	Medium	Superior
		Resilience (R)	Effective water treatment (BWQR)	Water Treatment based on circulated balance of nutrient salts	Existence	Unclear	Non-existence
	Landscape (LS)	*Diversity (D)*	Harmonization of area (BLSD)	Rate of artificial structural object	Low	Medium	High
		Vulnerability (V)	Community development & disaster prevention (BLSV)	Environs improvement (ex: coastal levee)	Active	Medium	Static
		Resilience (R)	Landscape conservation (BLSR)	Arrangement of landscape	Existence	Unclear	Non-existence
Productive	Population (PO)	*Diversity (D)*	Intergenerational balance (PdPOD)	Transition of the composition ratio of younger generation	$\geqq 20\%$	$10\%\sim20\%$	$\leqq10\%$
		Vulnerability (V)	Tsunami and disaster prevention (PdPOV)	Flood control, stock, emergency evacuation area, cooperation with citizen and company	Active	Unclear	Non-existence
		Resilience (R)	Community revitalization (PdPOR)	Achievement rate of community revitalization	High	Medium	Low
	Fishery (FP)	*Diversity (D)*	Multiproduct of Fish Species (PdFPD)	Number of pieces	Superior	Medium	Inferior
		Vulnerability (V)	Climate change (PdFPV)	Transition of seawater temperature	Inferior	Medium	Superior
		Resilience (R)	Succession planning of fishery (PdFPR)	Transition of new persons engaged	$\geqq 0\%$	$0\%\sim-10\%$	$\geqq-10\%$
Prosperous	Culture (CU)	*Diversity (D)*	Community traditional culture (PsCUD)	Participant of community traditional culture	Superior	Medium	Inferior
		Vulnerability (V)	Cultural succession (PsCUV)	Extinct cultural event	Inferior	Medium	Superior
		Resilience (R)	Movement to boost development (PsCUR)	Number of events, number of participants	Superior	Medium	Inferior
	Exchange (EX)	*Diversity (D)*	International exchange (PsEXD)	Inbound tourists	Superior	Medium	Inferior
		Vulnerability (V)	Traffic accessibility (PsEXV)	Access time by public transportation and automobile	Short	Medium	Long
		Resilience (R)	Community cultural exchange (PsEXR)	Budget of publicity	Superior	Medium	Inferior

TABLE 5.11 Sustainability Assessment Indicators Step 3 Will

	Criterion	Component of evaluation	Item of evaluation	Index	Evaluation of criterion		
					A	B	C
Clean/ Beautiful	Water Quality (WQ)	Management (M)	Institution of water quality (CWQM)	Achievement rate of water quality standard	High	Medium	Low
		Capability (C)	Participation of community residents (CWQC)	Number of participants	Superior	Medium	Inferior
		Social consensus (SC)	Acceptance of water quality standard (CWQSC)	Setting standards based on the area condition	Existence	Unclear	Non-existence
	Landscape (LS)	Management (M)	Institution of landscape conservation (CLSM)	Ordinance	Superior	Medium	Inferior
		Capability (C)	Activity of community residents (BLSC)	Recognition degree of institution of landscape conservation	High	Medium	Low
		Social consensus (SC)	Common basic values of landscape (BLSSC)	Landscape conservation	High	Medium	Low
Productive	Population (PO)	Management (M)	Population policy (PdPOM)	Ways and means of social increase in population	Existence	Unclear	Non-existence
		Capability (C)	Intraregional move-in and out of (PdPOC)	Ratio of change on move-out	Move-in excess	±5%	Move-out excess
		Social consensus (SC)	Future target of population (PdPOSC)	Transition of future population	$\geqq -10\%$	$-10\% \sim -25\%$	$\geqq -25\%$
	Fishery (FP)	Management (M)	Activity of fishermen's union (PdFPM)	Management of fishery's facilities, obtain certification	High	Medium	Low
		Capability (C)	Encouragement of fishermen's union (PdFPC)	Transition of new number of fishermen's union	$\geqq 0\%$	$0\% \sim -10\%$	$\geqq -10\%$
		Social consensus (SC)	Regional alliances (PdFPSC)	Set of regular consultative meetings	Existence (>4)	Existence (<3)	Non-existence
Prosperous	Culture (CU)	Management (M)	Public administration & tourist association (PsCUM)	Budget of activity/ number of staff	High (>3)	Medium (1–2 person)	Low/0
		Capability (C)	Cultural creation (PsCUC)	Train up of corporate actor	High	Medium	Low or nothing
		Social consensus (SC)	Common basic values of community value (PsCUSC)	Affection to area	High ($\geqq 70\%$)	Medium (35%~70%)	High ($\leq 35\%$)
	Exchange (EX)	Management (M)	Activity of common basic values of (PsEXM)	Power of transmission	High	Medium	Low
		Capability (C)	Encourage of exchange & human resources development (PsEXC)	Participation to community event	High	Medium	Low
		Social consensus (SC)	Value of exchange program (PsEXSC)	Social cognition of exchange program	High ($\geqq 70\%$)	Medium (35%~70%)	Low ($\leq 35\%$)

FIG. 5.20 The calculation procedure of dynamic sustainability assessment.

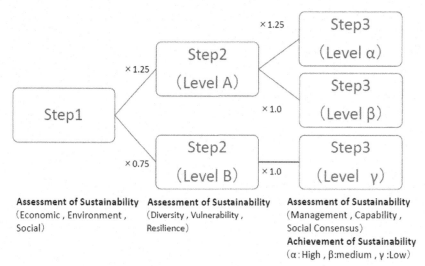

Assessment of Sustainability (Economic , Environment , Social)

Assessment of Sustainability (Diversity , Vulnerability , Resilience)

Assessment of Sustainability (Management , Capability , Social Consensus)

Achievement of Sustainability (α : High , β:medium , γ :Low)

The procedure is as follows. [Comprehensive evaluation criteria]

A+ (161–189)—Sustainability is realized and expected to be developed.
A (118–160)—Large potential for sustainability.
B (74–117)—Efforts to maintain current conditions are required.
C (45–73)—Difficulties realizing sustainability.

5.4.2 The Current Situation and Issues in the Target Areas

Japanese coastal zones have been changing greatly, along with natural and social conditions, especially the artificialization of coastal zones. Urbanization and industrialization have decreased tidal flats and seaweed beds significantly, degrading the fishing environment. In addition, the decrease of fishing catches has become prominent recently. The trend of depopulation and aging in coastal zones is very serious. The Fisheries Agency has pointed out that fishing villages have a higher aging rate than the national average, and the higher the fishery rate is, the higher the depopulation rate.

As a maritime nation, Japan formulated the Basic Act on Ocean Policy in 2007 and the Basic Plan on Ocean Policy in 2013, which aim for integrated coastal zone management. In spite of such planned activities to consider comprehensively coastal zones (not only from the perspective of fisheries), when the tsunami hit the Tohoku coast area after the Great East Japan Earthquake in 2011, many of the fishermen were victims and there was a catastrophic demolition of fishery facilities. Under such harsh circumstances, seeking sustainability of fisheries is an extremely difficult issue and a new challenge for revitalizing the community. It is not a simple case of resolving issues facing fishing villages. It also aims to revitalize the local community toward the creation of coastal zones in a broad sense, and not only offering ecosystem services in coastal zones.

In the research project, three areas were chosen as targets: the Seto Inland Sea as a representative of a closed sea area, the Sanriku coast area, including Shizugawa Bay as a representative of an open inner bay, and the Japan Sea coast area as a representative of an international closed sea area. In this document, we set Hinase Bay in Bizen City, Okayama Prefecture, Shizugawa Bay in Minamisanriku Town, Miyagi Prefecture, and Nanao Bay in Nanao City, Ishikawa Prefecture as our target areas to clarify their characteristics. The characteristics of the target areas are shown in Table 5.12 which is summarized in the following section. We organized statistical data of the three areas as follows. We labeled Hinase Bay in Bizen City, Okayama Prefecture as Bizen City; Shizugawa Bay in Minamisanriku Town, Miyagi Prefecture, as Minamisanriku Town; and Nanao Bay in Nanao City, Ishikawa Prefecture, as Nanao City; to clarify their characteristics as shown in Table 5.12.

(1) Population

In all three areas, the population has decreased significantly, almost 30% from 1985 to 2015: Bizen City (48,112 to 35,179), Minamisanriku Town (21,970 to 12,370), and Nanao City (69,915 to 55,325). In particular, Minamisanriku Town was affected directly by the Great East Japan Earthquake. The population has been rapidly aging in the last

TABLE 5.12 General Situations of the Target Areas

			Bizen City (※1)	Minamisanriku Town (※2)	Nanao City (※3)
Population	Total population (person)	1985	48,112	21,970	69,915
			14,086	5195	19,063
	Household		14.2	12.9	14.2
		2015	35,179	12,370	55,325
	Elderly person (older than 65 years) (%)		13,878	4041	20,855
			36.3	33.5	34.7
Economic	Industry	Manufactured goods shipment value (2014/10,000 yen)	25,546,652	2,065,312	5,531,613
	Fishery	Fish catches (2014/ton)	353	8485	8396
		Number of fishery management (2013)	105	472	283
Finance	Revenue of general account (1000 yen)		20,930,323	51,805,832	33,231,978
	Expenditure of general account (1000 yen)		19,750,140	45,671,462	32,893,229
	Total revenue of municipal tax (1000 yen)		5,029,380	1,096,006	7,989,531

※1 2005: Bizen City (old), Hinase Town, and Yoshinaga Town were merged to form Bizen City.
※2 2005: Shizugawa Town and Utatsu Town were marged to form Minamisanriku Town.
※3 2004: Nanao City (old), Tatsuruhama Town, Nakajima Town, and Notojima Town were merged to form Nanao City.

30 years as shown in the shift of the elderly ratio from 1985 to 2015: Bizen City (14.2% to 36.3%), Minamisanriku Town (12.9% to 33.5%), and Nanao City (14.2% to 34.7%).

(2) Economy

Product shipments in the target areas: Bizen City (25.546 billion yen), Minamisanriku Town (2.065 billion yen), and Nanao City (5.531 billion yen). Fish catches and fishery management entities, respectively: Bizen City (353 tons, 105 people), Minamisanriku Town (8485 tons, 472 people), and Nanao City (8396 tons, 283 people). The scale differs among the three areas: Bizen City has a high ratio of ceramic industries and a low dependence on fishing, while in Minamisanriku Town and Nanao City, fishing is an important industry.

(3) Finance

The revenue of the general accounts: Bizen City (20.9 billion yen), Minamisanriku Town (51.8 billion yen), and Nanao City (33.2 billion yen). Minamisanriku Town has experienced a large impact from construction investment because of reconstruction after the earthquake, which is a special situation for the region.

(4) Characteristics of the bay
    (1) Hinase Bay

The Hinase Islands (14 islands including Kakuijima Island and Kashirajima Island). The Hinase Bay is a part of the Seto Inland Sea. The depth of sea between the mainland and Kakui Island, the largest of the Hinase Islands, is largely <10 m, so the sea is basically calm except for typhoons and storm surges.

(2) Shizugawa Bay

The bay faces the Pacific Ocean and has an area of 47 km^2. The maximum depth is 54 m. Eight rivers flow into the bay, all of which originate in Minamisanriku Town.

(3) Nanao Bay

The bay faces the Japan Sea and is divided by Notojima Island (area: 47 km^2, population: about 3000 people) into three parts, Nanao-kita Bay, Nanao-nishi Bay, and Nanao-minami Bay. The maximum depth is 58 m. Tourism resources such as the Wakura hot spring area face Nanao-nishi Bay.

5.4.3 Comparison Analysis of Shizugawa Bay, Hinase Bay, and Nanao Bay

(1) Sustainability assessments of the target areas

Hinase Bay: High grades for fishing grounds conservation and regional culture education. Holistically "B" grade for diversity, vulnerability, and resilience. "A" grade for capability for clean; management of fish catches for productive; and social consensus of culture for prosperous.

Shizugawa Bay: High grades for fishing activities, regional communication promotion, and communication between regional communities and urban cities. "A" grade for vulnerability and resilience, but "C" grade for water quality and landscape. "A" grade for management of fish catches and social consensus for productive; and management of communication for prosperous.

Nanao Bay: High grades for tourism and the cultural economy. "A" grade for diversity in fish catches, culture and communication. "B" grade for all items of management, capability and social consensus.

(2) Comparison analysis of Shizugawa Bay, Hinase Bay and Nanao Bay

The comparison of the results of the sustainability assessment for the three areas is shown in Figs. 5.21–5.23.

Step 1 shows sustainability states for the economy, environment, and society. In the clean category, Nanao Bay has the highest grade, and in the productive category, Hinase Bay and Shizugawa Bay have high grades for fish catches, whereas Nanao Bay has a high grade for population. In the prosperous category, Shizugawa Bay has a high grade for communication.

Step 2 shows the abilities of diversity, vulnerability, resilience, etc. for sustainability. In the clean category, Nanao Bay has a high grade for landscapes; in the productive category, Shizugawa Bay has a high grade for population; and in the prosperous category, Nanao Bay has a slightly higher grade, but all three are almost the same.

Step 3 shows the will of management, capability, etc. for sustainability. In the clean category, Nanao Bay has a high grade for landscapes; in the productive category, Shizugawa Bay has a high grade for population; and in the prosperous category, Nanao Bay has a high grade.

(3) Comprehensive results

The comprehensive results of the three areas are shown in Table 5.13. Nanao Bay and Hinase Bay got "A" grades and Shizugawa Bay got a "B" grade for the comprehensive results. We assume that Nanao Bay was rated highly for its political efforts regarding conservation of Satoumi and its focus on tourism using Wakura hot springs and the sea. For Hinase Bay, the results of the sustainability assessment were improved because of the strong relationship among regional residents and people in the Kansai area as a result of the eelgrass beds restoration. And for Shizugawa

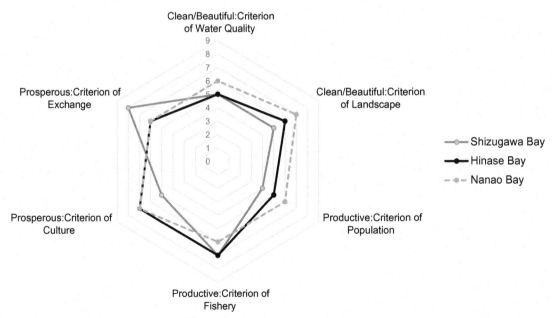

FIG. 5.21   Result of comprehensive assessment of Step 1 (State).

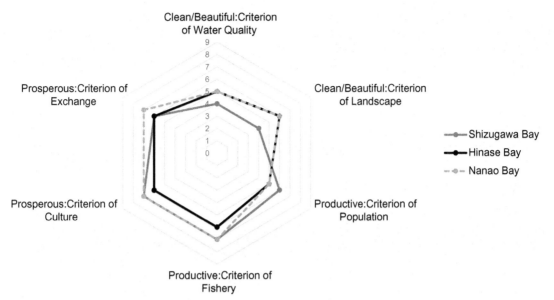

FIG. 5.22 Result of comprehensive assessment of Step 2 (Ability).

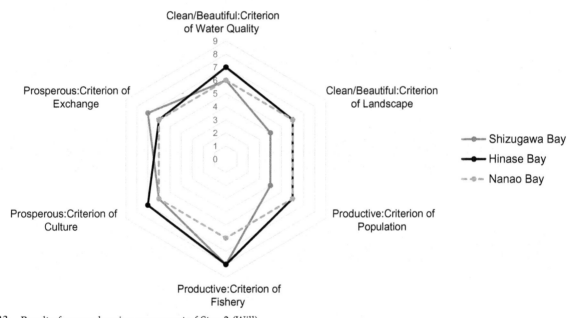

FIG. 5.23 Result of comprehensive assessment of Step 3 (Will).

TABLE 5.13 Comprehensive Assessment of Three Areas

	Step 1	GAP	Level	Adjusted value	Step 2	GAP	Level	Adjusted value	Step 3	Total score	Judgment
Shizugawa Bay	34	20	B	0.75	34	20	a	0.75	35	86	B
Hinase Bay	36	18	A	1.25	34	20	b	1	40	119	A
Nanao Bay	38	16	A	1.25	37	17	c	1.25	36	129.75	A

Bay, there were serious impacts from the Great East Japan Earthquake and tsunami on citizens and fishermen who were victims, and the annihilation of the natural coast because of the construction of coastal levees. From the perspective of sustainability assessments, we expect that sustainability will be improved because the stakeholders of the fisheries have a strong will to continue fishing (Nakagami et al., 2018).

5.5 COASTAL REGIONAL DEVELOPMENT AND THE FISHING INDUSTRY

N. Obata and T. Yoshioka

5.5.1 Current Situation Regarding the Coastal Fishing Industry

According to the Japan Fisheries Agency's White Paper on Fisheries (2017 edition), the production volume of the Japanese fishing and aquaculture industry peaked in 1984 (12.82 million tons) and then decreased rapidly until around 1995, after which it has continued to gradually decrease (Fisheries Agency of Japan, 1998). The production volume of the Japanese fishing and aquaculture industry in 2016 decreased by 270,000 tons (6%) year on year to 4.36 million tons. This amounts to just 34% of peak production volume. Within this, the coastal fishing industry produced 994,000 tons in 2016. The mariculture industry shrank slightly to 1,033,000 tons. Output by the mariculture industry, however, increased by 23.1 billion yen (5%) year on year to 509.7 yen billion. The number of fishing industry workers, which was one million at its peak, continued to decline, dropping to 153,000 in 2016. The average age of these workers was older than 60. The depopulation of fishing villages is progressing. This aging of fishing populations is not a recent phenomenon. It is the result of decades of ignoring the lack of new participants into the industry.

In the Seto Inland Sea area, production peaked at 372,000 tons in 1982 and since then it has continued to decline, reaching 157,400 tons in 2016. Aquaculture industry production also peaked in 1988 at 402,000 tons, but this also has declined predictably, reaching just 237,800 tons in 2018. Japan's fishing industries are declining in earnest. It either needs to restart production on a smaller scale or face extinction.

In Shizugawa Bay, Miyagi Prefecture's (surface area: 46.8km^2) first coho salmon fisheries were started in 1975. By the industry's peak in the late 1980s, coho salmon farmed in Miyagi made up 90% of Japan's domestic production. The growth of cheap Chilean imports, however, reduced coho salmon production volumes dramatically. Furthermore, the large tsunami caused by the 2011 earthquake caused catastrophic damage, including the destruction of all aquaculture facilities. Because of the efforts of various parties, turnover has been returning to levels close to what it was before the disaster, and in 2017 it exceeded predisaster levels as shown Fig. 5.24. One problem being faced is the slow recovery of ascidians, sea urchins, and other species because of factors including trade embargos and coastal denudation.

In this way, whether we look at catch volumes, catch earnings, or number of workers, Japan's coastal fishing industry peaked in the early 1980s and has seen no growth since. The coastal fishing industry is seeing a shift in direction from being an industry focused on catching fish, to one that increases and cultivates resources by taking eggs from fully grown fish in facilities on land, incubating these eggs, raising the juvenile fish to a certain size, and then releasing them into the sea.

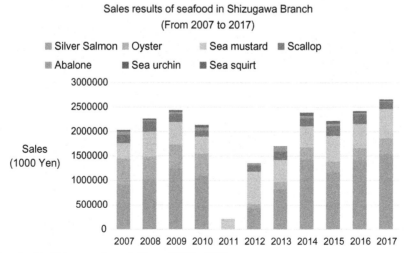

FIG. 5.24   Sales result of seafood in Shizugawa branch (From 2007 to 2017).

Looking at Japan alone, it would seem that the fishing industry has no future, but casting our eyes overseas reveals a different reality. The UN Food and Agriculture Organization (FAO) made forecasts regarding production by every country's fishing industry up to 2025, and it predicted a 17.4% increase in worldwide catch volumes. The global fishing industry continues to grow and remains profitable. The FAO forecasts also predicted a 13.4% decrease in catch volumes for Japan, which was the lowest growth ratio out of all countries with a significant fishing industry.

The decline of Japan's fishing industry is the result of decreasing fish stocks in the seas around Japan because of the lack of regulations on fishing (such as the establishment of maximum limits for catches or the protection of breeding grounds). Although it is widely known that the Japanese eel and tuna stocks are disappearing, other familiar fish stocks also are shrinking and being replaced with imported fish.

We also are seeing, however, the emergence of movements such as marine resource management, appropriate coastal management (the creation of Satoumi community coastlines), sustainable fishing business, and expected changes in the operation and function of fisheries cooperatives.

5.5.2 Development of Coastal Regions and the Deterioration of the Fishing Industry Environment

(1) Industrial Development in Coastal Regions and the Near-Death of the Seto Inland Sea

Toward the end of the Meiji period, the agricultural economist Inazo Nitobe described the Seto Inland Sea as a "jewel of the world." The vista created by its chain of 700 islands has charmed countless people, and it provides a habitat for about 3400 species of living creatures.

The area development vision included in the Comprehensive National Development Plan (1st CNDP) formulated in 1962 established many regional development centers in the Pacific belt region, and hubs for the regional development of the heavy and chemical industries were established in the Seto Inland Sea region. Many coastal areas were designated as concentrated industrial development zones because of the New Industrial City Construction Act of 1962 and the Special Industrial Development Region Act of 1964. This large-scale industrialization led to the reclamation of land occupied by mud flats and seagrass beds, completely transforming the scenery of the Seto Inland Sea. The inflow of pollutants caused eutrophication, red tides, sludge deposits, oxygen starvation in bottom waters, the appearance of fish with malformed spines, and other factors that contributed to the near death of the Seto Inland Sea.

On the coastlines, which are important locations for fishing industry production, the abundance of the sea is disappearing, as evidenced by the recent frequency of the color-loss phenomenon in aquafarms cultivating nori seaweed and shrinking catches for various fish and shellfish species. This poses a significant obstruction to the survival and development of the fishing and aquaculture industry and is one of the factors in the decline of the coastal fishing industry. The Ministry of the Environment is considering new environmental standards for water quality and is advancing initiatives aimed at conserving and regenerating coastal environments and recovering biological production, such as incorporating the creation of mud flats and seagrass beds in future plans for total water pollutant load reduction.

(2) Conserving the Seto Inland Sea's Water Quality and Aquatic Environment

In response to this crisis situation, the Act on Special Measures Concerning Conservation of the Environment of the Seto Inland Sea was established in 1973, and the volume of pollutants was reduced. First, regulations were established for industrial water discharge, and in 1979 regulations for total volumes, total pollutant load control, were introduced. In 1993, environmental standards for water quality for total nitrogen and total phosphorous were added, as well as new water quality conservation targets, and regulations for nitrogen and phosphorous in water discharge started. The fifth edition of total pollutant load control in 2001 added total nitrogen and total phosphorous as designated items. These regulations have largely been implemented as planned and chemical oxygen demand, nitrogen, phosphorous, and the burden they produce have been reduced on target. During this time, sewerage infrastructure also has been improved and the penetration rate, which was 24% in 1973, had risen to 63% in 2000.

According to the results of the 2014 Measurement of Water Quality for Public Waters, achievement rates regarding environmental standards for water quality for total nitrogen and total phosphorous for the Osaka Bay area and the Seto Inland Sea area, excluding Osaka Bay, were extremely high at 100% and 96.5% respectively. Compared to the respective achievements rates of 0% and 60% recorded in 1995, when the environmental standards first were applied, it appears improvement efforts have been successful. Achievement rates for chemical oxygen demand, however, were low, at 66.7% and 78.0%, respectively. Taking into consideration that these are similar levels to when the standards were applied, it would be difficult to say that the goal of achieving environmental standards for water quality has been met. The achievement rate for category C water bodies in particular was high at 100%, but in contrast, the achievement rate for category A bodies, which are expected to have good water quality, were extremely low in both Osaka Bay and the Seto Inland Sea as a whole. For example, in Hiroshima Bay, which is included in the Seto Inland Sea area, a lot of water bodies are designated as category A, and nearly all of these failed to achieve targets.

Along the Seto Inland Sea coastline, many of the mudflats and seagrass beds that provide habitats and breeding grounds for various living things have been lost. Between 1960 and 1989–90, about 70% of eelgrass beds were lost, and between 1898 and 2006, about 50% of mud flats disappeared. Between 1989–90 and 2006, however, a slight increase in the area reclaimed was reported (Seto Inland Sea Fisheries Coordination Office, 2018).

(3) Movement Toward Realizing Satoumi Community Seas

In 2015, a law was promulgated as part of partial revisions to the Act on Special Measures Concerning Conservation of the Environment of the Seto Inland Sea formulated during the 189th Session of the Diet (Act No. 78 of 2015, amended law from here on) and came into effect. The amended law took the current situation of the Seto Inland Sea into account and its purpose was to return the seas to abundance by promoting effective measures for conserving its environment. As previously mentioned, a comprehensive range of measures have been advanced to conserve the environment of the Seto Inland Sea in light of its special characteristics, including beautiful landscapes and valuable fishery stocks. These measures included the Act on Temporary Measures Concerning Conservation of the Environment of the Seto Inland Sea formulated in 1973, which was amended in 1978 to become the Act on Special Measures concerning Conservation of the Environment of the Seto Inland Sea, a permanent law containing new measures to counter eutrophication following damage from red tides and the like. As a result of this kind of initiative, we have seen a certain amount of success in improving water quality, but issues still need to be tackled in order to protect the diversity and productivity of life in the sea. Furthermore, attention has been drawn to the need for a detailed response in regard to each bay and sea area in each season.

The amended law clarifies the approach to "conserving the environment of the Seto Inland Sea" by maintaining good quality water conditions, as well as realizing an abundant sea by fully demonstrating the functions and value possessed by the sea, such as conserving the diversity and productivity of life. To further promote effective measures for conserving the sea's environment, it also establishes new guiding principles regarding the conservation of the Seto Inland Sea, revises basic plans and prefectural regulations regarding this conservation, and proposes additional actions that can be taken.

This amendment represents a significant shift in environmental policy and vision for the Seto Inland Sea, from aiming for a beautiful sea to an abundant sea. It also incorporates the concept of a Satoumi community sea, in which citizens play a part in increasing the variety and abundance of life. Conservation through regulation has not recovered catch sizes. Going forward, active intervention in accordance with the condition of seas in each area is required (Takata, 2017).

5.5.3 Overview of Japanese Fisheries Cooperatives and Current State of Activities

The Ministry of the Environment's Environment Research and Technology Development Fund (S-13) promotes research on the theme of sustainable management for coastal areas and Satoumi community coastlines, and the fund itself, which was launched in 2014, covers Japanese coastal areas that fall into the following three categories as shown in Table 5.13.

In this section, we will present an overview and the current state of activities of the Miyagi Fisheries Cooperative's Shizugawa Branch, the Hinase Fisheries Cooperative, and the Ishikawa Fisheries Cooperative's Nanaka Branch.

(1) Overview of the Miyagi Fisheries Cooperative's Shizugawa Branch

The Miyagi Fisheries Cooperative was formed by the merging of 33 of the prefecture's 35 fisheries cooperatives from 2007 to 2009. Until 2007, the town of Minamisanriku in Motoyoshi District, Miyagi Prefecture (formed in 2005 by the amalgamation of the towns of Shizugawa and Utatsu) was affiliated with the former Shizugawa Fisheries Association, but now it is under the jurisdiction of the Miyagi Fisheries Cooperative's Shizugawa Branch, which is affiliated with the Kesennuma General Branch. The branch has just under 700 members (almost 40% of whom are full members) and it registers annual sales of around 4.3 billion yen, making it one of the most profitable branches in Miyagi Prefecture.

The Shizugawa Branch rebuilt its main office in September 2016 in an area adjacent to the Minamisanriku district wholesale market, and it has a field office in the Tokura region on the southern side of Shizugawa Bay. The former Shizugawa Fisheries Association was established in 1949. In 1975, it started Miyagi Prefecture's first coho salmon mariculture facility in Shizugawa Bay and, at its peak in the late 1980s, Miyagi Prefecture as a whole was producing 90% of Japanese coho salmon. An increase in the volume of cheap Chilean imports, however, reduced production volumes dramatically, and the 2011 earthquake caused catastrophic damage to aquaculture facilities. However, because of the efforts of various parties, production of coho salmon, scallops, and other species is gradually recovering.

The branch's other main products include wakame seaweed, oysters, sea urchin, scallops, and ascidians. Wakame production restarted in 2011, after the earthquake, but factors such as the effects of coastal denudation mean that the production volume of sea urchins remains low. Furthermore, South Korea, the biggest buyer of ascidians, has retained a trade embargo on the product since the earthquake, resulting in many cases where fresh catches have had to be discarded.

As a result of the tsunami that followed the 2011 earthquake, about 1000 cages for aquafarming oysters from the Shizugawa Branch Tokura Field Office were lost completely. The use of 1000 cages had caused overcrowding, however, and it was recognized within the industry that the number needed to be reduced, even before the earthquake. Following the loss of the cages, a decision was made to reduce the number to 300 cages. As a result, the amount of time it takes an oyster to reach a marketable size has decreased from about three years to just under one year, and water quality is improving. This initiative has received positive recognition for improving the sustainability of the aquaculture industry (particularly from an environmental perspective) and resulted in the Shizugawa Branch receiving Japan's first Aquaculture Stewardship Council (ASC) certification.

(2) Overview of the Hinase Fisheries Cooperative

The Hinase Fisheries Cooperative is located in the Hinase region of the city of Bizen, Okayama Prefecture (formed in 2005 through the amalgamation of the towns of Bizen, Hinase, and Yoshinaga). Okayama Prefecture differs from Miyagi Prefecture and Ishikawa Prefecture in that it does not have a single unified prefectural fisheries cooperative.

The Hinase Fisheries Cooperative has about 150 members (more than half of whom are full members), and it is particularly noted for oyster aquaculture. In addition to having the largest production volume in Okayama Prefecture at 2000 tons per year (weight after shells are removed), it also operates Gomi-no-Ichi, its own direct sales facility, which attracts aabout 400,000 visitors a year. Considering that Bizen has a population of about 35,000, around 6900 of whom live in the former town of Hinase (both figures from the 2015 national census), this is an incredibly large number of visitors.

One of the most distinctive activities being carried out by the Hinase Fisheries Cooperative is the regeneration and conservation of eelgrass beds. Eelgrass is a species of seagrass that offers various functions, such as providing a spawning ground for various types of shellfish and improving water quality. From the mid-1940s to the mid-1950s, Hinase (in this case, referring to Hinase Bay and surrounding areas) had 590 ha of eelgrass beds, but Japan's postwar economic boom led to concrete seawalls being put in place. This meant that pollution from human wastewater was not removed effectively, resulting in the deterioration of the marine environment and a rapid decrease in eelgrass beds.

A survey carried out on Hinase's Kakui Island in 1985 confirmed that only a few natural eelgrass habitats remained, and leading the cooperative to begin to regenerate these habitats. Initial efforts did not go smoothly, because the condition of the sea floor was not conducive to the establishment of eelgrass beds, but it was discovered that oyster shells could be used as a material for improving conditions, and since 2009, the Hinase Seagrass Bed Creation Promotion Association (nicknamed the Eelgrass Club) a group with 83 members (as of 2016) who primarily work in the fishing industry, have been carrying out regeneration activities.

The regeneration of eelgrass beds, however, has been accompanied by detritus that separates from the plants during their main growth period from spring to summer and drifts into bays. This detritus has an impact on marine navigation, among other effects. As a result, since 2013, Hinase Fisheries Cooperative has cooperated with a local junior high school (Bizen Municipal Hinase Junior High School) to collect the detritus as part of an environmental education program, using boats provided by the cooperative. Currently, during their three years of junior high school, students not only collect eelgrass, but also take part in the planting and harvesting of cultivated oysters, as well as the washing process (while in the sea, barnacles and other things become attached to the oyster shells, which need to be detached before the oysters can be shipped). The students also take part in investigative learning that involves interviewing fishing industry workers and using these interviews to write a paper. This initiative is a leading example of environmental education and has been covered by a succession of media outlets.

The Okayama Cooperative Society (Okayama Co-op) also is supporting the regeneration of eelgrass beds, and together with Hinase Junior High School, is providing a valuable case study of nonfishing industry participation in tackling the issues of coastal management.

(3) Overview of the Ishikawa Fisheries Cooperative's Nanaka Branch

The Ishikawa Fisheries Cooperative was formed in 2006 by a merger of the prefecture's 27 fisheries cooperatives. The Nanaka Branch (formerly the Nanaka Fisheries Cooperative) has about 1100 members (>30% of whom are full members) and is Japan's second-largest single branch. The Nanaka Branch has its main office in Nanao's market area and a field office in Nanao Western Bay. The cooperative is engaged primarily in fixed shore net fishing, trawl fishing, and gillnetting, and the Nanao Western Bay Field Office is primarily responsible for oyster farming. The former Nanaka Fisheries Cooperative was formed in 1996 through the amalgamation of the Nanaokashima, Enome, and Nozaki fisheries cooperatives, and at that time it already belonged in the largest category for single organizations in Ishikawa Prefecture. Although many other regions have fishing port facilities and the like managed by the prefectural fisheries cooperative or municipal authority, the area under the jurisdiction of the Nanaka Branch has a history of management being undertaken by fishing industry workers themselves since the time of the former Nanaka Fisheries Cooperative, and, as a result, it is considered to be highly independent compared to Shizugawa and Hinase.

The Nanaka Branch's jurisdiction includes the city of Nanao (formed in 2004 through the amalgamation of the towns of Nanao, Tatsuruhama, Nakajima, and Notojima), which also contains the Nanao and Sazanami branches. These, however, are relatively small branches with around 100 members each, and together they have a fishing fleet of 134 boats, a small number compared to the Nanaka Branch's 830.

5.5.4 Fishing Industry Workers' Views Regarding Coastal Areas

In order to explore what was possible in terms of solving these issues, a questionnaire survey, among other methods, was carried out over several years involving fisheries cooperatives and branches in Minamisanriku, Bizen, and Nanao. In this section, we focus on the survey investigating the Shizugawa Branch in Minamisanriku.

To date, we have carried out two attitude surveys with the cooperation of Shizugawa Branch. They primarily focused on individuals holding official positions (such as the chairperson, deputy chairperson, and the heads of various subcommittees) and were carried out in February 2015 and January 2016. The first survey produced useable answers from 95 respondents, and the second from 89. The surveys were composed of a questionnaire sheet that was posted to respondents, onto which they recorded their answers.

The main results of the attitude surveys are as follows:

① Age range, etc.:

45 respondents were in their 60s, the largest group in a single-age range. This was followed by respondents in their 50s and 70s, meaning that a little >60% of the total were 60 or older. Eighty-five had been a member of the cooperative for >20 years; 92 were men.

② Primary or secondary occupation:

61 had fishing as their primary occupation.

③ Successor issue:

Although the majority of respondents were in official positions, 33 respondents, the largest group, said they already had a successor in place. Combined with those who answered that they had plans or were searching for a successor, 68 respondents, or close to 70%, were looking to pass on their business.

Statistics for the boat-based fishing industry as a whole show that 14% of businesses on average already have a successor secured.

④ Priorities regarding the conservation of coastal areas

Almost half of respondents answered that conservation should be carried out even if it reduced their own fishing activity, surpassing the number of respondents who answered that fishing industry activity should take priority.

Although 14 respondents answered that the situation should continue as it is now, these answers included the opinion that the fact it is a period of recovery and restoration following the 2011 earthquake means that things are fine as they are for the time being.

⑤ Satisfaction with the current state of fishing industry activities

The largest single group of respondents were the 35 (37% of the whole) who answered "neither satisfied nor unsatisfied." 36 respondents (38%) answered they were "extremely satisfied" or "fairly satisfied," while 23 (24%) answered they were "extremely unsatisfied" or "fairly unsatisfied."

⑥ Sense of priority regarding the conservation and management of coastal areas

45% answered that conservation and management was needed even if it meant the shrinking of the fishing industry, which surpassed the 32% who answered that fishing industry activity should take priority.

The amount that people were willing to pay toward conservation activities is outlined as follows. We asked the amount people were willing to pay regarding three categories: seagrass beds, coastal denudation, and activities concerning forests, rivers and the seas. The amount paid was supposed as an annual payment continued over a 10-year period. It is possible that this is the first time that fishing industry workers have been surveyed about how much they are willing to pay toward coastline conservation. An overview of the questions asked are as follows.

1. Seagrass beds have a crucial role as habitats, breeding grounds, and feeding sites for coastal organisms. Supposing that various environmental conservation activities were implemented to plant seagrasses in bays that have lost seagrass beds, then payments would go toward this.
2. The factors behind the loss of seaweeds that has triggered coastal denudation include changes to ocean currents, feeding damage by creatures that feed on seaweed (such as sea urchins), the large-scale influx of river water, and seawater pollution. Supposing that measures were implemented to counter coastal denudation, then a payment would be made every year toward these measures.
3. Rain water that soaks through leaf mulch in the forest contains nutrient elements, and it also picks up mineral elements from rocks in the ground as it flows into rivers. When the nutrients from this river water mix flow into the sea, it cultivates various seagrasses and shellfish. Supposing that an association was formed to consider this cyclical relationship between the forest, rivers, and sea, then a set payment would be made every year toward this association.

The results regarding the amount people were willing to pay toward conservation activities showed that about 80% of respondents were willing to pay for conserving forests in upstream areas, while more tha 90% were willing to pay to support coastal denudation countermeasures and seagrass bed planting. As for the options regarding payment amounts, in the level below 10,000 yen, earning levels did not make much of a difference regarding the amount chosen as shown in Fig. 5.25.

Although it is difficult to make a simple comparison, the results of other surveys regarding willingness to pay are shown in Table 5.14.

Supposing that all three measures for conserving the fisheries environment in Shizugawa Bay were carried out, the total cost would be 17,712 yen. Fishing industry workers seem to have a fairly strong attachment to the bay. This can be seen in the fact that 35% of responses considered the value of Shizugawa Bay to be more than five times the amount of revenue produced by the aquaculture industry, which shows they value the bay particularly highly and are strongly emotionally invested.

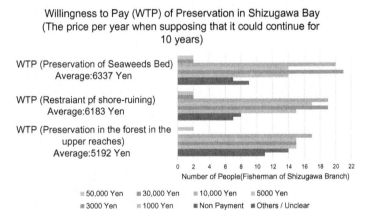

Willingness to Pay (WTP) of Preservation in Shizugawa Bay (The price per year when supposing that it could continue for 10 years)

FIG. 5.25 Willingness to pay (WTP) for preservation in Shizugawa Bay (The price per year when supposing that it could continue for 10 years).

TABLE 5.14 Case Comparison Regarding Willingness to Pay

Case name	Average amount	Median amount	Remarks
Japan-wide mudflat restoration	¥4431	¥2916	Survey date: 2014
Isahaya Bay mudflat conservation	¥6422	¥2908	Survey date: 2001
Seto Inland Sea natural environment (multiple conditions set)	¥4689–¥8899	(No data)	Survey date: 1998

Source: Ministry of the Environment (JAPAN).

Furthermore, the average citizen was willing to pay 316 yen per month (3702 yen per year) toward mudflats and seagrasses, about half that of people in the industry.

In order to make a comparison with fishing industry workers, we also surveyed members of the Minamisanriku Forestry Cooperative regarding their willingness to pay and other issues.

The Minamisanriku Forestry Cooperative was formed in 1988 through the merger of the Shizugawa and Utatsu forestry cooperatives; 77% of Minamisanriku's total area (163.4 km^2) is forested.

In fall 2015, environmentally friendly business activities in the town were assessed and relevant organizations, including the Minamisanriku Forestry Cooperative, were accredited by the Forest Stewardship Council (FSC). When the ASC accreditation gained in spring 2016 is added, Minamisanriku can be considered top class in terms of environmentally friendly forestry and fisheries.

The Fisheries Cooperative and Forestry Cooperative has deep ties and for >20 years it has cooperated to plant forests in the town's upstream areas. When the 2011 earthquake hit, the Forestry Cooperative allowed the Fisheries Cooperative to use part of its office as a temporary office. Talking about that time (in an interview), the Forestry Cooperative member responsible said, "As cooperatives in the same town, it is natural that we should work together." Eight rivers flow into Shizugawa Bay and all these rivers reach their end within Minamisanriku. Unlike the Kesennuma case, which straddles multiple municipalities, this falls into a rare category.

The survey carried out in March 2017 covered regional representatives and yielded 51 useable responses. None of the respondents was engaged in forestry as their primary occupation, and 20% were engaged in fishing as a secondary occupation.

When asked about the significance of considering forests, rivers, and sea in an integrated manner, 66% of all respondents said it was "of great significance," and when combined with respondents who answered "a certain amount of significance," this number reached 90%.

When asked about their impression regarding the FSC environmental accreditation gained in 2015, 60% had a favorable opinion. More than half of these respondents, however, also answered they felt uneasy as to whether this would limit their own activities. Also, a small but material number among those that answered "other" had the opinion that "restrictions posed by FSC standards may have a negative impact on the environment."

When asked about their impression of forest conservation efforts by fishing industry workers, just under 70% of respondents gave a positive response. About 30% of all the respondents answered, "It is good that they are planting trees, but management to maintain these trees afterwards is lacking." Forests cannot be cultivated by just planting trees, but also need appropriate management such as the trimming of undergrowth and branches. As a result, examples of activities that involved not only planting trees, but also emphasized forest maintenance and management emerged.

Regarding willingness to pay toward environmental conservation, we surveyed respondents from the perspective of realizing abundant seas, asking them about their willingness to pay (hypothetical annual payments paid over the course of 10 years) using the same questions as those given to the fishing industry workers. There was only a slight difference in the overall averages, with the fishing industry willing to pay 5192 yen and the forestry industry willing to pay 4957 yen. Also, the proportion of respondents who would not pay was 15.7% for the fishing industry but only 11.8% for the forestry industry. Therefore, although the questions called for planting activities that were for the direct benefit of the fishing industry, a greater proportion of forestry industry workers were willing to pay.

5.5.5 Trial Projects Aiming for a Sustainable Fishing Industry

According to future population estimates for municipalities across Japan released by the Japan Policy Council's subcommittee for considering the issue of Japan's shrinking population led by Hiroya Masuda, the rate of change (2010 → 2040) of the young female population (20–39 years old) in coastal areas covered by the investigation is as follows. Ranked in order of the size of population decrease were Minamisanriku, Miayagi Prefecture, 69.5% (highest in Miyagi Prefecture); Noto, Ishikawa Prefecture, 81.3% (highest in Ishikawa Prefecture); Nanao, Ishikawa Prefecture, 59.5% (seventh highest in Ishikawa Prefecture); Himi, Toyama Prefecture, 57.2% (third highest in Toyama Prefecture); and Bizen, Okayama, 60.1% (second highest in Okayama Prefecture). These occupy the highest population decreases in their respective prefectures, ranking first to third. Increasing the sustainability of these fishing regions is an extremely pressing issue.

In response to this issue, Minamisanriku has created the Minamisanriku Town Biomass Industry City Design (2013) shown in Fig. 5.26, which aims to make the town sustainable through the integrated management of forests, rivers, agricultural land, and sea.

According to the design, Minamisanriku's forest, agricultural land, and sea have all become relatively compact, but trying to engage in each of these areas separately would lead to less than ideal conditions in terms of scaling up and

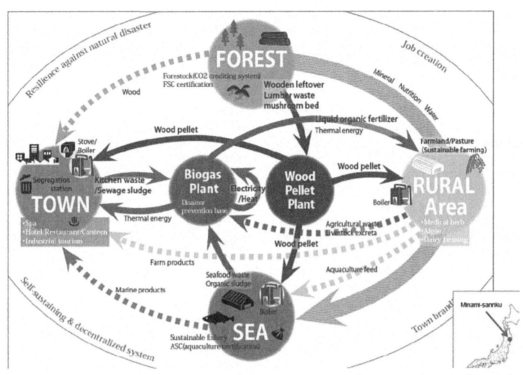

FIG. 5.26 Full image of the Minamisanriku Town Biomass Industry City Design. *(Source: The Minamisanriku Town Biomass Industry City Design).*

efficiency. Engaging in these areas as a composite formed through mutual relationships, however, will unlock the potential of the resources offered by all three, as well as the town itself.

Forest offcuts can become a stable energy source to support agriculture, the marine production industry, and citizens' everyday lives. A well-kept forest also produces water rich in minerals, which in turn creates abundant seas. Organic waste produced when processing marine produce and food waste from agricultural produce, such as skins, can be used as ingredients for biogas and fertilizers. Human waste excreted by town dwellers also can be used to feed methane bacteria at biogas facilities. The biogas process generates liquid fertilizer that can be reduced into an environmentally friendly organic fertilizer for agricultural land. Also, beautifully kept Satoyama community forests, agricultural land that conserves the environment, and abundant seas are all attractive to people looking to spend a fulfilling time (experiences/visits).

Using the sea as an example, before the 2011 earthquake there were about 1000 oyster cages in the southern part of Shizugawa Bay, causing overcrowded conditions that resulted in it taking two to three years for oysters to grow to a marketable size. After all of these cages were lost during the disaster and during the recovery process, the Tokura Field Office and the oyster producers' association decided to reduce the number of oyster cages to less than one-third. This has led to oysters only taking one year to grow to a marketable size and water quality has improved with the decrease in the number of oysters. As a result of this implementation of appropriate oyster aquaculture, it became the first area in Japan to receive accreditation by the international accreditation body the Aquaculture Stewardship Council (ASC) in March 2016.

This kind of sustainable fishing business is also one of the goals included in the UN Sustainable Development Goals (SDGs). Seventeen goals adopted by the UN in 2015 comprise the SDGs, and each goal contains targets, of which there are 169 in total. One of the goals includes conserving the abundance of seas and, in order to recover resources, it requires the effective regulation of harvesting and the implementation of science-based management plans by 2020.

If the fishing industry is too active, there is a danger that fish stocks will dwindle. Unless strict regulations are created and adhered to, the recovery of these stocks can be slowed. The fishing industry is tied closely to the natural environment. If the fishing industry also can tackle the SDGs, it will be of huge significance. Through aquaculture businesses, we must aim to prevent marine pollution, such as river water inflow containing agricultural chemicals, to create abundant seas that can produce a variety of life. An international accreditation system has been created that applies to fish catches in these seas. This fishing regulation considers ecosystems through a mechanism in which fish are accredited. EON Co., Ltd. will increase its purpose-built sales venues for accredited fish by 60% by 2020, and Panasonic Corporation is providing an accredited fish menu at all its employee cafeterias in Japan. With the danger of depletion looming, businesses are turning against overfishing and looking to leave marine resources for the next generation (reported in the Nikkei, May 13, 2018).

The representative accreditations systems are the ones created by the Marine Stewardship Council and by environmental organizations, including the World Wildlife Fund (WWF). Following a strict investigation, these systems award accreditation to fishing business that do not overfish and aquaculture businesses that do not pollute the sea through feeding. The accreditation symbol is also known as a marine ecolabel, and fishing industry participants, such as fishermen or fisheries cooperatives, undergo investigations to receive them. There is demand for adopting fishing methods that prevent overfishing and restore species with depleted stocks to create a fishing industry that conserves the structure and diversity of ecosystems.

The Aquaculture Stewardship Council (ASC), an NPO, operates an accreditation system for aquaculture businesses that engage in environmental conservation.

In order to use ASC accreditation for oyster production as an opportunity to drive further movement toward the sustainability of the Shizugawa area, it is important that the area's qualities are leveraged in ways such as mountain and sea tours, oyster smoking using wood chips, etc.

The environmental conservation efforts of the local forestry cooperative (Minamisanriku Forestry Cooperative) was assessed in fall 2015 and received accreditation from the Forest Stewardship Council (FSC).

It is important to tie together these trials and leverage the potential of forest, agricultural, and marine resources.

When asked about the effects on Minamisanriku exerted by efforts such as the conservation and restoration of seagrass beds in Shizugawa Bay and the integrated approach to forest, rivers, and seas, when considered from a sustainability perspective, 76 respondents (85% of the total) answered that they had "a large effect" or "a fairly large effect."

When members of the Minamisanriku Forestry Cooperative were asked the same question, 96% of the all respondents answered "a large effect" or "a fairly large effect."

Furthermore, the Japan Fisheries Science and Technology Association, which promotes research related to maintaining abundant marine production in coastal areas, has forwarded the following emergency proposals aimed at maintaining abundant marine production in coastal areas.

Proposal 1: Establish a safe and secure marine product supply framework. Maintain and increase the ratio of fish and seafood food products that are produced in Japan and communicate Japanese food culture to future generations in a way that protects coastal fishing industries.

Proposal 2: Maintain and improve fishing industry production by managing the nutrient salt cycle. Enlarge the ecological pyramid found in food chains and establish nutrient salt providers and production venues that use material recycling to effectively use both land and sea.

Proposal 3: Water quality standards aimed at realizing an abundant sea. Depart from total pollutant load controls and design environmental standards for water quality that are guided by living organisms.

Proposal 4: Measures for realizing an abundant sea. Enact detailed measures that use the unique characteristics of each bay and sea, enhance monitoring using scientific methods, and strengthen cooperation between government ministries and agencies.

Proposal 5: Strengthen investigative research and the like. Enhance the framework for measuring coastal environments and obtain scientific data.

These proposals were approached with the thinking that previous environmental standards for water quality in coastal areas were established based on the idea that it is better that the load and density of nutrient salts are as low as possible, with the aim of preventing water pollution. Healthy marine environments, however, are abundant habitats where a diverse range of life uses a diverse range of production grounds. Nutrient salts provided by the land are absorbed by these environments before they are removed from the ecosystem by the fishing industry, creating a form of material recycling, or, in other words, realizing an abundant sea. This means that nutrient loads are not removed in a uniform manner, leading to a design that can be thought to strongly embrace the Satoumi approach to maintaining abundant biological productivity.

These activities can be actively developed using "sustainability" as a keyword. Going forward, we will drive ahead to realize a sustainable fishing industry in which each coastal region cultivates a treasure of which to be proud. New developments will emerge in the form of "one bay, one treasure" (in the same way as agricultural villages have the "one village, one product" movement), with examples including the eelgrass beds, ASC-certified oysters, and use of oyster shells. Rather than transient measures, realizing sustainable, integrated management of forests, rivers, agricultural land, and seas will require the participation of all kinds of people.

5.6 LOCAL RESIDENTS' SENSE OF PLACE TOWARD COASTAL AREA AND POTENTIAL OF SATOUMI EDUCATION PROGRAM

Ryo Sakurai

5.6.1 Social Psychological Research About Local Residents' Willingness to Conserve Coastal Area

In order to sustainably conserve the natural resources of coastal areas, the participation of not only fishermen but also ordinary local residents is important. Local residents contribute to the conservation of coastal areas in various ways, such as recreation and consumption of fish catch. As the number of fishermen decreases and rural areas face depopulation and aging, it is important to grasp local residents' perceptions toward and willingness to engage in the conservation of coastal areas in order to perform coastal management effectively and sustainably. In this research, the author carried out a Web-based questionnaire survey of local residents of Shizugawa Bay, Miyagi Prefecture, and Hinase Bay, Okayama Prefecture, and clarified factors that affect their willingness to conserve coastal areas and their sense of place toward coastal areas.

5.6.1.1 Local Residents' Willingness to Conserve Coastal Area Around Shizugawa Bay

Shizugawa Bay is an area where sustainable fishing has been performed for many years, and is famous as a Satoumi site, where both the level of biodiversity and productivity have been improved through the positive interaction of people and the natural environment. In this study, a Web-based questionnaire survey was conducted on local residents aged 20 or older living within 100 km of Shizugawa Bay. Questionnaires were distributed from February to March 2015, and 1746 residents responded. The questionnaires included items that measured sense of place (23 items related to biophysical, psychological, sociocultural, and political-economic aspects based on Ardoin et al. (2012), time-related factors such as people's perceptions regarding future generations (4 items), sociodemographic factors such as gender, age, and educational level, and perceptions of coastal areas and experience of visiting these sites (13 items). Questions were asked with five-point answers (1 = disagree, 5 = agree) except sociodemographic items. In the analysis, two items of sense of place factors that measure willingness to conserve were used as dependent variables: "I am willing to invest my time and effort to make the coastal area a better place" and "I am willing to make financial sacrifices for the sake of the coastal area." The other 38 items were used as independent variables, and multiple regression analysis was conducted to identify factors that affect people's willingness to conserve. SPSS18 (IBM, Tokyo, Japan) was used for analysis and a stepwise method was used to identify the best-fit model.

As a result, 53% of respondents were male; 47% were female. Regarding age, 14% were in their 20s, 22% in their 30s, 26% in their 40s, 21% in their 50s, and 17% were older than 60. Among respondents, 13% answered that they would invest time and effort to conserve coastal areas; 16% answered that they would make financial sacrifices. Moreover, 13% of respondents felt attachment to coastal areas.

The best-fit model identified by the stepwise multiple regression analysis for explaining willingness to conserve the coastal area (investing time and effort) included 11 variables. The model had an explanatory power of 58%. The two items "I want to know more about the wildlife of the coastal area" ($B = 0.21$) and "I am strongly connected to the land and natural features that make up the coastal area" ($B = 0.15$) had the strongest influence on people's willingness to conserve. Regarding people's intention to make financial sacrifices, 14 items were identified as the best-fit model by the stepwise multiple regression analysis. The explanatory power was 51%. Especially, the two items "I want to know more about the wildlife of the coastal area" ($B = 0.13$) and "I like various outdoor activities that are available at the coastal area" ($B = 0.12$) strongly affected the dependent variable (Sakurai et al., 2016a).

Because "I want to know more about the wildlife of the coastal area" had the strongest effect on both willingness to invest time and effort, and to make financial sacrifices to conserve coastal areas, implementing outreach programs to increase wildlife lovers, for example, could be effective for encouraging public involvement in conservation. When conducting such outreach programs, participants' curiosity could be stimulated by providing information and asking about the ecology and characteristics of various species living in coastal areas. Moreover, because people's feeling of connection to coastal areas also affected respondents' willingness to conserve, such outreach activities that show the relationship between coastal areas and people's lives (e.g., a program in which participants could learn how the water used by people flows into the ocean) could be effective. To practice sustainable coastal management with the participation of local residents, it would be important to conduct a survey to understand factors affecting their perception as well as to suggest and implement outreach programs that reflect the results of such surveys.

For details about this study, please refer to the following research article: Sakurai et al. (2016a).

5.6.1.2 Distance Between Location of Residence and Coastal Area and Residents' Willingness to Conserve: Targeting Residents Around Hinase Bay

This research examined how the distance between the location of residence and coastal area affects residents' perceptions regarding the coast and conservation. Hinase town in Bizen city is in the southeastern part of Okayama Prefecture. To examine how many residents are expected to join in activities to conserve the coastal area, it is important to understand first the areas that are attractive (how broadly residents are asked to participate in the activities).

A questionnaire survey was conducted of residents living around Hinase Bay and the relationship between perception of Hinase Bay (e.g., willingness to support and sense of place) and distance between location of residence and the coast was examined. The method was similar to those explained in Section 5.6.1.1 and a Web-based questionnaire survey was conducted from February to March 2015, targeting residents aged 20 or older living within 60 km from Hinase Bay ($n = 1630$). Moreover, because those respondents who do not know Hinase Bay might not have any perception of this coastal area, samples who had heard about Hinase Bay ($n = 1623$) were included in the analysis. The questions were based on those of previous studies (Ardoin et al., 2012, Wakita et al., 2014, etc.).

The chi-square test was used to identify the relationship between sociodemographics/experience of visiting Hinase Bay and location of residence. Analysis of variance (ANOVA) (with multiple comparisons using Tukey's method) was conducted to identify the relationship between age and location of residence. In addition, ANOVA (with multiple comparisons using Tukey's method) was conducted to determine whether respondents' willingness to conserve Hinase Bay, sense of place, perceptions of wildlife, perceptions regarding future generations, and awareness of Satoumi differ depending on the distance of residence from Hinase Bay. SPSS 22 (IBM, Tokyo, Japan) was used for the analysis, with a statistically significant level of $p < 0.05$.

Among respondents ($n = 1263$), 53.6% were male and 46.4% were female. The average age was 47.9 years old. As for location of residence, 367 respondents (29.1%) lived within 20 km from Hinase Bay, 470 (37.2%) between 21 and 40 km, and 426 (33.7%) between 41 and 60 km. A majority of respondents (59.9%) answered that they had visited Hinase Bay. The chi-square test and ANOVA (multiple comparison using Tukey's method) found no significant difference between respondents' gender/age and location of residence. There was, however, a significant difference between respondents' experience of visiting coastal areas around Hinase Bay and location of residence ($X^2 = 19.03$, $P < .01$); those who lived near Hinase Bay tended to have visited the coastal area (0–20 km: 65.7%, 21–40 km: 63.0%, 41–60 km: 51.6%).

Regarding the relationship between location of residence and respondents' perceptions about Hinase Bay, their willingness to conserve (whether they were willing to invest time and effort) significantly decreased as location of residence exceeded 21 km from the bay. Meanwhile, respondents' willingness to make financial sacrifices for conservation significantly decreased as location of residence exceeded 41 km. Respondents' familiarity with Hinase Bay was significantly different between those living within 20 km and those living 21–40 km from the bay, and also between those living 21–40 km and those living 41–60 km from the bay. Their sense of familiarity with Hinase Bay significantly decreased as the location of residence was farther from the bay. Respondents' sense of place toward Hinase Bay was significantly higher for those living within 20 km than those living >21 km from the bay. As for respondents' attitudes toward future generations, "I hope that future generations will feel attached to Hinase Bay" and "I want to preserve Hinase Bay in a good condition for future generations," there was a significant difference between those living within 20 km and those living farther than 41 km from the bay (Table 5.15) (Sakurai et al., 2016b).

When performing outreach programs regarding Satoumi or promoting participatory conservation activities, few studies have tried to identify target residents and area based on the results of social research. Stakeholders such as local governments, NPOs/NGOs, and fishery associations in charge of implementing outreach programs could use their budgets more effectively if they knew to whom (residents living in which area) they should convey the message. As the number of fishermen decreases, various people including citizens are expected to participate in the conservation of coastal areas. Identifying areas to which residents feel a sense of attachment and are willing to conserve coastal areas is the first step to realizing participatory coastal management.

For details about this study, please refer to the following research article: Sakurai et al. (2016b).

5.6.1.3 Local Residents' Sense of Place Toward Coastal Areas

This section describes the result of analyzing the relationship between sense of place toward coastal areas and perceptions regarding future generations based on the survey of residents around Shizugawa Bay and Hinase Bay described in the previous sections.

This study analyzed data of all samples who answered the questionnaire (around Shizugawa Bay: $n = 1746$, around Hinase Bay: $n = 2851$). Structural equation modeling (SEM) was conducted using respondents' attitudes toward future

TABLE 5.15 Respondents' ($n = 1263$) Perceptions About the Hinase Bay Based on Location of Their Residence; Distance From the Hinase Bay (There Was No Significant Difference for the Average Score Surrounded by the Same Dotted Line Based on Multiple Comparison With Tukey Method (Sakurai et al., 2016b)

Items	Average score / Standard deviation		
	0-20km (Bizen city, Ako city)	**21-40km** (Okayama city, Himeji city)	**41-60km** (Kurashiki city, Kakogawa city)
I am willing to invest my time and effort to make the Hinase Bay a better place	2.45 / 1.08	2.24 / 1.03	2.18 / 0.97
I am willing to make financial sacrifices for the sake of the Hinase Bay	2.41 / 1.06	2.28 / 1.02	2.11 / 0.95
Hinase Bay is familiar to me	3.29 / 1.22	2.93 / 1.19	2.56 / 1.13
I am very attached to the Hinase Bay	2.62 / 1.16	2.37 / 1.06	2.13 / 1.08
I want to know more abou the wildlife of the Hinase Bay	2.60 / 1.11	2.51 / 1.11	2.43 / 1.07
I think the wildife of the Hinase Bay is facinating	2.90 / 1.11	2.78 / 1.11	2.74 / 1.11
I hope that the future generation will feel attached to the Hinase Bay	3.02 / 1.12	2.86 / 1.08	2.78 / 1.14
I want to preserve the Hinase Bay in a good condition for the future generation	2.96 / 1.21	2.95 / 1.16	2.77 / 1.21
I heard about "Satoumi" before	1.48 / 0.72	1.38 / 0.64	1.38 / 0.63

generations as the dependent variable and four factors regarding sense of place (biophysical, sociocultural, psychological, and political-economic factors explained in Section 5.6.1.1) as independent variables. The analysis also compared the regions (Shizugawa Bay and Hinase Bay) in analyzing the relationship between sense of place and attitudes toward future generations. SPSS18 (IBM, Tokyo) and AMOS16 (IBM, Tokyo) were used for the analysis.

A reliability analysis testing whether items that compose the latent constructs had internal reliability revealed that all four factors (biophysical, sociocultural, psychological, and political-economic factors) and attitudes toward future generations had a Cronbach alpha score of >0.7. This implied that these items had enough internal correlation to compose the latent constructs that the author had hypothesized.

Based on the exploratory model of SEM that tests all possible correlations and associations between latent constructs, biophysical factors, and psychological factors significantly affected respondents' attitudes ($P < .01$) in both Shizugawa Bay and Hinase Bay (Figs. 5.27 and 5.28). In Shizugawa Bay, biophysical factors had three times more influence than psychological factors on attitudes toward future generations. Moreover, political-economic factors had a negative relationship with attitudes toward future generations. For example, the residents who were willing to make financial sacrifices were less likely to want to preserve the area for future generations. The overall SEM had a satisfactory level of fitness at both sites as RMSEA (root mean square error of approximation) was <0.1 and GFI (goodness of fit index) was nearly 1.0. Models of both study sites had an explanatory power of >0.65, meaning that the model with the sense of place variables explained the majority of the attitudes toward future generations regarding conservation of coastal areas (Sakurai et al., 2017).

This research is one of the first studies in Asian countries to test the validity of the sense of place model (Ardoin et al., 2012) developed in the United States. The study proved that the model is valid and a reliable psychological theory even in regions that are culturally and socially different. The study is also the first to clarify the relationship between people's sense of place and their attitudes toward future generations: Whether residents have a sense of place toward their area affects their willingness to conserve coastal areas for future generations.

Based on the results of this study, strategic outreach/participatory conservation activities targeting that foster people's sense of place toward an area need to be implemented to increase residents' willingness to conserve coastal areas. For example, activities that connect people's lifestyle (hobbies, jobs, etc.) with the local area are important. Programs that involve local companies (e.g., employees participating in volunteer activities in the local area) or activities by residents' associations that families could join would be effective. Such outreach activities should be conducted to foster residents' sense of place toward the target area and their willingness to conserve the coastal area for future generations.

For details about this study, please refer to the following research article: Sakurai et al. (2017).

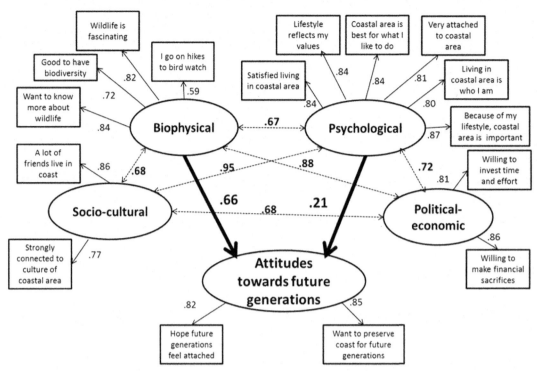

FIG. 5.27 Results of structural equation modeling at Shizugawa identified the Exploring Model Identification analysis ($n = 1746$, $P < .01$, GFI = 0.89, RMSEA = 0.09; all scores represent standardized coefficients [all relationships were significant at 1%]) (Sakurai et al., 2017).

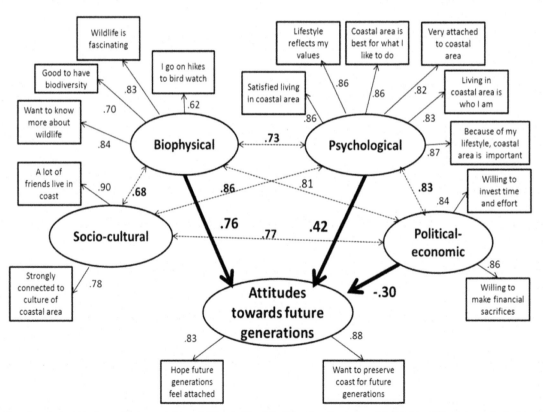

FIG. 5.28 Results of structural equation modeling at Hinase identified by the Exploring Model Identification analysis ($n = 2851$, $P < .01$, GFI = 0.90, RMSEA = 0.09; all scores represent standardized coefficients [all relationships were significant at 1%]) (Sakurai et al., 2017).

5.6.2 Satoumi Educational Program at Hinase Junior High School at Bizen City, Okayama Prefecture, and Effectiveness of the Program

Because many coastal areas in Japan face various issues such as pollution and decrease of fishermen, it is important to provide education to help increase people's understanding of and interest in Satoumi in order to manage the oceans sustainably. It is also important to provide marine education to children in coastal areas who are likely to be responsible for managing Satoumi in the future. In practice, however, schools do not have enough teachers and educational materials to do so. Therefore, efforts must be made to develop a marine education curriculum and implement research to evaluate its effectiveness. Local fishermen at Hinase town, Bizen city, Okayama Prefecture, have been actively involved in conserving the ocean, and students of all grades in the local Hinase Junior High School receive marine education through collaboration with a fishing association using the Period for Integrated Studies. Actual activities include attaching oyster seeds to scallop shells, sorting and sowing eelgrass seeds, washing, processing, and tasting the oysters, and conducting interviews with local fishermen.

In this study, we conducted semistructured interviews with the students [$n = 108$; 36 students in each grade (seventh-, eighth-, and ninth-graders)] at the school in April 2017 to understand the effectiveness of the education program. Based on a qualitative and quantitative analysis, it was revealed that the program appears to have changed students' perceptions, such as recognizing the importance of the sea and eelgrass, as well as their behavior such as no longer throwing waste into the sea. The higher the grade level was, the more that students felt close to and were willing to care for the sea. Our study suggests that the program has helped to develop individuals who are knowledgeable about the fishing community of Hinase, fishermen's roles, and activities that would contribute to biodiversity conservation and who are motivated to conserve Satoumi in the future.

The marine education program at the school provides opportunities for interaction among various local stakeholders, connecting both the ocean and people as well as people with each other, and contributing to creating sustainable local communities through collaboration with fishermen. This program, which could be called Satoumi Education, is different from traditional ocean literacy programs and is a model example of realizing the sustainable development of coastal areas.

For details about this study, please refer to the following research article: Sakurai et al. (2018) and Sakurai, R. 2018. Effectiveness of marine education program on junior high school students with as specific focus on Satoumi. Jpn. J. Environ. Educ. 28(1), 12–22 (in Japanese with English abstract).

5.7 MULTI-LEVEL MANAGEMENT SYSTEM OF THE COASTAL SEA

Takeshi Hidaka

Half a century has passed since the concept of the coastal area in Japan's national policy was submitted by the Third Comprehensive National Development Plan in 1977. The importance of coastal areas has not decreased, and the need to manage coastal areas in an integrated manner is stated in the latest basic plan for the oceans. A legal system for integrated coastal management (ICM), however, has not yet been established. Instead, there are individual legal systems such as the Coast Act and projects such as for revitalizing Tokyo Bay and Osaka Bay, which are similar in character to the ICM. In addition, many activities to preserve coastal areas by local governments and inhabitants called voluntary ICM are performed. These activities are sometimes called Satoumi, and their number has been increasing. It is important to effectively integrate these individual systems and activities and to establish a mechanism for managing and using coastal areas effectively and continuously overall. Otherwise, errors in composition (overall management of coastal areas might not be successful, even if individual actions are good) could occur. This section aims to clarify ways to avoid errors in composition and achieve smart use of coastal areas as a whole.

Hidaka (2014) suggested a two-level method for managing coastal areas: the jurisdiction of the prefectural government and the areas handled by local governments. Based on this proposition, through analyzing various activities concerning coastal management and organizing the layer structure of governance, I will suggest a system of multilevel management for managing coastal areas integrally. I believe that network governance is effective for organizing this mechanism well and building the analysis framework based on it. Then, through a preliminary example analysis, I indicate the effectiveness and problems of the multilevel system of managing coastal areas by network governance.

5.7.1 Task of Coastal Management

Coastal areas consist of land areas and ocean spaces that should be managed as a unit across a shoreline. Land areas are administrative boundaries of coastal municipalities, and ocean spaces are territorial waters, according to the Japan Coastal Association. In this chapter, however, the target area is limited to the nearshore waters including territorial waters and coasts as coastal waters, in consideration of the complexity of land areas.

Coastal waters have various uses, which can be divided into three types: resources, nature, and space. According to Hidaka (2002), three values acquired from coastal waters are identified: biological environmental value, economic value, and life culture value. In order to extract maximum value from coastal waters, consistent, proper management is needed so that multifaceted uses and creation of value can be realized.

When considering the management of coastal waters, the legal system treats sea surfaces and many coasts as national property with no administrative law. That is, they are public goods. Many natural resources in nearshore waters, however, are bona vacantia, which means there is no owner. Although I noted above that there is no single method of administration, many methods of administration with different individual purposes exist (such as the Port and Harbor Law and the Coast Law). Although the range of territorial waters is designated, the jurisdictional limits and responsibilities of local governments are not clear. This complicated legal system hinders the management of coastal waters, including the introduction of Satoumi, which are defined as "coastal sea waters which have high biodiversity and biological productivity through human intervention" (Yanagi, 2006). Accordingly, many efforts have been made to construct an integrated coastal management plan and to establish ICM laws to overcome this challenge, but nothing concrete has been introduced yet.

5.7.2 Dynamism and Multilevel Management System of the Coastal Sea

Instead of instituting ICM, many projects similar in character to ICM and activities called Satoumi exist in Japan (Hidaka and Yoshida, 2015). I hypothesize that the management system should differ depending on the combination of spatial scale and complexity of use, and that such projects and activities can be plotted on a graph with spatial scale on the horizontal axis and complexity of use on the vertical axis (Hidaka, 2016b), as shown in Fig. 5.29. The cases explained later are based on this figure.

Fishing-ground management based on a common fishery right or the right of demarcated fishery goes into the lower-left corner of the figure. This means that the management of fishing grounds and marine resources within

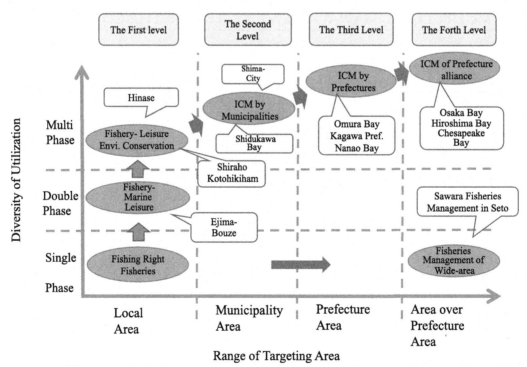

FIG. 5.29   Dynamics of the ICM.

an area by fishing communities is a primitive form of Satoumi and ICM. Many activities called Satoumi or voluntary ICM are carried out in the area bordering communities and the territory of local governments. Although there are few instances of ICM within the territory of prefectural governments, some prefectural governments such as Nagasaki Prefecture and Kagawa Prefecture have a management plan for their area of jurisdiction. Moreover, in wider areas beyond prefectural territories, projects to revitalize a huge bay such as Osaka Bay and Hiroshima Bay have been carried out in cooperation with the national government and some prefectural governments.

A detailed analysis of each instance reveals that the management system, such as organizational structure and management process, differs depending on the space. Moreover, the management body varies greatly, including local residents (fishermen, fishermen's cooperative associations, NPOs, etc.), local governments, prefectural governments, and the national government. If the sequence is taken in advance, the structure of the coastal management system does not develop into the upper-right corner from the lower-left corner of the figure; rather, different structures overlap as the space spreads. That is, the structure of the management system of a certain stage is formed by combining the system of the stage and the previous stage. Thus, the whole management system of the whole area consists of four types of management. This is the core feature of the multilevel management system of coastal waters. Furthermore, the principle for moving the system successfully as a whole is network governance, which is described later. Next, I explain the features of the structure of each step.

5.7.2.1 First Level

The structure of the management in which the fishing community formed in a coastal area manages the nearshore waters based on convention or customary law goes into the lower-left corner. This structure had been legislated as the exclusive fishing right by the Meiji Fishing Law, and has transformed into the common fishery right of the current fishing law.

The situation, however, has changed. Although fishing communities consisting of fishery operators and the use of nearshore waters concerned only fishing at the outset when this legal system was enacted, inhabitants other than fishery operators have gradually increased in number in fishing communities, and use other than for fishing has increased in nearshore waters as well, particularly since the 1970s. This coincides with the time when the importance of coastal areas was raised by the national policy.

If the use of nearshore waters other than for fishing, such as for ocean recreation, is simple then those involved should just discuss a solution directly. As the types of use increase and the participants become more diverse, however, the participants gather and form a committee to draw up common rules. Fig. 5.30 shows the structure of the committee that is advancing ICM focusing on the growth of turtle grass in Hinase, Bizen city (Hidaka, 2016b). The point is that some fishermen began to grow turtle grass as a measure by the fishermen's cooperative association, and various stakeholders in the community of Hinase participated in the movement. I call this process the whole-of-region approach.

This structure is a type of network. Members do not belong exclusively to the committee, and can come and go freely; there is a flat relationship among members with no hierarchical order. Furthermore, the representative of the committee has no government authority.

As the organization is not hierarchical and participation is easy, the organizational structure can react to changes in the situation by taking the whole-of-region approach. This is the framework of the system for managing the Satoumi of nearshore waters.

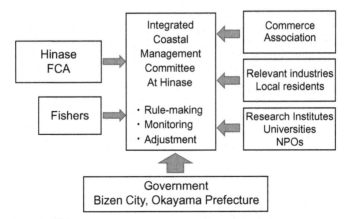

FIG. 5.30 Organizational arrangement in Hinase.

5.7.2.2 Second Level

The management system of the first level is effective within limited nearshore waters where the geographical and socioeconomic conditions are similar. When the conditions change, however, it becomes impossible to use this management system. Regarding the nearshore waters of local governments, various types of coastal waters have different conditions as usual. If some Satoumi according to the conditions of each coastal water is formed and linked as a network, the network can cover the coastal waters of the local government. It is necessary to form some Satoumi at coastal waters that are important in terms of biology or use, and to network them together effectively. Such an instance is observed in Shima city, where there are three types of coastal waters having different natural and social conditions: Ago Bay, Matoya Bay, and the Pacific coast. Under the Satoumi Creation Plan of the Shima city government, three Satoumi are formed according to each condition and are networked by the promoting committee, a network I call the Satoumi network.

5.7.2.3 Third Level

Although such Satoumi and Satoumi networks involve the creation of Satoumi by people involved in the nearshore waters of the district, issues such as the water quality of nearshore waters, environmental protection, and conservation regulations are the jurisdiction of the national and prefectural governments, which have budgets and authority concerning coastal waters. I call these the coastal infrastructure, in the sense of the most fundamental and important social and environmental base of coastal waters.

As the areas of coastal infrastructure are diverse in terms of environment, aquatic engineering, fishery, etc., the jurisdictional divisions of the national and prefectural governments are also diverse. Accordingly, when introducing ICM, the vertical division of relevant government divisions can become an issue easily and cause a wicked problem. To solve this vertical division and to provide consistent coastal infrastructure, related divisions need to cooperate beyond vertical administrative boundaries. I call such activity the whole-of-government approach (Bevir, 2012).

In Omura Bay, the Nagasaki Prefectural Government established the Omura Bay Environmental Protection and Revitalizing Action Plan in 2003 and has been tackling the bay's management (Hidaka, 2018). The management organization consists of an executive board chaired by the vice governor and with directors-general as members, and a management board of division managers. The management board brings together the measures related to Omura Bay and adjusts them to promote environmental protection and activation effectively. As the related measures by different divisions are adjusted through these processes, I consider this to be the whole-of-government approach.

5.7.2.4 Fourth Level

The fourth level is the management system targeting the wider ocean space exceeding the prefecture's jurisdiction. In this case, the national government takes the lead, collects all related prefectures, and devises necessary measures. In the coastal region administrative plan based on the Grand Design of the Country in the 21st Century (1998, National Land Agency), the whole country is divided into eight coastal region categories, and the committees consisting of related prefectures manage the coastal regions. The national government serves as facilitator and intermediary, and organizes all related prefectures.

With regard to coastal management by a group of multiple governments, it is important to provide coastal infrastructure in common among the governments. This requires the implementation of common regulations and common conservation measures. Because the Satoumi and Satoumi network are formed in response to the conditions of each coastal region, it is difficult to unify them among all related prefectures. Instead, all these prefectures need to unify and raise the level of providing common infrastructure.

The Osaka Bay Revitalizing Committee was formed in 2003 by five national agencies, six prefectures, four major cities, and two related organizations with the national agency acting as secretariat. The Osaka Bay Revitalizing Action Plan was drawn up by the committee, and is about to be carried out. In this program, measures, evaluation indicators, and participants are designated, and the management is carried out systematically, with periodical evaluations. This program will be used to tackle revitalizing projects such as Tokyo Bay, Ise Bay, and Hiroshima Bay. Moreover, there is an example of reaching an agreement in the United States among related state governments bordering Chesapeake Bay.

5.7.3 Layer Structure of Governance

According to the legal system for coastal areas, because coastal waters and many coasts are public property, the national government has primary responsibility. Specific uses, disaster prevention, and environmental protection are the responsibility of prefectural governments. Local governments, however, undertake some public works, support voluntary activities by local residents and collaborate with them. Thus, each body plays different roles in coastal management. The roles of each body are shown in Table 5.8.1 as the governance layer structure.

The national government designs and constructs institutions for the whole country and acts as an intermediary among prefectures. The national government has a particularly significant role in the case of wider ocean spaces exceeding prefectural jurisdiction.

The prefectural governments have authority supported by law and resources concerning coastal infrastructure to take the lead among management bodies. Prefectural governments provide coastal infrastructure, such as disaster prevention institutions, environmental protection institutions and measures, and regulations for resource conservation.

In the narrow nearshore waters, however, local inhabitants form the Satoumi voluntarily and local governments facilitate and support these activities. Because the Satoumi created in nearshore waters is small and cannot cover the whole ocean space of the prefecture, several Satoumi are connected to form a Satoumi network. This function is handled mainly by local governments. These activities are called the supporting approach.

The roles of various management bodies are shown in Fig. 5.31. This indicates the layer structure of governance and the decision-making level concerning ICM. In this figure, three decision-making levels are plotted; the top row is the selection of institution, the middle row is the formulation of an overall plan, and the bottom row is the creation of on-site rules. Furthermore, the figure is divided by a diagonal line: the upper-left part is assigned to the government and the lower-right part to users (local residents, etc.).

By combining Fig. 5.31 and Table 5.16, the outline of the governance structure of ICM is explained in Fig. 5.32, which plots the major activities. The provision of coastal infrastructure by prefectural governments goes into the upper-left part. Here, the point is whether the related divisions of prefectural governments can provide coastal infrastructure

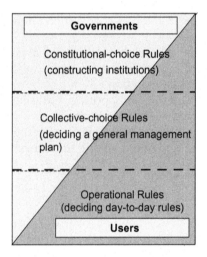

FIG. 5.31 Governance structure box.

TABLE 5.16 Contents of Each Layer for ICM

Management body	Contents of governance	Management type
National government	Adjustment among prefectures	Institutional design
	Establishment of institutions	Intermediation
Prefecture government	Environment conservation	Top-down and centralized management
	Land conservation	
	Disaster measures	
Local government	Adjustment of users and gov.	Voluntary management
	Management of local resources	Co-management
Local Residents	Day-to-day management of using, environment creation, adjustment of use	Voluntary management
		Community-based management

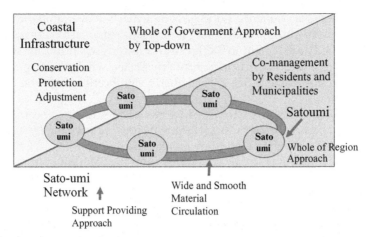

FIG. 5.32 Structure of multilevel management system.

cooperatively and properly, that is, the whole-of-government approach. The Satoumi and the Satoumi network by local residents and local governments are located in the lower-right part. Here, the point is whether the Satoumi can involve the stakeholders in the region by the whole-of-region approach and whether the Satoumi network is supported by local governments. This plot also explains the relationships among management bodies as well as the relationships among the different levels. This figure illustrates the global image of the multilevel managerial system.

The method of horizontal cooperation involves the whole-of-government approach and the whole-of-region approach. Regarding the wider ocean space, it involves cooperation among related prefectures to make the coastal infrastructure consistent.

Vertical cooperation is done through role-sharing among the national government, prefectural governments, local governments, and regional inhabitants. Furthermore, it is a structured relationship among the Satoumi, Satoumi networks, coastal infrastructure, and prefectural cooperation.

The supporting approach plays an important role in supporting and facilitating horizontal and vertical cooperation. To achieve a deflection, an intermediate agency sometimes is formed, or a structure of advancing cooperation is built. These are called strategic bridging or bridging-type social capital, according to environmental governance theory, and are equivalent to the process of promoting cooperation among the actors constituting a network.

5.7.4 Network-Governance

To manage the networks explained, the theory of network governance is useful. The easiest definition of governance in public administration is "government activities that use power" (Yamamoto, 2005). With the changing social situation in recent years, however, new governance has arisen to replace the old governance, which was characterized by hierarchy in the national government or local governments. The new governance is "the aspect of the administration which musters the stakeholders having resources required for government in the form of a network, and which plans to solve problems through cooperation among those actors" (Togawa, 2011).

Environmental governance theory also recognizes governance as the process of cooperation among actors. The environmental governance theory was proposed to cope with environmental problems that are complex, diverse, and stratified. It is explained as the process of solving problems by unifying governance from the top (government) and self-governance from the bottom (civil society), and actively involving related stakeholders while making use of various efforts to create a sustainable society (Matsushita and Ono, 2007).

According to both theories, the governance that should be applied to ICM solves problems through a process of collaboration among the entities forming a network. Coastal waters are managed by a network of entities such as various stakeholders on an equal footing, instead of the pyramid-type top-down governance. Problems are solved through a cooperative process of exchanging and combining resources among stakeholders.

The characteristics of environmental governance theory are that goal-setting is called for in alignment with the postulate of environmental sustainability, and that governance becomes stratified at the local, national, regional, and global levels. Sustainability is necessary when environmental issues become the target of management. Stratified governance might require a management system that connects every different spatial scale vertically from an environmental viewpoint.

TABLE 5.17 Checklist for Multilevel Management System of Coastal Waters

Function	Content	Subject
Alliance of prefectures	◆ Management of wider ocean area over prefecture jurisdiction	National government
	◆ Unifying coastal infrastructures provided by prefecture governments	
	◆ Intermediation of related prefectures	
Coastal infrastructure	◆ Regulations such as water quality control, use regulations	Prefecture governments
	◆ Projects such as coast conservation, port construction	
Satoumi network	◆ Networking plural Satoumi created at different areas	Prefecture and local governments
Satoumi	◆ Creating Satoumi to conserve and use coastal resources	Local governments
	◆ Creating economic, social values	Local residents

TABLE 5.18 Checklist for Network-Governance Factors of the Multilevel Management System

Category	Standard
Horizontal alliance	□ Are there any measures beyond vertical dividing by the whole of government approach?
	□ Are there any activities in which whole regional stakeholders participate by the whole-of-region approach?
Vertical alliance	□ Is there the a structured relationship among Satoumi, Satoumi network, coastal infrastructure and alliance among related prefectures?
	□ Is there any organization in which the national government, prefecture governments, local governments and regional residents are unified?
Unifying as a whole	□ Are there any targets and indicators that embody in sustainability postulate?
	□ Are there proper organization and process that promote PDCA cycle based on indicators?
Supporting	□ Are there any supporting activities to facilitate horizontal collaboration?
	□ Are there any supporting activities to facilitate vertical alliance?

Four factors are extracted from the network governance theory and environmental governance theory: process of cooperation among constituents, equal relations with the government, stratification connecting the different spatial scales vertically, and the target of the sustainability postulate. Because these factors respond to the framework of the multilevel management system, the whole-of-region approach and the whole-of-government approach, the multilevel management system for coastal waters provides a skeleton for network governance. To check the system from the viewpoint of network governance, we can check these four factors of the system.

I created two checklists as a test framework for checking these factors. The first list checks the contents of Satoumi, Satoumi networks, coastal infrastructure, and supporting activities (Table 5.17), and the second checks the four factors from the viewpoint of network governance (Table 5.18). By checking the current system against these two lists, it is clear whether the management structure and actions are based on network governance.

I conducted case studies at Hinase, Shizukawa Bay, Osaka Bay, Hiroshima Bay, and Chesapeake Bay. From the results, I could evaluate whether these cases are multilevel management by network governance, what is missing, and the problems. These matters will be examined in another book of this project.

5.7.5 Result and Policy Proposal

5.7.5.1 A Transduction of a Multilevel Management System

Because many individual laws and Satoumi already have been created in Japan, it is difficult to unify the structure of new management as an integrated system. Networking them into a multilevel network organization seems more appropriate than unifying them as a hierarchical organization. A network organization means that various individual management laws and Satoumi are accumulated with nesting and networked toward the same purpose: the sustainable and effective use of coastal waters. The multilevel management system consists of Satoumi, Satoumi network, coastal infrastructure, and an alliance of related prefectures, in accordance with different spatial scales. They are accumulated and unified as a whole based on the network governance principle. Through theoretical examination and some case studies, the multilevel management system seems to be effective, with network governance as the governance principle.

I propose that this system be introduced for coastal management in Japan. The framework for checking the structure and functions to ensure that this system works well as a network organization is shown in Tables 5.17 and 5.18. This framework makes it immediately clear what is missing and any problems of the current system. This system could be applied to other environmental resources and countries.

5.7.5.2 Policy Support

In order to introduce the multilevel management system as a coastal waters management system within a prefecture's jurisdiction in Japan, policy support by the government is necessary. First, a principal ordinance for coastal waters management needs to be established by the prefectural government to introduce the whole-of-government approach. Second, a principal ordinance for coastal waters management also should be established by the local government, and coastal waters management should be specified in its general plan to promote the whole-of-region approach. Both these principal ordinances should prescribe support for the approach. It is important that they determine the framework as the principal ordinance so that coastal waters management can be carried out continuously. On a country level, a measure to promote the establishment of principal ordinances by prefectures and local governments is necessary. It also is necessary to build a system of subsidies and grants to secure the feasibility of the measure based on the principal ordinances.

Concerning a wider area beyond the jurisdiction of a prefecture, a national agency should serve as secretariat as a mediator and promotor, and related prefectures should get together and cooperate as in the case of Osaka Bay. It would be useful for related prefectural governments to draw up cooperative agreements to set common goals and ensure cooperation on activities for coastal infrastructure, like the Chesapeake Bay Watershed Agreement.

5.8 COLLABORATIVE OCEANOGRAPHY—MULTISECTORAL DESIGNATION OF MARINE PROTECTED AREAS IN TSUSHIMA

Satoquo Seino

Multisectoral collaborative scientific research contributes to the conservation and sustainable use of biological diversity by eliciting local knowledge through collaborative observation of oceanographic drift buoys deployed along the coast of the Tsushima Islands (Fig. 5.33).

(1) Participation and collaboration from the survey phase

Marine Protected Areas (MPAs) are attracting attention not only for protecting biological diversity and conserving the natural environment, but also for promoting sustainable fisheries and tourism on the local level. Bottom-up style decision-making that respects each stakeholder's independence is needed to manage a protected area in a busy part of the sea that is subject various uses, and it promotes harmony among stakeholders.

Information about various weather phenomena known for generations by local fishermen and coastal residents, through traditions, experience, wisdom, historical documents, and local history, fused with scientific knowledge and the latest technology can result in marine management methods well adapted to local conditions.

Surveys were conducted in Tsushima City, Nagasaki prefecture, and its adjoining seas, where a marine protected area policy has been in force since 2010. As part of our consideration of "Collaborative oceanographic monitoring grounded in local knowledge," we checked the correspondence between good fishing grounds (marine areas known to be ecologically important by fishermen) and physical oceanographic findings (Tsushima Marine Protected Areas Science Committee, 2014).

A drift buoy (ZTB-R6-P3 (S)) was thrown from a boat into the sea, and its progress was followed from July 29 to August 7, 2015. Information about its location was relayed through a satellite channel and remotely received by email every hour. Each record of the buoy's location and track was made public and discussed through the researcher's SNS account (Facebook).

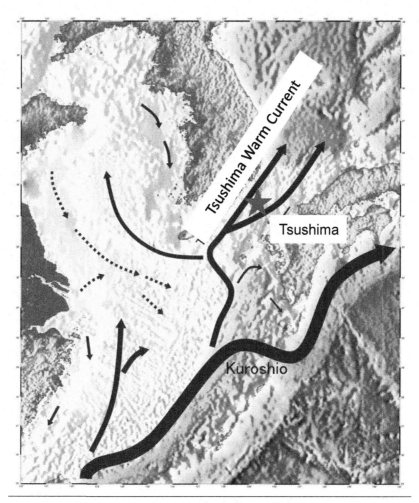

FIG. 5.33   Tsushima Warm Current and Tsushima Island.

(2) Drift buoy observation results

The buoy was dropped into the sea over the Tsushima Trench offshore of Saosaki in northwest Tsushima, and it moved north along the trench for the next 16h. It then entered the marine area off the northeast end of the island and followed a gentle northbound track while tracing a circle approximately every 4 days. The weather continued to be calm and no typhoons or other heavy weather hit during this time. Observations made by the same kind of device the previous year (deployed off Tsushima's northeast coast on May 29, 2014, recovered near Izuhara on June 9) recorded the same kind of circuitous route.

These circular tracks are what is called the Tsushima eddy, and suggest the possibility that this swirling current moves northward together with the entire water mass of the Tsushima Warm Current. It also is thought that the current moving directly northeast along the Tsushima Trench is a topographic phenomenon (Fig. 5.34).

(3) Knowledge for MPA design and management

The marine area where we observed the eddy approximately coincides with a good fishing ground where the fishermen of northern Tsushima enjoy abundant catches of 33 diverse species (Fig. 5.34). The fishermen independently suggested that an area between this eddy and the island be protected under a Nagasaki prefectural plan in order to restore tilefish resources; it seems likely that this area is a nutrient repository. In future, MPA design and management will have to consider not just the Tsushima Warm Current, but also the characteristics of the currents occurring between the coasts and offshore on a variety of spatiotemporal scales.

(4) Expansion of collaborative oceanography

The area for observation and the contents of the study were chosen through discussions with fishermen. By sharing the survey results through SNS with citizens, fishermen, and experts, new perspectives were obtained about fishing

FIG. 5.34 Collaborative oceanographic research in Tsushima. Trace of Tsushima Eddy observed a floating buoy. Location of protected area of red tile bream by Nagasaki fishery resource autonomic management.

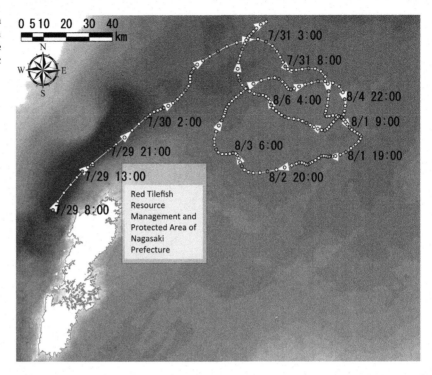

grounds, marine drift litter collection, harbor topography, and other matters relating to local knowledge and oceanography. Looking at the study as a drift buoy experiment, an important point in physical oceanography is what agents of geomorphic change are being used to measure the current; another issue is the need to improve the compliance of the device with the surface current. As in the case of our study, however, when looking at biota as well as floating and washed-up debris, an important factor is where debris washes up as a result of the action of a variety of forces. Collaborative oceanography is a new way to use drift buoy surveys, an established oceanographic observation method (Seino, 2016).

References

Abson, D.J., von Wehrden, H., Baumgärtner, S., Fischer, J., Hanspach, J., Härdtle, W., … Walmsley, D., 2014. Ecosystem services as a boundary object for sustainability. Ecol. Econ. 103, 29–37.

Anderies, J.M., Folke, C., Walker, B., Ostrom, E., 2013. Aligning key concepts for global change policy: robustness, resilience, and sustainability. Ecol. Soc. 18 (2), 8.

Andersson, E., McPhearson, T., Kremer, P., Gomez-Baggethun, E., Haase, D., Tuvendal, M., Wurster, D., 2015. Scale and context dependence of ecosystem service providing units. Ecosyst. Services 12, 157–164.

Ardoin, N.M., Schuh, J.S., Gould, R.K., 2012. Exploring the dimensions of place: a confirmatory factor analysis of data from three ecoregional sites. Environ. Educ. Res. 18 (5), 583–607.

Arrow, K.J., Dasgupta, P., Mäler, K.G., 2003. Evaluating projects and assessing sustainable development in imperfect economies. Environ. Resour. Econ. 26 (4), 647–685.

Arrow, K.J., Dasgupta, P., Goulder, L.H., Mumford, K.J., Oleson, K., 2012. Sustainability and the measurement of wealth. Environ. Dev. Econ. 17 (3), 317–353.

Ballinger, R., Pickaver, A., Lymbery, G., Ferreria, M., 2010. An evaluation of the implementation of the European ICZM principles. Ocean Coast. Manag. 53 (12), 738–749.

Banzhaf, H.S., Boyd, J., 2012. The architecture and measurement of an ecosystem services index. Sustainability 4 (4), 430–461.

Barbier, E.B., 2012. Ecosystem services and wealth accounting. In: UNU-IHDP and UNEP, Inclusive Wealth Report 2012: Measuring Progress Toward Sustainability. Cambridge University Press, Cambridge.

Barbier, E.B., Markandya, A., 2013. A New Blueprint for a Green Economy. Routledge, Oxford.

Bastian, O., Syrbe, R.U., Rosenberg, M., Rahe, D., Grunewald, K., 2013. The five pillar EPPS framework for quantifying, mapping and managing ecosystem services. Ecosyst. Services 4, 15–24.

Bennett, E.M., Cramer, W., Begossi, A., Cundill, G., Díaz, S., Egoh, B.N., … Woodward, G., 2015. Linking biodiversity, ecosystem services, and human well-being: three challenges for designing research for sustainability. Curr. Opin. Environ. Sustain. 14, 76–85.

Berque, J., Matsuda, O., 2013. Coastal biodiversity management in Japanese satoumi. Mar. Policy 39, 191–200.

Bevir, M., 2012. Governance: A Very Short Introduction. Oxford University Press, London, p. 152.

Cabinet Office, Government of Japan. 2015. Prefectural Economic Accounting 2012. http://www.esri.cao.go.jp/jp/sna/data/data_list/kenmin/files/contents/pdf/gaiyou.pdf (Accessed 25 September 2016).

Cameron, T.A., Quiggin, J., 1994. Estimation using contingent valuation data from a "dichotomous choice with follow-up" questionnaire. J. Environ. Econ. Manag. 27 (3), 218–234.

Central Environment Council. 2012. State of Future Course Which Should Aim at the Future in Setonaikai and Environment–Conservation and Restoration (in Japanese).

Central Environmental Council. 2012. The Desired Future States and Environmental Conservation and Restoration in the Seto Inland Sea. Reports Submitted to the Ministry of the Environment Government of Japan.

Chan, K.M., Guerry, A.D., Balvanera, P., Klain, S., Satterfield, T., Basurto, X., … Hannahs, N., 2012. Where are cultural and social in ecosystem services? A framework for constructive engagement. Bioscience 62 (8), 744–756.

Chan, K.M., Balvanera, P., Benessaiah, K., Chapman, M., Díaz, S., Gómez-Baggethun, E., … Luck, G.W., 2016. Opinion: why protect nature? Rethinking values and the environment. Proc. Natl. Acad. Sci. U. S. A. 113 (6), 1462–1465.

Common, M., Stagl, S., 2005. Ecological Economics: An Introduction. Cambridge University Press, Cambridge.

Costanza, R., 2015. Ecosystem services in theory and practice. In: Figgis, P., Mackey, B., Fitzsimons, J., Irving, J., Clark, P. (Eds.), Valuing Nature: Protected Areas and Ecosystem Services. Australian Committee for IUCN, Sydney, p. 6.

Costanza, R., d'Arge, R., de Groot, R., Farberk, S., Grasso, M., Hannon, B., Limburg, K., Naeem, S., O'Neill, R.V., Paruelo, J., Raskin, R.G., Sutton, P., van den Belt, M., 1997. The value of the world's ecosystem services and natural capital. Nature 387, 253–260.

Costanza, R., de Groot, R., Sutton, P., van der Ploeg, S., Anderson, S.J., Kubiszewski, I., … Turner, R.K., 2014. Changes in the global value of ecosystem services. Glob. Environ. Chang. 26, 152–158.

Cummins, V., McKenna, J., 2010. The potential role of sustainability science in coastal zone management. Ocean Coast. Manag. 53 (12), 796–804.

Daniel, T.C., Muhar, A., Arnberger, A., Aznar, O., Boyd, J.W., Chan, K.M., … von der Dunk, A., 2012. Contributions of cultural services to the ecosystem services agenda. Proc. Natl. Acad. Sci. U. S. A. 109 (23), 8812–8819.

Dasgupta, P., Mäler, K.G., 2004. The economics of non-convex ecosystems: introduction. In: Dasgupta, P., Mäler, K.G. (Eds.), The Economics of Non-Convex Ecosystems. Springer, Netherlands, pp. 1–27.

de Groot, R., Brander, L., van der Ploeg, S., Costanza, R., Bernard, F., Braat, L., Christie, M., Crossman, N., Ghermandi, A., Hein, L., Hussain, S., Kumar, P., McVittie, A., Portela, R., Rodriguez, L.C., ten Brink, P., van Beukering, P., 2012. Global estimates of the value of ecosystems and their services in monetary units. Ecosyst. Serv. 1(1), 50–61.

Derissen, S., Quaas, M.F., Baumgärtner, S., 2011. The relationship between resilience and sustainability of ecological-economic systems. Ecol. Econ. 70 (6), 1121–1128.

Duraiappah, A.K., Nakamura, K., Takeuchi, K., Watanabe, M., Nishi, M., 2012. Satoyama—Satoumi Ecosystems and Human Well-Being. United Nations University Press, Tokyo.

Enclosed Sea Bodies, Water Environment Division, Water and Air Environment Department, Ministry of the Environment 2010. Heisa sei kaiiki chyuchyoki vision sakutei ni kakawaru konndannkai.

Eppink, F.V., van den Bergh, J.C.J.M., 2007. Ecological theories and indicators in economic models of biodiversity loss and conservation: a critical review. Ecol. Econ. 61 (2–3), 284–293.

European Commission. n.d.. Integrated Coastal Management. http://ec.europa.eu/environment/iczm/home.htm (Accessed 25 September 2016).

Fischer, J., Gardner, T.A., Bennett, E.M., Balvanera, P., Biggs, R., Carpenter, S., … Tenhunen, J., 2015. Advancing sustainability through mainstreaming a social–ecological systems perspective. Curr. Opin. Environ. Sustain. 14, 144–149.

Fisheries Agency of Japan. 1998.

Fisheries Agency of Japan and Fisheries Research Agency. 2014. The Resource Evaluation of the Seto Inland Sea Japanese Spanish Mackerel. http://abchan.job.affrc.go.jp/digests26/details/2654.pdf (Accessed 25 September 2016) (in Japanese).

Fu, B., Jones, K.B. (Eds.), 2013. Landscape Ecology for Sustainable Environment and Culture. Springer, Netherlands.

Fu, B., Su, C., Lü, Y., 2013. Coupling landscape patterns and ecological processes. In: Fu,, Jones (Eds.), Landscape Ecology for Sustainable Environment and Culture. Springer, Netherlands, pp. 3–20.

Fukken Co., Ltd. 2015. Hiroshima wan syuhen no saiteki kaki yousyokuryo tou no suiteikekka gyomu houkokusyo.

Gómez-Baggethun, E., de Groot, R., Lomas, P.L., Montes, C., 2010. The history of ecosystem services in economic theory and practice: from early notions to markets and payment schemes. Ecol. Econ. 69 (6), 1209–1218.

Gordon, H.S., 1953. An economic approach to the optimum utilization of fishery resource. J. Fish. Res. Board Can. 10 (7), 442–457.

Governors and Mayors' Conference on the Environmental Conservation of the Seto Inland Sea. 2007. Measurements for the Realization of Bounty and Beautiful Seto Inland Sea: Regeneration of Satoumi.

Graham, M., 1935. Modern theory of exploiting a fishery, and application to North Sea trawling. J. Cons. Int. Explor. Mer 10, 264–274.

Gu, H., Subramanian, S.M., 2014. Drivers of change in socio-ecological production landscapes: implications for better management. Ecol. Soc. 19 (1), 1–13.

Haines-Young, R., Potschin, M., 2010. The links between biodiversity, ecosystem services and human well-being. In: Raffaelli, D.G., Frid, C.L.J. (Eds.), Ecosystem Ecology: A New Synthesis. Cambridge University Press, Cambridge, pp. 110–139.

Harris, M., Pearson, L.J., 2004. Using inclusive wealth to measure and model sustainable development in Australia: a working example. In: 2004 Conference (48th), February 11-13, 2004, Melbourne, Australia (No. 58457), Australian Agricultural and Resource Economics Society.

Hattam, C., Atkins, J.P., Beaumont, N., Börger, T., Böhnke-Henrichs, A., Burdon, D., … Austen, M.C., 2015. Marine ecosystem services: linking indicators to their classification. Ecol. Indic. 49, 61–75.

Hein, L., Obst, C., Edens, B., Remme, R.P., 2015. Progress and challenges in the development of ecosystem accounting as a tool to analyse ecosystem capital. Curr. Opin. Environ. Sustain. 14, 86–92.

Henocque, Y., 2013. Enhancing social capital for sustainable coastal development: is satoumi the answer? Estuar. Coast. Shelf Sci. 116, 66–73.

Hidaka, T., 2002. Urban and Fishery; Coastal Utilization and Exchange. Seizando, Tokyo, p. 194.

Hidaka, T., 2014. A study on the management system of integrated coastal management: effectiveness of dual management systems and network governance. J. Jpn. Soc. Ocean Policy 4, 61–72.

Hidaka, T., 2016a. Satoumi and Coastal Management; Managing Satoumi. Association of Agriculture & Forestry Statistics, Tokyo (In Japanese).

Hidaka, T., 2016b. Satoumi and Coastal Management; Managing Satoumi. Agriculture, Forestry and Statistics Association, Tokyo, p. 173.

Hidaka, T., 2018. Case study of the regional ICM system introduces voluntarily by the prefectural government in Omura Bay, Japan. In: Guillotreau, P., Bundy, A., Perry, R.I. (Eds.), Global Change in Marine Systems: Integrating Natural, Social and Governing Responses. Routledge, New York, pp. 135–147.

Hidaka, T., Yoshida, M., 2015. A study of the structure and function of the Sato-umi management organization: preliminary analysis through questioner surveys. J. Coast. Zone Stud. 28 (3), 107–118.

Hinase Junior High School and Network for Coexistence with Nature, 2016. Marine Education Through People and Ocean: The Seagrass Bed Restoration Project at Hinase Junior High School. Hinase Junior High School, Okayama (In Japanese).

Hiroshima Prefecture. 2014b. Heisei 26 nendo Hiroshima kaki seisan syukka sisin.

Hiroshima Prefecture. 2014a. Kokyo yousui iki suishitsu sokutei kekka hyo.

IDEA Consultants, Inc., 2015. Oosaka wan no kaki toumeido to teisou DO no saigen gyoumu nikakawaru itakugyoumu houkokusyo.

Innami, T., 2015. The sixth industrialization and Sawara in Hinase. In: Sci. Forum of the Seto Inland Sea. vol. 70, pp. 67–70 (in Japanese).

Komatsu, T., Yanagi, T., 2015. Sato-umi: an integrated approach for sustainable use of coastal waters, lessons from human-nature interactions during the Edo period of eighteenth-century Japan. In: Marine Productivity: Perturbations and Resilience of Socio-Ecosystems. Springer, Netherlands, pp. 283–290.

Kulig, A., Kolfoort, H., Hoekstra, R., 2010. The case for the hybrid capital approach for the measurement of the welfare and sustainability. Ecol. Indic. 10 (2), 118–128.

Kuriyama, K., 1998. Kankyo no kachi to hyoka shyhou. Hokaido University Press.

Liquete, C., Piroddi, C., Drakou, E.G., Gurney, L., Katsanevakis, S., Charef, A., Egoh, B., 2013. Current status and future prospects for the assessment of marine and coastal ecosystem services: a systematic review. PLoS One. 8(7). e67737.

Maccarone, V., et al., 2014. The ICZM balanced scorecard: a tool for putting integrated coastal zone management into action. Mar. Policy 44, 321–334.

Managi, S.. 2009. Capacity Output and Possibility of Cost Reduction: Fishery Management in Japan, DPRIETI Discussion Paper Series 09-E-040 (http://www.rieti.go.jp/en/).

Matsuda, O. 2013. A Vital Role of Satoumi in the Implementation of ICM in Japan.

Matsushita, K., Ono, T., 2007. A neo development of an environmental governance theory. In: Kazuo Matsushita, K. (Ed.), Environmental Governance Theory. Kyoto-University Press, Kyoto, pp. 3–31.

Millennium Ecosystem Assessment, 2003. Ecosystems and Human Well-Being: A Framework for Assessment. Island Press, Washington, DC.

Minamisanriku Town Biomass Industry City Design. 2013. The Nikkei: Sustainable Fish Consumption Spreads; International Accreditation Prevents Overfishing and Water Pollution; Aeon Increases Sales Venues by 60%; Used in all Panasonic Cafeterias—May 13, 2018.

Ministry of Agriculture, Forestry and Fisheries of the Government of Japan, 2006. Survey on Conservation Activities. http://www.jfa.maff.go.jp/j/kikaku/tamenteki/sankou/index.html.

Ministry of the Environment Government of Japan. n.d.-a Sato-umi Net. https://www.env.go.jp/water/heisa/satoumi/en/01_e.html (Accessed 25 September 2016).

Ministry of the Environment Government of Japan. n.d.-b. Setouchi Net. Retrieved from https://www.env.go.jp/water/heisa/heisa_net/setouchi-Net/seto/index.html (Accessed 25 September 2016) (in Japanese).

Ministry of the Environment Japan, 2014. About the results of the economic valuation of biodiversity (CVM) using quetionnaire survey. http://www.env.go.jp/press/press.php?serial=18158.

Miyagawa, M., Masui, T., Akai, N., Suenaga, Y., Ishizuka, M., 2015. Nutrient management by control operation in sewage treatment process. Kaiyo Seibutsu 37 (3), 261–273 (in Japanese).

Nakagami, K., Yoshioka, T., Tomeno, R., Obata, N., 2018. Dynamic sustainability assessment towards the integrated coastal zone management. J. Policy Sci.. 12.

National Institute of Population and Social Security Research. 2013. Population Projections for Japan by Regions (March 2013). http://www.ipss.go.jp/pp-shicyoson/j/shicyoson13/2gaiyo_hyo/gaiyo.asp (Accessed 25 September 2016).

Naveh, Z., 2000. What is holistic landscape ecology? A conceptual introduction. Landsc. Urban Plan. 50 (1), 7–26.

Z. Naveh and A.S. Lieberman, 2013.

Nguyen, L., Nguyen, T.B., 2008. Assessment of Tonkin Gulf Fishery, Vietnam, Based on Bioeconomic Models. Paper presented to the Fourteenth Biennial Conference of the International Institute of Fisheries Economics and Trade, Nha Trang, Vietnam.

Ostrom, E., 2010. A multi-scale approach to coping with climate change and other collective action problems. Solutions 1 (2), 27–36.

Panel for the Visions of Seagrass Beds and Mudflats, 2015. A Preliminary Report. The Ministry of the Environment, Government of Japan (in Japanese).

Paredes, C., 2008. La industria anchovetera Peruana: Costos y Beneficios. In: Un Análisis de su Evolución Reciente y de los Retos par el Futuro. Estudio preparado por encargo del Banco Mundial al Instituto del Perú de la Universidad de San Martin de Porres. Trabalho en processo.

Pearson, L.J., Biggs, R., Harris, M., Walker, B., 2013. Measuring sustainable development: the promise and difficulties of implementing inclusive wealth in the Goulburn-Broken Catchment, Australia. Sustain. Sci. Pract. Policy 9 (1), 16. At the applied science and engineering universities.

Reis, J., Stojanovic, T., Smith, H., 2014. Relevance of systems approaches for implementing integrated coastal zone management principles in Europe. Mar. Policy 43, 3–12.

Reyers, B., Biggs, R., Cumming, G.S., Elmqvist, T., Hejnowicz, A.P., Polasky, S., 2013. Getting the measure of ecosystem services: a social-ecological approach. Front. Ecol. Environ. 11 (5), 268–273.

Rosenberger, R.S., Loomis, J.B., 2003. Benefit transfer. In: Champ, P.A., Boyle, K.J., Brown, T.C. (Eds.), A Primer on Nonmarket Valuation. Kluwer Academic Publishers, Berlin, pp. 445–482.

Sakurai, R., Ota, T., Uehara, T., Nakagami, K., 2016a. Factors affecting residents' behavioral intentions for coastal conservation: case study at Shizugawa Bay, Miyagi, Japan. Mar. Policy 67, 1–9.

Sakurai, R., Ota, T., Uehara, T., Nakagami, K., 2016b. Public perceptions of a coastal area among residents around Hinase Town of Okayama Prefecture: analysis based on location of residence. People Environ. 42 (3), 18–26 (in Japanese).

Sakurai, R., Ota, T., Uehara, T., 2017. Sense of place and attitudes towards future generations for conservation of coastal areas in the Satoumi of Japan. Biol. Conserv. 209, 332–340.

Sakurai, R., Uehara, T., Yoshioka, T., 2018. Students' perceptions of a marine education program at a junior high school in Japan with a specific focus on Satoumi. Environ. Educ. Res. https://doi.org/10.1080/13504622.2018.1436698.

Sala, S., Ciuffo, B., Nijkamp, P., 2015. A systemic framework for sustainability assessment. Ecol. Econ. 119, 314–325.

Schlueter, M., McAllister, R., Arlinghaus, R., Bunnefeld, N., Eisenack, K., Hoelker, F., Quaas, M., 2012. New horizons for managing the environment: a review of coupled social-ecological systems modeling. Nat. Resour. Model. 25 (1), 219–272.

Seino, S., 2016. Planning, management and sustainable use of marine protected areas. J. Environ. Conserv. Eng. 45 (3), 138–145 (in Japanese).

Seto Inland Sea Fisheries Coordination Office. 2018. Trends in Marine Fishery Catch Volumes in the Seto Inland Sea, Setouchi Net, etc.

Setonaikai Fisheries Coordination Office. 2012. Fishery in the Seto Inland Sea. http://www.jfa.maff.go.jp/setouti/tokei/pdf/24setonaikainogyogyo. pdf (Accessed 25 September 2016) (in Japanese).

Setonaikai Research Conference, 2007. Setonaikaiwo satoumini aratana siten niyoru saisei housokau. Kouseisya kouseikaku, Tokyo.

Shimizu, N., et al., 2014. Connectivity of Hills, Humans and Oceans: Challenge to Improvement of Watershed and Coastal Environments. Kyoto University Press, Kyoto.

Shirahata, Y., 1999. In: Goda, K. (Ed.), Srtonaikaino bunka to kankyo. ShinSetonaikai bunka Series, vol. 2. Setonaikai kankyo hozen kyokai, Kobe.

Singh, R.K., Murty, H.R., Gupta, S.K., Dikshit, A.K., 2012. An overview of sustainability assessment methodologies. Ecol. Indic. 15 (1), 281–299.

Sumaila, U.R., Marsden, D.A., 2007. Case Study of the Namibian Hake Fishery. Prepared for the FAO/World Bank rent drain study. Fisheries Economics Research Unit, Fisheries Centre, University of British Columbia, Vancouver.

Syme, G.J., Dzidic, P., Dambacher, J.M., 2012. Enhancing science in coastal management through understanding its role in the decision making network. Ocean Coast. Manag. 69, 92–101.

Y. Takata. 2017. Citizen Awareness Regarding Shallow Seas and Their Regeneration, and Future Initiatives.

Tanda, M., Harada, K., 2011. The example of measures to oligotrophy and a future problem. J. Jpn. Soc. Water Environ. 34 (2), 54–58 (in Japanese).

Terawaki, T., 2000. Research Note Limdep Programs for Estimating Parametric Models in Double-Bounded Dichotomous Choice CVM. vol. 33. Koube University nougyo keizai, pp. 101–112.

The Association for the Environmental Conservation of the Seto Inland Sea. 2014. The Conservation of the Seto Inland Sea 2013 (in Japanese).

The Association for the Environmental Conservation of the Seto Inland Sea. 2015. Environmental Conservation in the Seto Inland Sea (in Japanese).

The Fishery Agency of Japan. 2016. Fishery White Paper, Heisei 26, in Japanese.

Togawa, S., 2011. Networked governance and network-like NPM: a case study on hospital PFI. J. Soc. Sci. 31, 47–88.

Tsushima Marine Protected Areas Science Committee. 2014. Report of Tsushima Marine Protected Areas Science Committee, Tsushima City. http://www.city.tsushima.nagasaki.jp/web/updata/houkokusho.pdf (Accessed 25 April 2019) (in Japanese).

Turner, K., Schaafsma, M., Elliott, M., Burdon, D., Atkins, J., Jickells, T., Tett, P., Mee, L., van Leeuwen, S., Barnard, S., Luisetti, T., Paltriguera, L., Palmieri, G., Andrews, J. 2014. UK.

Uehara, T., 2013. Ecological threshold and ecological economic threshold: implications from an ecological economic model with adaptation. Ecol. Econ. 93, 374–384.

Uehara, T., Mineo, K., 2016. The database construction and guidelines for economic valuation studies on the Japanese coastal ecosystem services. Seisakukagaku 23 (2), 57–65 (in Japanese with English abstract).

UNU-IHDP, UNEP, 2012. Inclusive Wealth Report 2012. Measuring Progress Toward Sustainability. Cambridge University Press, Cambridge.

UNU-IHDP, UNEP, 2014. Inclusive Wealth Report 2014. Measuring Progress Toward Sustainability. Cambridge University Press, Cambridge.

van Oudenhoven, A.P.E., Petz, K., Alkemade, R., Hein, L., de Groot, R.S., 2012. Framework for systematic indicator selection to assess effects of land management on ecosystem services. Ecol. Indic. 21, 110–122.

Varian, H., 2014. Intermediate Microeconomics—A Modern Approach, eighth ed. W. W. Norton & Co., New York and London.

Wakita, K., Shen, Z., Oishi, T., Yagi, N., Kurokura, H., Furuya, K., 2014. Human utility of marine ecosystem services and behavioural intentions for marine conservation in Japan. Mar. Policy 46, 53–60.

World Bank and FAO, 2009. The Sunken Billions: The Economic Justification for Fisheries Reform. Agricultural and Rural Development, World Bank/FAO, Washington, DC/Rome.

Yamamoto, H., 2005. Civil society and country, and governance. J. Public Policy Stud. 5, 68–84.

Yamamoto, T., 2015. Seto nai kai no eiyouen kanri—jizokutekina kaisou yousyokuni mukete Hajimeni. Kaiyo seibutsu 37 (3), 207–208.

Yanagi, T., 2006. Satoumi Ron. Koseisha-Koseikaku, Tokyo, p. 102.

Yanagi, T., 2008. Establishment of Sato-umi in the coastal sea. J. Jpn. Soc. Water Environ. 21, 703 (in Japanese).

Yanagi, T., 2012. Japanese Commons in the Coastal Seas: How the Satoumi Concept Harmonizes Human Activity in Coastal Seas With High Productivity and Diversity. Springer, Netherlands.

Yoshida, K., Kinoshita, J., Egawa, A., 1997. Valuing economic benefits of agricultural landscape by double-bounded dichotomous choice CVM: a case study of nose-town, Osaka prefecture. Nouson keikaku gakkaishi 16 (3), 205–215 (in Japanese).

Zonneveld, I., 1989. The land unit—a fundamental concept in landscape ecology, and its applications. Landsc. Ecol. 3 (2), 67.

6

Integrated Numerical Model of the Coastal Sea

Tetsuo Yanagi*, H. Yamamoto*, T. Kasamo[†], Xinyu Guo[‡], T. Mano[§],
Katsumi Takayama[¶], and Takafumi Yoshida[∥]

*International EMECS Center, Kobe, Japan
[†]Researcher, ECOH Consulting Company Limited, Japan
[‡]Center for Marine Environmental Studies, Ehime University, Matsuyama, Japan
[§]Graduate School of Science and Engineering, Ehime University, Matsuyama, Japan
[¶]Ocean Modeling Group, Center for Oceanic and Atmospheric Research, Research Institute for Applied Mechanics,
Kyushu University, Kasuga, Japan
[∥]Regional Activity Center, Northwest Pacific Region Environmental Cooperation Center, Toyam City, Japan

6.1 DESIGN OF INTEGRATED NUMERICAL MODEL

Tetsuo Yanagi

We have to develop an integrated numerical model to support policymaking for coastal environment development and conservation. In this context, integration means the integration of land, coastal sea, open sea, sea bottom, and air, and the integration of the results from studies of natural, social, and human sciences.

Environmental changes in the coastal sea are affected by open sea, air, sea bottom, and land as shown in Fig. 6.1. To evaluate the policy for environmental conservation in coastal seas, we must include these effects in our numerical model.

The effects from the air and sea bottom are included in our integrated numerical model by boundary conditions expressed by an empirical formula or observed data, e.g. the dry or wet fault from the air, and the release and denitrification fluxes from the sea bottom will be based on the observed and/or experimental results. The flux sinking to the bottom, released flux from the bottom, and the oxygen demand by the surface sediment are calculated by an empirical formula depending on the water temperature, the concentration of organic matter in the surface sediment, and so on. The effect of open ocean variability will be obtained by the coupling of the results calculated by JCOPE2 (Miyazawa et al., 2009).

As for the effect from the land, which might have the greatest effect on the coastal sea, we take the mutual interactions between land and coastal sea into consideration and integrate both effects. For example, we calculate the

© 2019 Elsevier Inc. All rights reserved.

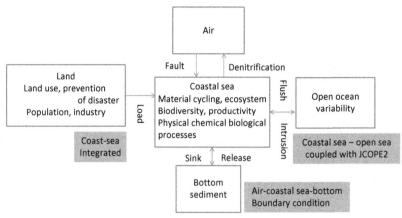

FIG. 6.1   Integrated numerical model.

change in the marine environment by examining the change in land use or land cover, that is, how much fresh water, nutrients, organic matter, heavy metals such as iron are changed by the change in areas of flat-leaf trees and needle-leaf trees. Moreover, we calculate the necessary nutrients and iron loads from land required for achieving a desirable primary production in the target coastal sea. We compare the results of both calculations and design the most desirable land use for the development and conservation of the marine environment of the target coastal sea.

Based on the results of these calculations using our integrated numerical model, we are able to propose the best policy for the development and conservation of the coastal marine environment.

In this project, integrated numerical models have been developed in Shizukawa Bay and the Seto Inland Sea, including Osaka Bay and Toyama Bay (Fig. 1.3). Shizukawa Bay is a typical small-scale open character coastal sea, which is affected mainly by open ocean variability from the viewpoint of primary production, and results from the study will be applied to Oohunato Bay, Ago Bay, Sukumo Bay, and elsewhere in Japan. Toyama Bay is a typical mid-scale, open character coastal sea, whose marine environment is affected mainly by the Tsushima Current, and results from the study will be applied to Karatsu Bay, Wakasa Bay, and elsewhere in Japan. The Seto Inland Sea is a typical semi-enclosed coastal sea, which is affected mainly by changes in land use, and the small bays and basins that are connected to it. Results will be applicable to Tokyo Bay, Ise Bay, and elsewhere in Japan.

We asked environmental consulting companies to perform the development of integrated numerical model in this project S-13 because most of the previous models for environmental policy have been carried out by consulting companies in Japan, not by universities (Yanagi, 2018). We prohibited them from using their most usual expression: "The calculated results nearly reproduce the observed results." Instead of this expression, we asked them to provide a quantitative evaluation of the degree of reproduction using a quantitative index.

Three kinds of indexes have been proposed by the three environmental consulting companies that took part into this project.

The first example from company A is the Taylor diagram (Fig. 6.2A, Taylor, 2001), where the length along the transverse denotes the average variance normalized by the standard deviation, and the plot near 1.0 means good reproduction. The angle denotes the correlation coefficient between observation and calculation, and transverse means the correlation is 1.0. The length of the plot from (1, 0) denotes the root mean squared error (RMSE) between observation and calculation. Therefore, the plot is located at (1, 0) when the calculation completely agrees with the observation.

RMSE and relative root mean squared error (RR) were proposed by company B. RMSE is defined as:

$$RMSE = \frac{1}{B_{obsAv}} \sqrt{\frac{1}{n} \sum_{i=1}^{n} (B_{cal} - B_{obs})^2} \tag{6.1}$$

where B_{obsAv} means an average of observed values, B_{cal} calculated values, and B_{obs} observed values. RR is obtained by dividing RMSE by the average value of the observed one, as

$$RR = \sqrt{\frac{1}{n} \sum_{i=1}^{n} \frac{(B_{cal}(i) - B_{obs}(i))^2}{B_{obs}(i)^2}} \tag{6.2}$$

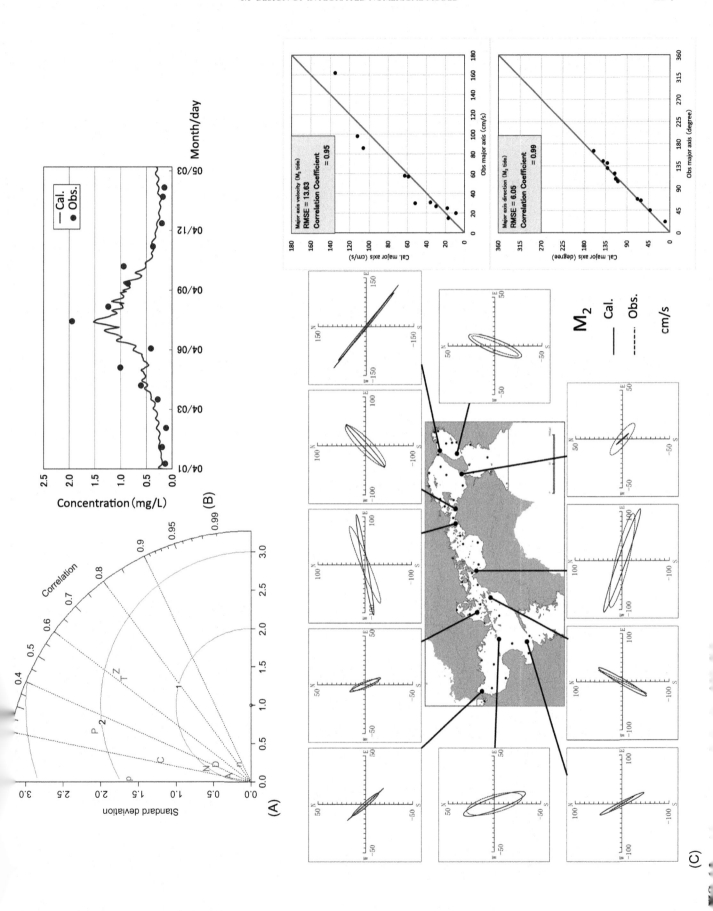

Both values are considered to be good when they take small values. RMSE (0.41 in Fig. 6.2B) is appropriate in cases where the initial stage is important, such as the blooming of phytoplankton, and RR (0.59 in Fig. 6.2B) is appropriate when the average value is important (Fujiwara et al., 2003).

A quantitative expression for the reproduction degree of tidal ellipse was proposed by company C (Fig. 6.2C).

It is regrettable that we are unable to report the response of policymakers about how to use such quantitative expressions of the degree of reproduction shown by the results calculated by the numerical model. We will continue this study about the relationship between policymaking and the expression of the quantitative correctness of the results obtained by the numerical model.

Another challenge for this project related to the numerical model is how to show visually the results of the numerical model calculation to ordinary people. The committee in Shizukawa Bay succeeded in making pictures of the results of the calculations easy to understand (Komatsu et al., 2018).

6.2 INTEGRATED NUMERICAL MODEL OF SHIZUKAWA BAY

Tetsuo Yanagi, H. Yamamoto

Shizukawa Bay is along the southern part of the Sanriku Coast, in the northeast of Japan, and faces the open Pacific Ocean. All rivers that empty into the bay belong to the local government of Minamisanriku Town. Therefore, the environmental management of their catchment area from the mountains to the coastal sea is somewhat easier compared to other bays in Japan.

In 2014, domestic production in Minamisanriku amounted to 746 billion yen with the primary industries accounting for 32 billion; secondary industries, 490 billion; and tertiary industries, 225 billion. In the primary sector, the fisheries industry came in at 27 billion yen, compared to 3 billion for agriculture and 2 billion for forestry.

The landing frame of the fishing catch amounted to 21 billion yen, and that of aquaculture was 27 billion, meaning that aquaculture is more important financially than the fishing catch in Shizukawa Bay (Minami-Sanriku-Town, 2018).

The fishermen who cultivate oysters in Shizukawa Bay lost all their facilities and equipment as a result of the devastating tsunami following the Great Tohoku Earthquake of March, 2011. They consequently decided to reduce the number of oyster rafts to just one-third during the rehabilitation period after the tsunami (Fig. 6.3) because they believed that the oyster culture before the tsunami was overcrowded (based on their experience or innate knowledge). At the committee meeting that was established in 2014 by the fishermen, local government officers and scientists (i.e., ourselves), the fishermen asked us to clarify scientifically the effect of the decrease in oyster rafts (in other words, combining their local knowledge with our scientific knowledge).

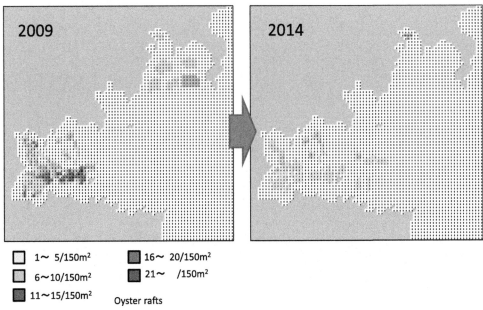

☐ 1~ 5/150m²	▦ 16~ 20/150m²
▨ 6~10/150m²	▩ 21~ /150m²
▦ 11~15/150m²	Oyster rafts

FIG. 6.3 Sites of oyster culture in 2009 and 2014.

The results of the integrated numerical model (the reproduction of current and lower trophic level ecosystems already has been confirmed, Yamamoto et al., 2017a) and the results of numerical experiments show that:

(1) The growth rate of cultured oyster roughly doubles because of increases in phytoplankton concentration as a result of the reduction in oyster rafts (Fig. 6.4) and the time needed for selling oysters shrinks from about 2 years to 1 year.

(2) The sinking organic matter decreases because of the decrease in oyster rafts and the increase in younger oysters (organic content in the fecal pellets increases as oysters age). That reduction decreases the amount of oxygen consumed for decomposition of organic matter and results in the disappearance of hypoxia in the bottom layer during the summer (Fig. 6.5).

The first Aquaculture Stewardship Council (ASC) citation in Japan was awarded to cultured oysters in Shizukawa Bay in March, 2016, and it is likely that cultured oysters from the bay will be on the menu at the Olympic Games village in Tokyo in 2020.

As for "wakame," (*Undaria pinnatifida*, a popular seaweed in Japan) culture in Shizukawa Bay, the results of integrated numerical model experiments show that:

(3) The decrease in cultured wakame biomass to 75% results in an increase in the growth rate of cultured wakame, and the harvest by, and income for, the fishermen will increase (Fig. 6.6).

In addition,

(4) The main seasons for the growth of oysters and wakame are different, and the change in their cultured biomass does not affect each other (Fig. 6.7).

In Shizukawa Bav, the limiting nutrients for primary production are not phosphorus and iron but nitrogen (Fig. 6.8), and the nitrogen budget before and after the tsunami is shown in Fig. 6.9. The bay is supplied mainly with dissolved inorganic nitrogen (DIN) from the open ocean.

The differences in the primary production and the recycling speed of nutrients before and after the earthquake are shown in Fig. 6.10, where the primary production (phytoplankton + wakame) increased by about 1.4 times after the earthquake and the recycling speed increased by about 4.5 times. DIN was recycled smoothly after the tsunami because it is assimilated to phytoplankton, grazed by cultured oysters, discharged as fecal pellets, and decomposed

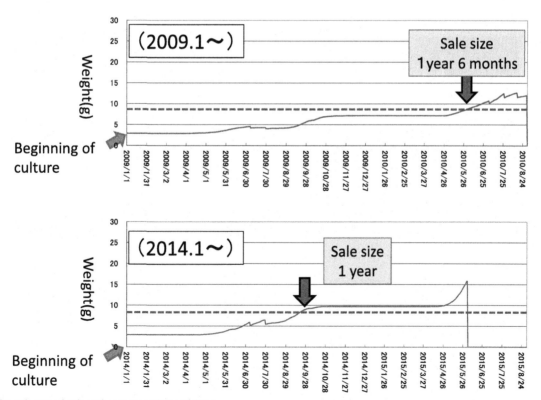

FIG. 6.4 Growth rate of cultured oyster in 2009 and 2014.

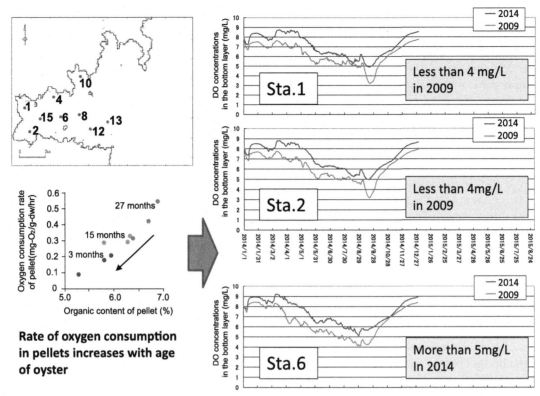

FIG. 6.5   Dissolved oxygen concentration 1 m above the bottom in 2009 and 2014.

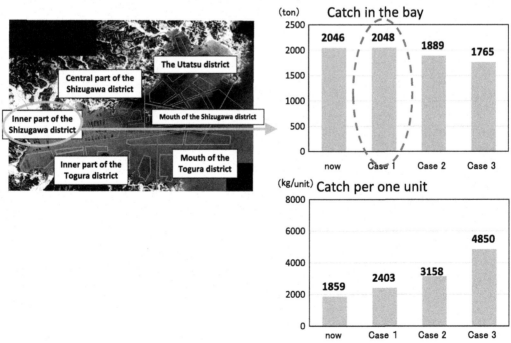

FIG. 6.6   Change in seaweed catch depending on the cultured biomass.

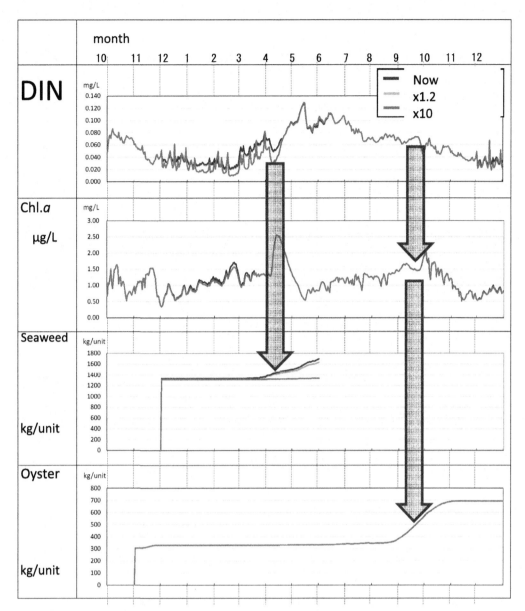

FIG. 6.7   Growth seasons of cultured seaweed and oyster.

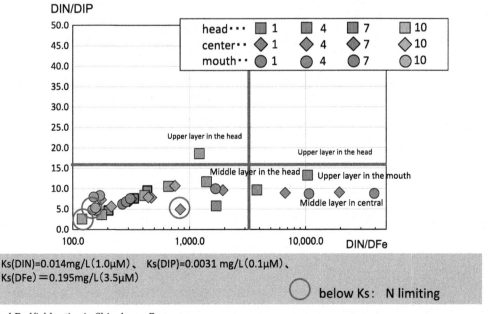

FIG. 6.8   Expanded Redfield ratios in Shizukawa Bay.

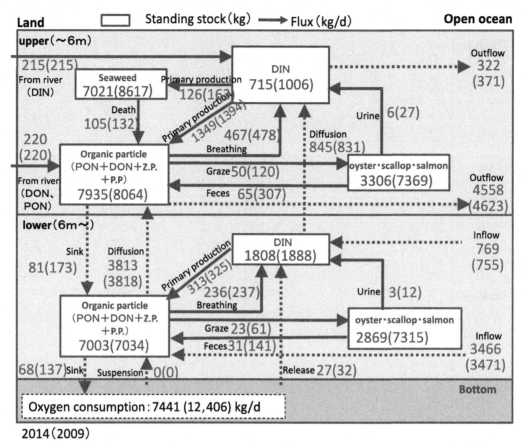

FIG. 6.9 Nitrogen budget in Shizukawa Bay.

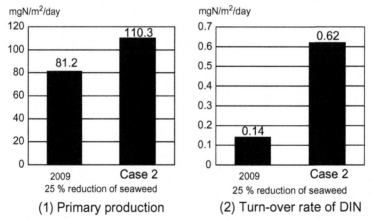

FIG. 6.10 Primary production and turnover rate of DIN in Shizukawa Bay during 2009 and 2014.

back to nutrients under suitable oyster culture. The appropriate amount of cultured oysters increased the recycling speed of DIN, which resulted in increased primary production (Yamamoto et al., 2017b).

These results suggest that appropriate human interaction increased biodiversity (plankton→plankton+ sea algae +bivalve) and productivity (primary production) in Shizukawa Bay. In other words, we created Satoumi (Fig. 6.11).

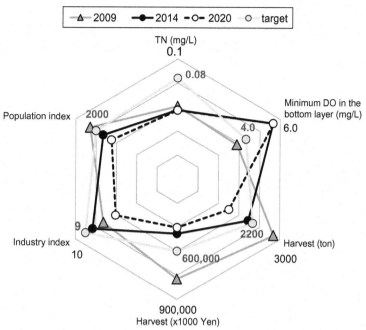

FIG. 6.11 Integrated indexes of clean, productive, prosperous, and sustainable Shizukawa Bay.

6.3 INTEGRATED NUMERICAL MODEL OF OSAKA BAY

T. Yanagi and T. Kasamo

The fish catch in the Seto Inland Sea during the eutrophication period is larger than that during oligotrophication period with the same transparency, and the mean trophic level of fish caught during the eutrophication period is smaller than during the oligotrophication period. In other words, hysteresis exists (Figs. 1.12 and 1.13).

Such hysteresis is generated by a process in which a large biomass of phytoplankton results in a high fish catch of phytoplankton feeder fish such as anchovies, but the large biomass of phytoplankton also generates anoxic water mass in the bottom layer during the summer of a eutrophication period. The anoxic water mass kills the eggs of zooplankton and benthos, which results in a decrease in transfer efficiency of organic matter to higher trophic level biota in the food web, and the mean trophic level of fish caught decreases. The main reason of such hysteresis is generated by the mutual interaction between water quality and sediment quality (Takeoka and Murao, 1997; Yanagi, 2015).

We developed an integrated numerical ecosystem model for Osaka Bay to clarify the relationship between the mutual interaction of hysteresis and sediment. The integrated numerical model we developed for Osaka Bay is shown in Fig. 6.12. Year-to-year variations in TN and TP load in the bay are shown from 1920 to 2020, while variations in TN and TP concentrations in Harima-Nada and the Kii Channel are the results of observations. Variations in water quality

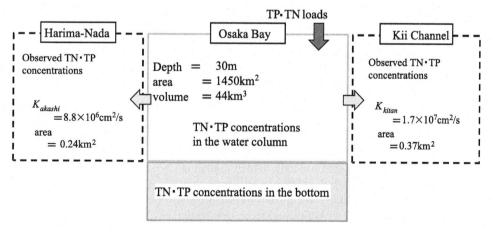

FIG. 6.12 Integrated box model of Osaka Bay.

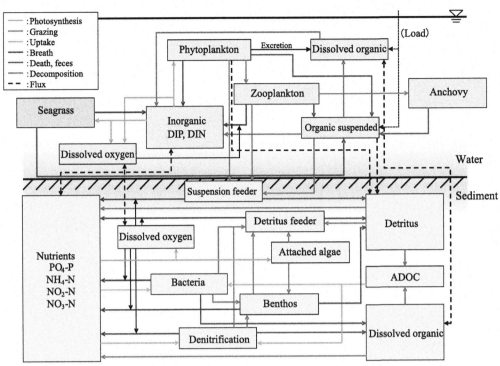

FIG. 6.13   Coupled numerical model of water quality and sediment quality for Osaka Bay.

TABLE 6.1   Contents of Integrated Numerical Model for Osaka Bay

Item	Contents
Water quality model	Lower trophic level ecosystem model (Nakata, 1993) One box model of Osaka Bay
Sediment quality model	Sediment quality model (Port Research Institute, 2012) Water-sediment combined model 10-layer sediment model (0.2, 1, 5, 10, 20, 40, 70, 100, 150, 200 cm)
Time step	External mode: 120 s (water quality and ecosystem) Internal mode: 5 s (sediment)
Calculation period	1920.1.1～2020.12.31 (100 years) (pre calculation of 15 years and the steady state was confirmed)
Physical field	Water temperature and salinity are given based on the observed data (one-month average) by the Osaka Prefectural Fisheries Experimental Station
Weather conditions	Observed values (every time) by the Osaka District Weather Observatory are given
Initial and boundary conditions	Initial condition during the winter of 1972 is given based on the observed values Boundary conditions at Akashi and Kitan Straits are given based on the observed values
Sediment conditions	Initial conditions are given based on the manual (Port Research Institute, 2012) Water content is given as the same value
N·P loads	T·N, T·P loads are given based on observations

and sediment quality in the bay are calculated by applying a simple ecosystem model of water and sediment as shown in Fig. 6.13 (Port Research Institute, 2012). The dispersion coefficients at the Akashi Strait, which connects Harima-nada and Osaka Bay, and at the Tomogashima Strait, which connects Osaka Bay and the Kii Channel are based on past study results (Yanagi et al., 1985). Details of this model are shown in Table 6.1.

Year-to-year variations in TN and TP loads are shown in Fig. 6.14, where TN and TP loads are based on their proportion to COD load, the data for which were obtained from the Ministry of Environment after 1978 (The Ministry of Environment, 1981, 1986, 1991, 1992, 2016, 2017) and from the Ministry of Land, Infrastructure, Transport, and

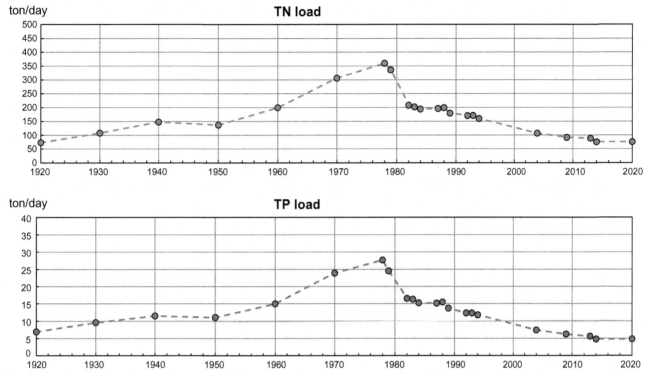

FIG. 6.14 Year-to-year variations in TN and TP loads to Osaka Bay.

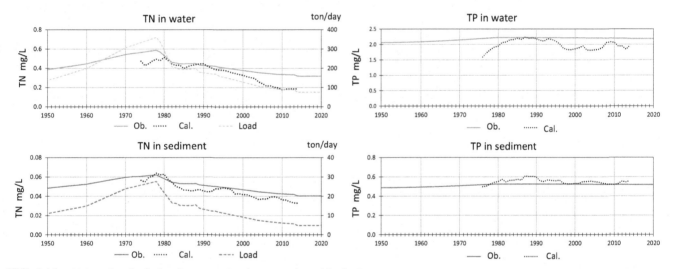

FIG. 6.15 Observed and calculated water and sediment quality of Osaka Bay.

Tourism from 1920 until 1977 (The Ministry of Infrastructure and Transport, 2010). TN and TP loads increased until 1977 and then decreased after 1978.

Observed and calculated TN and TP concentrations in the water column and surface sediment are shown in Fig. 6.15. Year-to-year variations in calculated TN and TP concentrations reproduce the observed TN and TP concentration in the water column of Osaka Bay, which are obtained from the literature (Idea Consulting Company, 2016; Osaka Prefecture, 1971–1982, 1983–2012), though the short-time scale variations have not been calculated.

The hysteresis of year-to-year variations in TN and TP concentrations of the water column and sediment are shown in Fig. 6.16, where the peak values in sediment lags behind those in the water column by about 10 years. The route during the oligtrophication period differs from that during eutrophication period, as shown.

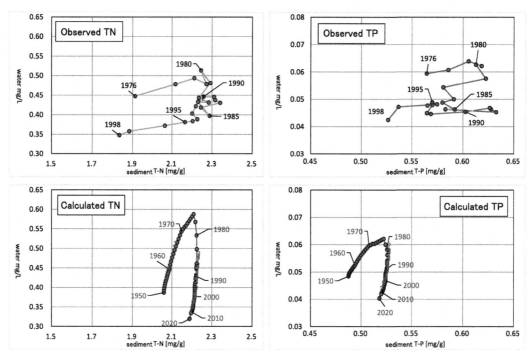

FIG. 6.16   Hysteresis of year-to-year variations in TN and TP concentrations of water column and surface sediment for Osaka Bay.

Such hysteresis is considered to be generated by the following process: The sinking flux of suspended organic matter increases to a volume larger than the release of flux from the bottom, and organic matter accumulates in the bottom during the eutrophication period. Thereafter, the TN and TP concentrations in the surface sediment become largest after time in the water column (largest when the TN and TP load are at their greatest). Suspended organic matter in the sediment decomposes into a dissolved state and moves into the pore water. The released flux of the TN and TP from the bottom to the water column becomes larger during the oligtrophication period than the eutrophication period because it is proportional to the concentration gradient between pore water and bottom water. The TN and TP concentrations in the water column during the oligotrophication period are larger than during the eutrophication period with the same TN and TP loads because the released flux from the bottom is larger during the oligotrophication period because of the accumulated organic matter in the sediment. The time lag between the worst water quality and the worst sediment quality depends on the rate of decomposition of organic matter in the sediment, in other words, a larger decomposition rate results in shorter time lag.

In the near future, TN and TP loads flowing into Osaka Bay from the land will have nearly the same values, and TN and TP concentrations in the water column and sediment will approach some quasi-steady state when the sinking flux and the released flux will have the same value. However, we cannot forecast absolute steady state values now.

Part of this paper already has been published (Kasamo et al., 2016).

6.4 INTEGRATED NUMERICAL MODEL FOR TOYAMA BAY

Xinyu Guo, T. Mano, Katsumi Takayama, and Takafumi Yoshida

Toyama Bay is an opening to the Tsushima Warm Current that flows along its offshore area. Consequently, the bay is under the influence of both the Tsushima Warm Current and rivers and groundwater. For an integrated model of Toyama Bay, we constructed a low-trophic ecosystem model based on a three-dimensional hydrodynamic model. We used this integrated model to examine the response of the nutrients and phytoplankton in the bay to the temporal changes in the nutrient loads from the land.

6.4.1 Hydrodynamic Model for the Toyama Bay

The base of the model is DREAMS_I (Nakada et al., 2014) of the Research Institute for Applied Mechanics, Kyushu University. The model domain is 136.5°E–138.5°E and 36.6°N–38.3°N. The horizontal interval of the grid point is 1/60° in the easterly direction, and 1/75° in the northerly direction. There are 36 vertical layers.

The driving forces (wind stresses, heat flux, freshwater flux) for the model are calculated using the results of the Meteorological Agency GPV–MSM (Meso–Scale Model). The lateral condition on the open boundary uses the results of DREAMS_M, also of the Research Institute for Applied Mechanics, Kyushu University. For the purpose of reproducing seasonal variations, the calculation period, including the spin-up time, was set as March 1, 2006, to December 31, 2007.

According to the surface flow and sea surface temperature (figure not shown) of February, May, August, and October produced by the model calculation, the inflow from the Japan Sea into the Toyama Bay is strong in May and August, and weak in February and November, indicating an apparent seasonal variation. The surface water temperature inside the bay decreases in the winter and increases in the summer. The difference in water temperature between Toyama Bay and the Japan Sea also has a remarkable seasonal variation: large in winter, small in summer.

6.4.2 Low-Trophic Ecosystem Model for Toyama Bay

A low-trophic ecosystem model was constructed for Toyama Bay by combining the NPZD model used in DREAMS_M (Section 4.1) with the hydrodynamic model. Only one type of nutrient (DIN) is used in the original low-trophic ecosystem model. In our study, we added DIP to this NPDZ model to consider the phosphorus limitation near the mouths of rivers and the phosphorus cycle in this system. With two types of nutrients in the model, we can calculate the N/P ratio, and therefore know where there are nitrogen and phosphorus limitations.

We used the output of DREAMS_M for the initial condition and the open boundary condition of the low-trophic ecosystem model for the bay. Because there is no DIP in the output of DREAMS_M, 1/16 of the DIN concentration was given to the bay model as DIP concentration. The calculation period is set in the same way as the hydrodynamic model. The results of calculating nutrient (DIN, DIP), phytoplankton, zooplankton, detritus, and dissolved oxygen (DO) were saved every hour.

Five primary rivers flow into the bay: the Oyabe River, Sho River, Jinzu River, Joganji River, and Kurobe River. Discharges from these rivers are applied to the model from the grid point at each estuary. Daily river discharges are available from the hydrological water quality database provided by the Ministry of Land, Infrastructure, Transport, and Tourism. We determined the nutrient concentrations in the river water by referring to reports by Tsugimoto (2012) and the Toyama Prefecture Fisheries Experimental Station (2007). Their values are shown in Table 6.2.

The discharge of submarine groundwater into Toyama Bay has been reported as about 25% river water, but the loads of nutrients from submarine groundwater is comparable to those of rivers (Hatta et al., 2005). Therefore, in order to accurately reproduce the low-trophic ecosystem of the bay, it is necessary to consider nutrient supply from submarine groundwater. However, observational data on the exact discharge of the submarine groundwater and the spout location are poor. Instead of explicitly including the discharge of the submarine groundwater, our calculation doubled the nutrient concentration of the river water. This method means the addition of the same nutrient loads from the submarine groundwater as the river water, which is consistent with the report by Hatta et al. (2005).

To evaluate the effects of submarine groundwater, we designed two cases for our calculations. Case 1 corresponds to the idea that the concentration of nutrients in the river water was doubled include the contribution of submarine groundwater; Case 2 corresponds to the idea that the source of nutrients from land is only river water. The comparison of the two cases can tell us what changes occur in the low-trophic ecosystem of the bay because of the presence of submarine groundwater.

TABLE 6.2 The Concentrations of DIN and DIP in the Oyabe River, Sho River, Jinzu River, Joganji River, and Kurobe River

	DIN (μM)	DIP (μM)	DIN/DIP
Oyabe River	80.0	5.0	16.0
Sho River	7.8	2.2	3.5
Jinzu River	80.0	2.5	32.0
Joganji River	39.0	2.3	17.0
Kurobe River	7.8	2.2	3.5

6.4.3 Results

Fig. 6.17 shows horizontal distributions of the monthly averaged currents, nutrient (DIN) concentration, N/P ratio, phytoplankton, and zooplankton at a depth of 1 meter in January, April, July, and October 2007, as obtained from the calculation of the model in Case 1. In the surface layer, the nutrient concentration is high in the inner part of Toyama Bay, where water flows from five rivers. Consequently, the phytoplankton and zooplankton also show a high concentration. The offshore range of water with such high nutrient concentration was greatest in July. Phytoplankton reaches a peak concentration over the entire bay in April, followed by a reduction in May and June. It shows a region of high concentration in the inner part of the bay again in July. In general, the phytoplankton has a slightly high concentration in the inner part of the bay. Following the phytoplankton bloom in April, zooplankton has a high concentration over a large area of the bay in April and May. Then, in July, it has a high concentration only in the inner part of the bay.

The N/P ratio was much higher than the Redfield ratio of 16 in the inner part of the bay. This suggests that the growth of phytoplankton is limited by phosphorus. The results of this model show that the high N/P ratio is about 50, which agrees with the results of observations by Tsugimoto (2012), who explained that phosphorus limitation is caused by excessive nitrogen supply from river water.

Fig. 6.18 shows the vertical distribution of nutrient (DIN) concentration, phytoplankton, and zooplankton, from a depth of 0 to 200 m in a section from the mouth of the Jinzu River to 30 km offshore. The high nutrient concentration originating from the river is distributed in the upper layer with a thickness of 10 m from the estuary to 10 to 15 km

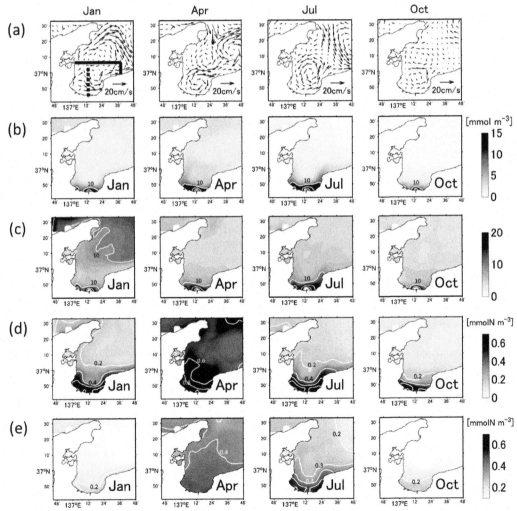

FIG. 6.17   Horizontal distribution of the monthly averaged (a) current velocity, (b) nutrient (DIN) concentration, (c) N/P ratio, (d) phytoplankton, (e) zooplankton, at 1 m depth in January, April, July and October, 2007. The solid line in the left panel (A) shows the range of the Toyama Bay area defined for Figs. 6.19–6.21; the dotted line shows the cross-section position used in Fig. 6.18.

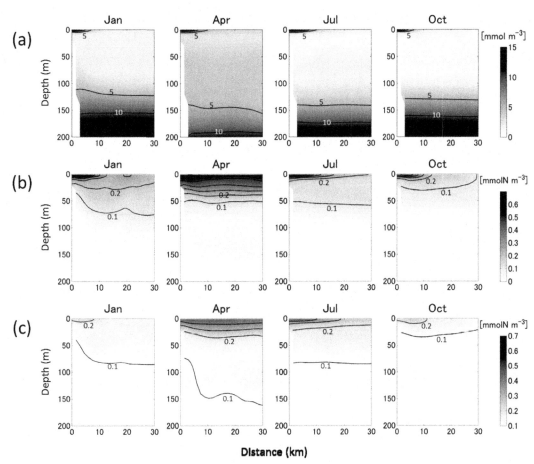

FIG. 6.18 Vertical distribution of the monthly averaged (a) nutrient (DIN) concentration, (b) phytoplankton, (c) zooplankton in January, April, July, and October, 2007. The vertical axis is water depth (m), and the horizontal axis is the distance (km) to the north from the mouth of the Jinzu River. The upper layer from 0 m to 200 m is presented here.

offshore. Although the amount of nutrients supplied from rivers varies seasonally with the river discharge, the rivers continuously supply nutrients to the bay, and consequently there is always a certain amount of phytoplankton near the estuary. From the winter to the following spring, vertical mixing also supplies nutrients from the depths to the upper layer, which is necessary for the phytoplankton bloom in April over the entire bay. In addition, coastal upwelling also supplies nutrients from the lower layer to the upper layer near the coast.

The total amount of DIN, DIP, phytoplankton, and zooplankton in the region of Toyama Bay as defined in Fig. 6.17(a) in the layer shallower than 100 m was calculated for both Case 1 (with submarine groundwater) and Case 2 (without submarine groundwater). A time series with a one-week running mean is shown in Fig. 6.19. Case 1 and Case 2 show almost the same seasonal variation. The amount of nutrients is one order of magnitude larger in DIN than in DIP. The tendency for seasonal variation, however, is almost the same for DIN and DIP. They start to increase from around December, become largest in early April, decrease from the middle of April, and remain at a low level in summer and autumn. There was a temporary increase in the amount of nutrients in July, which is likely because of massive flooding from the river.

The seasonal variation in phytoplankton is almost the same as in nutrients, but the peak time was a little earlier in late March. Zooplankton peaks in early May, which is more than one month behind the peak of phytoplankton. We calculated the ratio of Case 2 to Case 1 for each variable (Fig. 6.20). As a result, the difference, depending on the presence or absence of submarine groundwater, was largest during the summer and autumn (July–December) when the absolute amount of nutrient is small. In the absence of submarine groundwater, DIN decreased by about 19%, DIP by about 11%, phytoplankton by about 17%, and zooplankton by about 11%.

We also calculated the nutrient (DIN) flux in the Toyama Bay area defined in Fig. 6.17(a) by dividing it into loads from rivers and submarine groundwater (hereafter referred to as land origin fluxes), horizontal fluxes, and vertical fluxes along with currents. In this calculation, we changed the water column depth to 20 m, 50 m, 100 m, 150 m,

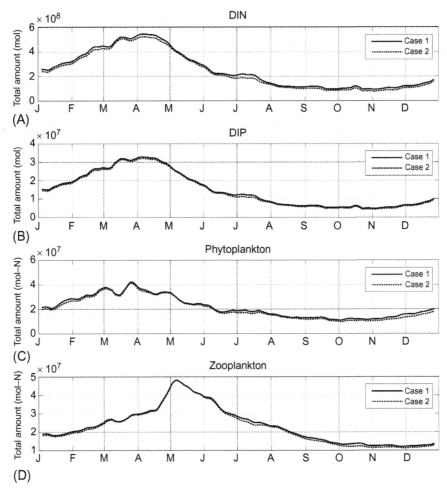

FIG. 6.19 Time series of total amount of (A) DIN, (B) DIP, (C) phytoplankton, (D) zooplankton in 2007 from sea surface to 100 m depth in the Toyama Bay area defined in Fig. 6.18. The solid line is for Case 1 (river water + submarine groundwater); the dotted line is for Case 2 (river water only); both are results of a running mean of one week. Phytoplankton and zooplankton are converted to nitrogen.

and 200 m. As an example, Fig. 6.21 shows the time series of nutrient fluxes in the area with a water depth less than 100 m. Although the results for the water column with the other depths are not shown, we describe their results below.

In areas with a water depth of less than 20 m, the flux originating from the land is the dominant source of nutrients in all seasons. In areas shallower than 50 m, the advection exceeds the flux originating from the land from February to April. This advection is because of vertical flux. In other months, however, the flux originating from the land is still a source of nutrient supply to the area. In areas shallower than 100 m, advection flux increases from October to March, exceeding flux originating from the land—notably 3 to 4 times over the period January to March (Fig. 6.21). From summer to autumn when the advection flux decreases, however, flux originating from the land becomes the main source of nutrients in the area. In areas with a depth less than 150 m, the advection and flux originating from the land are at the same order only in May; in most months, the advection flux greatly exceeds the flux originating from the land of nutrients. In areas with a depth of less than 200 m, the nutrient flux by advection is always larger than flux originating from the land. In the advection of nutrient fluxes, the vertical flux contributes to most nutrient input. The process responsible for this is the upwelling from the lower layer.

From these results, we conclude that the nutrient loads from rivers and submarine groundwater are an important source for the nutrient inventory in the upper 100 m layer in Toyama Bay in summer and autumn. In winter and spring, as well as for the deeper layers, the supply of nutrients from the deep layer increases because of vertical upwelling, and the proportion of nutrient supply of terrestrial origin is relatively small.

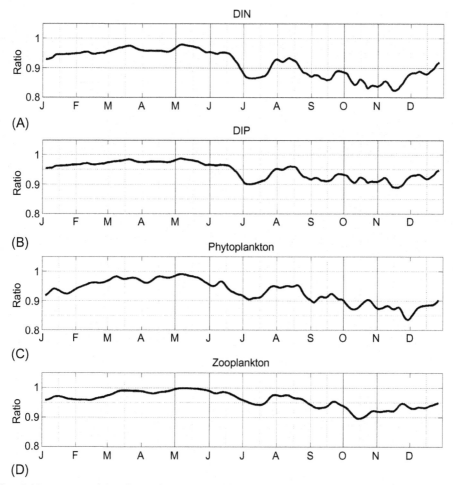

FIG. 6.20 Ratio of Case 2 (river water only) to Case 1 (river water + submarine groundwater) of the total amount of (A) DIN, (B) DIP, (C) phytoplankton, (D) zooplankton in 2007 from sea surface to 100 m depth in the Toyama Bay area. The result here is after a running mean of one week.

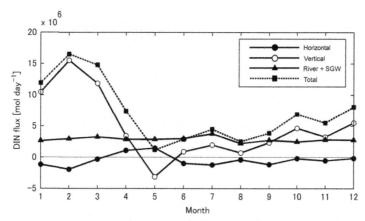

FIG. 6.21 Contents of nutrient (DIN) flux flowing into and out of the area of 0–100 m in the Toyama Bay area. The results are monthly averaged values in 2007. The inflow into the area is defined as a positive value. The solid line with black circles is nutrient flux by horizontal advection; the solid line with white circles is nutrient flux by vertical advection; the solid line with triangles is nutrient flux because of river and submarine groundwater; the dotted line with squares is the sum of all nutrient flux.

6.4.4 Predictions

As introduced in the previous section, nutrients from river waters and submarine groundwater have a remarkable influence on low-trophic ecosystems in Toyama Bay, especially along the coastal area. In the future, it is expected that the usage of groundwater will change because of human activities. In addition, the time when snow melts and the amount of melted water also can be expected to change because of climate change, such as global warming. Such long-term variations will naturally change the seasonal variation and the magnitude of terrestrial nutrient load, which will have a certain impact on the low-trophic ecosystem of Toyama Bay. Therefore, we are planning to clarify the characteristics of the response of the low-trophic ecosystem of the bay to the change in the nutrient load with terrestrial origins using the integrated numerical model for Toyama Bay developed in this study.

References

Fujiwara, Y., Tanakamaru, H., Hata, T., Tada, A., 2003. Error function for the determination of the best constant in tank model. J. Civil Eng. Agric. 225, 137–149.

Hatta, M., Zhang, J., Satake, H., Ishizaka, J., Nakaguchi, Y., 2005. Water mass structure and fresh water fluxes (riverine and SGD's) into Toyama Bay. Chikyukagaku (Geochemistry) 39, 157–164. in Japanese with English abstract.

Idea Consulting Company. 2016. Report of Environmental Investigation in the Seto Inland Sea for Achievement of Productive Coastal Sea.

Kasamo, T., Shiraki, Y., Shibaki, H., Yanagi, T., 2016. Model of anchovy during spring in Osaka Bay. J. Japan Civil Eng. Soc. B2 (Coast. Eng.) 72 (2), 1381–1386.

Komatsu, T., Sasa, S., Montani, S., Yoshimura, C., Fujii, M., Natsuike, M., Nishimura, O., Sakamaki, T., Yanagi, T., 2018. Studies on a coastal environment management method for an open-type bay: the case of Shizukawa Bay in southern Sanriku coast. Coast. Oceanogr. 56 (1), 21–29.

Minami-Sanriku-Town, 2018. Statistics of Minami-Sanriku-Town. 71 p.

Miyazawa, Y., Zhang, R., Guo, X., Tamura, H., Ambe, D., Lee, J.-S., Okuno, A., Yoshinari, H., Setou, T., Komatsu, K., 2009. Water mass variability in the western North Pacific detected in a 15-year eddy resolving ocean reanalysis. J. Oceanogr. 65, 737–756.

Nakada, S., Hirose, H., Senjyu, T., Fukudome, K., Tsuji, T., Okei, N., 2014. Operational ocean prediction experiments for smart coastal fishing. Prog. Oceanogr. 121, 125–140.

Nakata, K., 1993. Estimation of parameters in ecosystem model. J. Adv. Mar. Tech. Conf. 8, 99–138.

Osaka Prefecture. 1971–1982. Report of Environmental Problems in Osaka.

Osaka Prefecture. 1983–2012. Report of Environmental Problems in Osaka.

Port Research Institute. 2012. Manual for Ise Bay Simulator Version 1.2.

Takeoka, H., Murao, H., 1997. Response of water quality to the change of nitrogen and phosphorus loads. Coast. Oceanogr. 34 (2), 183–190 (in Japanese).

Taylor, K., 2001. Summarizing multiple aspects of model performance in a single diagram. J. Geophys. Res. 106, 7183–7192.

The Ministry of Environment. 1981. Estimation of Nitrogen and Phosphorus Loads in 1980.

The Ministry of Environment. 1986. Estimation of Nitrogen and Phosphorus Loads in 1985.

The Ministry of Environment. 1991. Estimation of Nitrogen and Phosphorus Loads in 1990.

The Ministry of Environment. 1992. Estimation of Nitrogen and Phosphorus Loads in 1991.

The Ministry of Environment. 2016. Estimation of Nitrogen and Phosphorus Loads to the Seto Inland Sea in 2015.

The Ministry of Environment. 2017. Estimation of Nitrogen and Phosphorus Loads to the Seto Inland Sea in 2016.

The Ministry of Infrastructure and Transport. 2010. Ecosystem Service of Ports in Japan in 2009.

Toyama Prefecture Fisheries Experimental Station. 2007. Fisheries Ground in Toyama Bay; Water Quality, Bottom Sediment Quality, Seaweed and Food. Report on Toyama Bay Fisheries Environmental Survey in 2006 (in Japanese).

Tsugimoto, R., 2012. Seasonal variation in nutrient concentration and phytoplankton biomass in the head of Toyama Bay. Coast. Oceanogr. 49 (2), 127–137 (in Japanese with English abstract).

Yamamoto, H., Yoshiki, K., Komatsu, T., Sasa, S., Hamana, M., Murata, Y., Yanagi, T., 2017a. Comparison numerical model experiments on the carrying capacity for oyster culture in Shizukawa Bay; before the earthquake and after the earthquake. J. Japan Civil Eng. Soc. B2 (Coast. Eng.), 73 (2), 1339–1344.

Yamamoto, H., Yoshiki, K., Komatsu, T., Sasa, S., Yanagi, T., 2017b. The best aquaculture method in Shizukawa Bay: from the results of integrated numerical model calculation. J. Japan Civil Eng. Soc. B2 (Coast. Eng.) 73–74.

Yanagi, T., 2015. Present situation and future in water environment under the oligotrophication. Environ. Technol. 44 (3), 147–153 (in Japanese).

Yanagi, T., 2018. Outline of the special project "development of management for the sustainable coastal sea". Coast. Oceanogr. 56 (1), 3–12. (in Japanese with English abstract).

Yanagi, T., Shibaki, H., Takeoka, H., 1985. Budgets of salt, nitrogen and phosphorus in Harima-Nada and Osaka Bay. Coast. Oceanogr. 22 (2), 159–164.

Further Reading

The Ministry of Land, Infrastructure, Transport and Tourism. 2010. Report of Ecosystem Services by Port.

7

What Can We Learn From Satoumi to Guide International Ocean Policies?

Yoshitaka Ota* and Wilf Swartz†

*School of Marine and Environmental Affairs, University of Washington and Nippon Foundation Nereus Program
†Institute for Oceans and Fisheries, University of British Columbia, Vancouver, BC, Canada

Satoumi is a Japanese concept describing a mosaic of marine ecosystems and coastal human communities. As a model for coastal and fisheries management that includes the integration of traditional ecological knowledge, Satoumi recently has garnered interest in international ocean policy discourse. The idea of Satoumi, however, cannot be fully appreciated under the existing framework of marine conservation initiatives, which traditionally have focused primarily on ecological attributes and states. Moreover, while some have suggested that Satoumi is a win-win approach that aligns fishing objectives with conservation ideals, this perspective fails to capture the diverse and complex relationships among fishers, their communities, and the surrounding marine environment that underpins the seascape encapsulated by Satoumi.

This contribution examines the discrepancies between Satoumi and internationally predominant approaches to ocean management. It identifies locality and relationships between fishers and marine environment as the critical attributes of Satoumi. By defining Satoumi as seascape for coastal livelihood, in contrast to the concept of socioecological system, it discusses what Satoumi can contribute to the international ocean policy discourses.

7.1 SATOUMI AMONG INTERNATIONAL OCEAN GOVERNANCE

It is not suitable to view Satoumi simply as an ocean management strategy, given that the three commonly recognized models of marine management do not adequately apply to the framework of Satoumi. Although it might be inappropriate to generalize, emerging trends in marine management over past several decades can be described as: Spatial management that ascribes ocean in three-dimensional space to be allocated to different user groups; ocean governance approach that emphasizes stakeholder involvement and self-regulation; and socioecological system perspective that considers the environment as an integration of social (human) and natural systems. Various interpretations of these three models can be prescribed to the policies promoted by the marine conservation groups in North America and Europe, fisheries management programs in development, particularly as co-management and other community-based management programs, and in the sustainable development goals promoted by the international community

Integrated Coastal Management in the Japanese Satoumi
https://doi.org/10.1016/B978-0-12-813060-5.00007-9

© 2019 Elsevier Inc. All rights reserved.

With respect to the marine conservation model, Satoumi is described in the context of the ecosystem-based approach (Kakuma et al., 2018). An ecosystem-based approach to resource management, as opposed to the traditional management approach that considers a single species or stock in isolation, recognizes the full array of interactions within an ecosystem that aims to manage fisheries with considerations for impacts beyond the targeted stock while integrating trophic dynamics in assessment of stocks. At the same time, an ecosystem-based approach shifts our perception of fisheries from a singular object of resource management to a broader potential driver of ecosystem change and biodiversity loss. The ecosystem-based approach inevitably requires management to extend into regulation of ocean space, including allocation of areas explicitly for conservation of marine ecosystem integrity and essential habitats. In extreme cases, spatial protection is viewed as the only appropriate means for conservation of marine ecosystems. Thus, the increased popularity of ecosystem-based approach leads to the emergence of spatial management (point 1).

To some extent, the ecosystem-based approach also spurred the emergence of ocean governance. The ecosystem-based approach and spatial protection that the approach inspired is the most common area where marine conservation objectives and fisheries objectives come into conflict. As marine protected areas gain support from international communities and enshrined into global ocean management commitments and beyond (e.g., UNFCCC, SDGs), the size-based targets that these commitments demand have minimized the function of marine protected areas as conservation of endemic or sensitive areas. Rather, the emphasis on the total areas protected as a proportion of national waters (i.e., exclusive economic zones) has driven implementation of these areas based on quantity over quality (Peter, 2014). This drive for bigger MPAs led to a wave of very large MPAs (e.g., Papahānaumokuākea Marine National Monument in 2006 by US President George W. Bush) as a political performance, enshrined as a national monument rather than a component of broader ocean management strategy. Implementation of similar very large MPAs in Palau, the Seychelles, and Kiribati, often in partnership with US environmental NGOS, are now promoted as a marine management goal in itself.

These measures, however, often are incompatible with the needs of local stakeholders or with scientific understanding of local marine ecosystems. The Chagos Marine Protected Areas and Rapa Nui Marine Park, for example, have been criticized as ocean grabbing (Bennett et al., 2017) in spite of the local consultations, and when viewed in the context of social seascape, exclusions or restrictions of local fisheries can lead to further marginalization of these indigenous and minority communities. Moreover, although some studies suggest correlations between management benefits and sizes of protected areas (e.g., management cost efficiency, McCrea-Strub et al., 2011), others have indicated that such benefits might be marginal if social costs to affected coastal communities are to be considered (Peter, 2014). With the increased recognition of such social impacts (and their variability from one case to another), there are heightened demands for better governance framework that explicitly consider the social impacts even among the conservation NGOs promoting spatial protection (Christie et al., 2017).

The theory of governance that currently prevails in marine conservation and fisheries management policies has been the principal of self-governance based on economic optimization. Carlisle and Grudy, for example, considers community leadership as the central function of the co-management governance, and therefore the management is sustained by the community governance structure. (Carlisle and Gruby, n.d.) Nevertheless, proponents of self-governance in ocean management fail to address the broader concerns of accountability, transparency, and legitimacy, and remain constrained to resource management rather governance of socioenvironmental communities (Schlager and Ostrom, 1999; Scott, 1993). It is critical that analyses of fisheries governance consider the process of decision-making, particularly: where and how accountability for management is ensured; how are transparency for policy decisions about resource allocation and access rights secured; and how are consensus among stakeholders achieved with adequate consideration heterogeneity among them.

Given that the current discourse about self-governance and fisheries management continue to view economic optimization as the outcome of environmental conservation and resource management, discussion falls short of becoming more comprehensive about the common and social impacts. The case of territorial use right fisheries (TURF) in Chile, for example, is put forward routinely as the showcase of self-governance; yet, some scholars argue that this model of collective action might be unsuitable in addressing local social issues such as aging populations or shortage of successors (Tam et al., 2018).

7.2 PARALLELS OF SATOUMI AND OCEAN MANAGEMENT

Before expanding on the third emerging attribute of ocean management—socioecological systems perspective—let us consider the parallels between Satoumi the previously discussed attributes. First, spatial management can be viewed as a derivative of the scale-driven ecosystem-based approach promoted under Satoumi as a smooth material cycle of nutrients and other ocean properties. Given Satoumi's emphasis on ecosystem linkages, species- or stock-based approaches of more conventional fisheries management would be inadequate in capturing the nature of Satoumi. Kakuma (2018), for example, argues that Satoumi as a concept based on multifaceted interactions between human communities and the sea must be perceived as a comprehensive spatial management system that fully integrates regulations of demersal and pelagic systems. Kakuma (2018) goes on to describes MPAs as a tool for Satoumi. Unlike the politically motivated implementations of very large MPAs or ocean grabbing, however, the delineation of ocean space under Satoumi is not driven by the need for differentiation based on user purposes but by the zoning driven by fishers and other stakeholders to manage and conserve ocean resources and environment. Therefore, there are fundamental differences in both objectives and approaches between spatial management and Satoumi.

Attempts to characterize Satoumi as a model of co-governance also are inappropriate because they fail to fully appreciate the historic context and social dynamics involved in the decision-making process of Satoumi. Specifically, unlike the self-governance framework, which limits scope to collective decision-making by the stakeholder community, the scope of Satoumi extends beyond ocean policy making, as a greater, regional governance framework. If Satoumi is to be considered as a social system rather than spatial system, however, it is possible to describe it without mirroring the existing coastal management approach. For example, unlike protected areas or fishing access management outside of Japan, Satoumi is based explicitly on information sharing among social actors, both administrative and economic. Through a system of information sharing, Satoumi facilitates management consensus. Oceans are freed from the unitization as production system, and mosaics of management units are removed. The case in Okayama, for example, restoration of seagrass was implemented by local fishers, however, other stakeholders such as recreational fishers and local school students were integrated into the management system as new stakeholders, allowing continued support from local public administrative bodies (Yanagi, 2018). This network of local actors differs from the role of leadership as described by the self-governance theory, in that they are not organized hierarchically and the emphasis is on local coordination. Moreover, such structure acts to moderate the likelihood of concentration of power to certain stakeholders in the distribution of resource access (Tsurita, 2015). This cross-sectoral participation of various actors in Satoumi is reflected in the coastal management in Japan, involving Fisheries Cooperatives, local administration, and academia.

From the perspective of ocean governance, discussions about the roles of social capital in formation and maintenance of Satoumi have been limited. In Japanese coastal regions, social capital is not viewed as a mechanism for regional social integration or regional norms, but as social intervention in a natural system to enhance its productivity. Here, social intervention primarily denotes habitat and ecosystem restoration by fishers and represents the local network and support for fishers in the community. Tanaka (2018), for example, argues that the system of localized fishing rights based on historical context form the social context of Satoumi, and provide the central role for fishers. More broadly, the belief that fishers must be the stewards and protectors of the oceans that define their role and, thus objectives of local coastal governance and marine conservation efforts. Administratively speaking, Japan's Fisheries Act was the institutionalization of the historic system of localized fisheries governance and provided rights to local coastal communities. Collective fishing rights that are allocated to fishers and fisheries cooperatives are responsible for managing their adjacent waters. Satoumi, therefore, can be considered to be a system of social contract between fishers and local communities founded upon the awareness of complex interactions between the ocean and humans. Nevertheless, Satoumi is not a concept that is universally recognized or defined, but rather a concept that is applied dynamically on a case-by-case basis. Ocean management that is described as Satoumi does not implement a specific set of measures, but rather, it is an academically defined cultural category retroactively applied to local management systems. Cultural categories are founded upon cross-disciplinary frameworks integrating both natural and social sciences. Although Satoumi's key functions are described as productivity enhancement, conservation of environment, promotion of communication, and cultural succession, it is not necessary for all these functions to be identified explicitly in a system described as Satoumi. Cultural succession, for example, is not an explicit objective of coastal management, and its function is not generally discussed.

The report about Satoumi published by the Conservation on Biological Diversity (CBD) (United Nations University Institute of Advanced Studies Operating Unit Ishikawa & Kanazawa, 2011) argues that the cultural aspect of Satoumi makes it an effective model for integration of traditional knowledge. This view is supported by Crosby (Crosbby, 2018) in his assessment of the locally managed marine areas (LMMA) in Fiji, and stems from the differences in the attitudes about environment and ecosystems between North America and Europe and Asia. Although Crosby stressed this philosophical perspective about the interactions and co-existence between people and environment, in his assessment of Satoumi, in practice, it is clearly driven by the idea of benefit optimization through management of environment. Under the Satoumi model, conservation is a means to improve fisheries productivity and mobilize social networks for a conservation effort with the understanding that such effort supports fisheries. If the function of Satoumi as cultural succession is to be strengthened, the argument that ocean management not only benefits economically but can sustain the fisheries culture of coastal communities and that its fishing rights regime can sustain social network of the fisheries communities must be emphasized.

In conclusion, the concept of Satoumi differs from marine spatial planning or ocean management based on self-governance in its function, historical context, and social structure. This is because Satoumi is founded upon the social contracts under which fisheries and fishers operate differs from more abstract principles of conservation or marine governance that underpins the latter programs. Fishing grounds in the context of Satoumi are not ocean space or use but it is a function, with rights and responsibilities, that is embedded in sato or community. Moreover, fisheries are identified as both economic and social activity under the governance framework of Satoumi, and as such, Satoumi contains core governance functions such as accountability, transparency, and legitimacy.

If Satoumi is to serve a cultural function, then the culture that is referred to here is the culture of fisheries that is different from the culture of conservation that underpins marine protection and other ocean management programs. Satoumi is a culture of the coastal communities in which the community manages ocean resources as a foundation of the community livelihood. Under such culture, the ocean is not perceived as a natural system distinct from the community, but one embedded in the community. (Yanagi, 2018) recalls the criticisms from foreign colleagues to his position about Satoumi as being in the service of fishers. He argues against such comments by stressing Satoumi's relevance in the context of more mainstream management programs such as ecosystem-based approaches, integrated coastal management, and marine spatial planning. It is possible that Satoumi can serve both marine conservation and fisheries management and sustainability objectives, and that Satoumi can contribute to international ocean governance discussions and analyses. If Satoumi is to be considered a form of social contract that transcends conservation and fisheries aspirations, however, it is critical that further discussions must focus on oceans as an integral part of the coastal community, rather than environment that the coastal community occupies and uses.

7.3 SATO AND SOCIAL ECOLOGICAL SYSTEM

It is clear that sato is the central concept in perceiving Satoumi as a cultural system. At the same time, interpreting Satoumi from the cultural lens rather than as spatial management or self-governance lens highlights its dual characteristics defined by sato (society) and oceans (nature). Yet, the concept of sato is vague, and its interpretation from anthropological or cultural studies perspectives is limited. Hirakawa (2003), for example, has traced sato as an administrative unit to the eighth century, and expands its definition to include production and religious unit. Hirakawa (2003) also views sato as a unit of animistic worship, with a shrine dedicated to a local deity. In addition to administrative and religious interpretation, it also is defined as a location of social interactions for people living off the land.

Such historical concepts of sato evolved in the 1990s to include the concept of natural environment, as a keyword for Japan's conservation objectives. More specifically, satoyama began to circulate internationally as a Japanese model for conservation, landscape, and biodiversity. Under this revised interpretation, sato became the symbol of co-existence between human communities and their livelihood and natural environment. From this expanded interpretation of sato, the term Satoumi emerged as coastal regions where productivity and biodiversity increased through human involvement (Nakamura and Honda, 2010). Since its origin, therefore, Satoumi has encompassed both fisheries and marine conservation as its objective.

More recently, this dual nature of Satoumi has been integrated in the field of sustainability studies and in the concept of social ecological system that balances historical and cultural aspects of societies with environmental aspects such as biodiversity and biological productivity. Even in discussions about Satoumi, social ecological system has become a keyword to describe its previously ambiguous concept of community-supported ecosystem.

The concept of social ecological system was proposed by Elinor Ostrom, a key proponent of self-governance of environmental systems. According to Ostrom (2007), social ecological system denotes linkages between social and natural systems, encapsulating the complexity and adaptability of such linkages. This concept has been interpreted to include much broader concept of vulnerability, resilience, and sustainability, particularly in relationship to global environmental change, such as climate change. In spite of its wide use, criticisms include absence of linkage between social and natural systems, or lack of considerations for political or cultural aspects of social systems. Others view a social ecological system as merely an expansion of the system boundary and failed to analyze the relationship between the two systems. In the context of ocean management, Stojanovic et al. (2016) noted that although most models of social ecological systems consider cultural identity as its key index, most fail to adequately assess the cultural implications of environmental and social changes.

As interpretations of social and natural systems evolve and the boundaries expand, it is critical that how and by whom these new boundaries are defined and what groups representing society are made transparent. For example, some scholars highlight traditional ecological knowledge as a system by which indigenous communities understand and adapt to their environments (Folke et al., 2003). A more holistic system can emerge through integration of nature and culture, society and environment, or science and briefs. Traditional ecological knowledge is evolving through internal diversity or social relationships (Ingold, 2002) and is not a product of a community's interaction with environment. Traditional ecological knowledge is defined through reinterpretation of culture and history, and can be distorted by political, social, and cultural objectives of the time. In the case of ocean management, how traditional ecological knowledge is defined often has economic incentives of influencing resource distribution.

For Satoumi, the actors involved are fishers, administrators, researchers, and coastal communities, and their dynamics continue to evolve. At its core, however, the objective of enhancing fisheries productivity through active human engagement is transparent and consistent. Satoumi is not constrained by social ecological systems considerations such as spatial management or governance, but simply is an interaction between a community and its ocean environment and is therefore a cultural system. By interpreting Satoumi through spatial management or self-governance framework, it is perceived as an administrative mechanism rather than multidirectional relationships between sato (society) and umi (ocean). Consequently, the social ecological system fails to fully appreciate Satoumi, particularly in its interpretation of sato as a cultural concept. Social ecological systems retain the hierarchical relationship between society and its ecosystem, whereas Satoumi does not differentiate between the two.

7.4 SATOUMI AS ECOLOGY OF LIVABILITY

This contribution examined Satoumi in the context of key developments in international ocean governance and critiqued attempts to frame Satoumi as a model for spatial planning, self-governance, or social ecological system. Specifically, it examined the shortcomings of spatial planning (e.g., ocean grabbing and marketization of marine space) and ocean governance (e.g., limited considerations of social diversity and power dynamics) and cautioned against applying such frameworks to interpret Satoumi. The trend to view Satoumi as a form of social ecological system blurs its core objective, which is to enhance natural productivity through active human interventions. This contribution, therefore, argues for the need to view Satoumi through the cultural lens, as a focus of human activities instead of analyzing Satoumi as an ocean system and to analyze it using social science tools to fully understand it from the political ecology lens.

I conclude by posing the concept of ecology of livability as a mean for interpreting Satoumi. If sato is to be the focal point of human activities, then to connect ecology of ocean ecosystems with livelihood of coastal communities through explicit expression of agencies, power relations, and discourse would enable more complete understanding of the social aspects that are increasingly emphasized in international ocean governance discussions.

References

Bennett, N.J., et al., 2017. An appeal for a code of conduct for marine conservation. Mar. Policy 81, 411–418.

Carlisle, K., and Gruby, R.L. n.d. From Community-Based to Polycentric: Fishery Governance in the Northern Reef of Palau. In Press. Environmental Policy and Governance.

Christie, P., et al., 2017. Why people matter in ocean governance: incorporating human dimensions into large-scale marine protected areas. Mar. Policy 84, 273–284.

M. Crosbby (2018) Effects of Satoumi concept to the world—symbiosis of mankind and marine ecosystem. In. "Satoumi-Gaku" ed. by S. Kakuma, T. Yanagi, and T. Sato, Bensei-Shuppan, 30-48, (In Japanese).

Folke, C., Colding, J., Berkes, F., 2003. Synthesis: building resilience and adaptive capacity in social-ecological systems. In: Navigating Social-Ecological Systems: Building Resilience of Complexity and Change.pp. 352–387.

Hirakawa, M., 2003. Sato and village in the ancient time—materials and analysis. In: Research Report of National Museum for Estonic Culture.Vol. 108. (in Japanese).

Ingold, T., 2002. The Perception of the Environment: Essays on Livelihood, Dwelling and Skill. Routledge.

S. Kakuma (2018) What is Satoumi? In. "Satoumi-Gaku" ed. By S. Kakuma, T. Yanagi, and T. Sato, Bensei-Shuppan, pp. 9-29, (In Japanese).

Kakuma, S., Yanagi, T., Sato, T. (Eds.), 2018. Satoumi-Gaku. Bensei-Shuppan, Tokyo (in Japanese).

McCrea-Strub, A., et al., 2011. Understanding the cost of establishing marine protected areas. Mar. Policy 35 (1), 1–9.

Nakamura, T., Honda, H., 2010. Transition of satoyama and satoumi concepts. In: Research Report of Chiba Biodiversity Research Center.Vol. 2, pp. 13–20.

Ostrom, E.. 2007. Sustainable Social-Ecological Systems: An Impossibility?.

Peter, J.S., 2014. Governing Marine Protected Areas: Resilience Through Diversity. Routledge.

Schlager, E., Ostrom, E., 1999. Property rights regimes and coastal fisheries: an empirical analysis. In: Polycentric Governance and Development: Readings From the Workshop in Political Theory and Policy Analysis. Vol. 87. University of Michigan Press, Ann Arbor.

Scott, A., 1993. Obstacles to fishery self-government. Mar. Resour. Econ. 8 (3), 187–199.

Stojanovic, T., et al., 2016. The social aspect of social-ecological systems: a critique of analytical frameworks and findings from a multisite study of coastal sustainability. Ecol. Soc. 21(3).

Tam, J., et al., 2018. Gone fishing? Intergenerational cultural shifts can undermine common property co-managed fisheries. Mar. Policy 90, 1–5.

T. Tanaka (2018) Satoumi and Fisheries Right. Aqua-Net, 21–51, (in Japanese).

I. Tsurita. 2015. Marine Conservation Activities by Sea-Grass Rehabilitation in Hinase. Doctoral Thesis in University of Tokyo, (in Japanese).

United Nations University Institute of Advanced Studies Operating Unit Ishikawa & Kanazawa. 2011. Biological and Cultural Diversity in Coastal Communities, Exploring the Potential of Satoumi for Implementing the Ecosystem Approach in the Japanese Archipelago. Secretariat of the Convention on Biological Diversity, Montreal, Technical Series No. 61.

Yanagi, T., 2006. Satoumi. Koseisha-Koseikaku, Tokyo, p. 102.

T. Yanagi (2018) History of Satoumi creation. In. "Satoumi-Gaku" ed. by S. Kakuma, T. Yanagi, and T. Sato, Bensei-Shuppan, pp. 48-71, (In Japanese).

8

Conclusions

Tetsuo Yanagi* and Kenichi Nakagami[†]

*International EMECS Center, Kobe, Japan
[†]Professor Emeritus, Policy Science, Ritsumeikan University, Ibaraki, Osaka, Japan

8.1 PROSPEROUS AND SUSTAINABLE COASTAL SEA

Tetsuo Yanagi

In the S-13 project, prosperous means that the people in a coastal region who use Satoumi can thrive and pass on a good life to their children and grandchildren. In Theme 2, the cleanness of Shizukawa Bay is expressed by dissolved oxygen (DO) concentration in the bottom layer, and the productivity is expressed by dissolved inorganic nitrogen (DIN) concentration in the upper layer, which determines the phytoplankton growth rate. To express the prosperity, the catch, and income of aquaculture, the population trend and the activity of fishery industries are used for their indexes. The example of expression of indexes for cleanness, productivity, prosperity, and sustainability for Shizukawa Bay is shown in Fig. 8.1. The change in indexes before and after the great tsunami (2007 and 2014, respectively) and the future forecast for 2020 are shown in this figure, which helps us understand the possibility of sustainability. The members of committee discussed the possibility of sustainability of local society of Shizukawa based on this figure. If they find any problems related to sustainability, they have to take some countermeasures to sustain their society. They have to solve the population and industry problems on land area of Shizukawa Bay in the future to achieve sustainability.

8.2 INTEGRATED COASTAL MANAGEMENT METHOD TO REALIZE THE SUSTAINABLE COASTAL SEA

The proposed coastal management method for achieving Satoumi by the S-13 project is shown in Fig. 8.2.

We cannot directly manage the environmental indexes such as transparency, TN and TP concentrations, DO, average TL of caught fish, and so on because they are decided by the result of material cycling in the ecosystem of

© 2019 Elsevier Inc. All rights reserved.

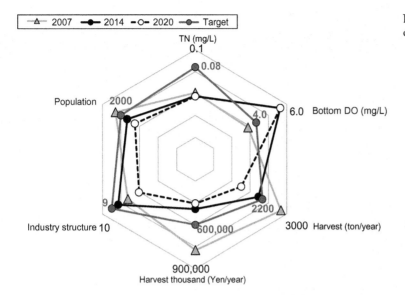

FIG. 8.1 Material for judging clean, productive, prosperous and sustainable Shizukawa Bay.

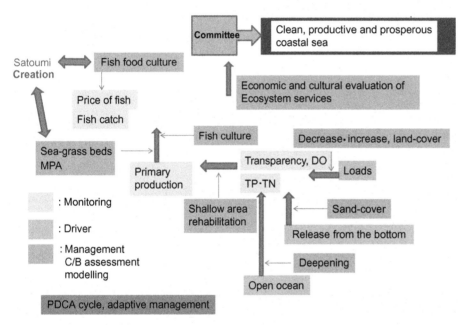

FIG. 8.2 Management method for achieving clean, productive, prosperous and sustainable coastal sea.

target coastal sea. We already understand the process by which such indexes are changed, that is, TP and/or TN concentrations mainly change by the effects of loads from land, bottom release, and environmental fluctuation in the open sea. We also understand the qualitative effects of mankind activities. We quantitatively forecast the necessary activities that change the environmental indexes using the integrated numerical ecosystem model and decide the suitability of such activity after the C/B (cost to benefit) calculation. Of course, such activity will not always succeed, and we might have to carry out adaptive management applying the PDCA (plan, do, check, and action) cycle.

This process can be applied not only to water quality, but also to marine habitat, aquaculture, fish food expansion, and so on. The results of our proposals are assessed by integrated numerical model experiments and then presented to the committee, including stakeholders such as fishermen, navigator, tourism, NPO, and officers to obtain the agreement for the proposed activity.

The S-13 project was carried out as one of the SIMSEA (Sustainability Initiative in the Marginal Seas of South and East Asia) projects by International Council for Science, Regional Office for Asia and the Pacific, under the umbrella of Future Earth project. The aim of Future Earth is to establish the cooperative design of future coastal areas with the collaboration of all stakeholders and scientists in those areas.

We propose the necessary policy plan for achieving the sustainable coastal area in the Seto Inland Sea, Shizukawa Bay, and the coastal areas of the Japan Sea based on our (scientists) close discussion among fishermen, other stakeholders, and government officers, as shown in the next section.

8.3 ENVIRONMENTAL POLICY PROPOSALS FOR THE INTEGRATED MANAGEMENT OF COASTAL AREAS

Kenichi Nakagami and Tetsuo Yanagi

The Headquarters for Ocean Policy was established in order to promote marine-related policy in a concentrated and comprehensive manner as legally established in the Basic Act on Ocean Policy (2007). To date, it has facilitated vigorous discussions that have produced tangible results, such as the Basic Plan on Ocean Policy (2013) and the Basic Policy for Conserving Isolated, Inhabited Border Island Regions and Maintaining Communities in Designated Inhabited Border Island Regions (2016). Safeguarding marine areas has become a particularly pressing concern in recent years, and the situation requires more effective policy implementation. Over the last few decades, fishing industry activity, particularly by fishing industry participants active in more peaceful coastal areas, has seen abrupt changes such as dwindling catches and a rapid decline in fishing industry workers. The reality is that the sustainable management of coastal areas has become difficult.

Rather than viewing coastal areas simply as fishing grounds, we need to focus on the ecological services provided by coastal environments and position them as the common property of the Japanese people, and pursue management techniques based on the value they provide. Therefore, we need to develop a deeper understanding of the ecological services provided by coastal environments and a regional revitalization policy for handling them. We also need to establish sustainable, integrated methods for managing the three elements that are needed to realize Satoumi community coasts—cleanliness, abundance, and vibrancy—which are focused on the ecological services provided by coastal environments.

The Strategic Plan for Biodiversity 2011–20 and the Aichi Biodiversity Targets that were adopted at the COP 10 Convention on Biological Diversity that was held in Nagoya in October 2010 contained new global targets that went into effect in 2011. These highlight the importance of conserving coastal environments as evidenced by Strategic Goal B: "Reduce the direct pressures on biodiversity and promote sustainable use," which includes target six, calling for sustainable management of all fish species to avoid overfishing and realize recovery, and Strategic Goal D: "Enhance the benefits to all from ecosystem services," which includes target 14, calling for the restoration and safeguarding of ecosystems that provide essential services for people. The 65th session of the UN General Assembly held in December 2010 looked to contribute to the achievement of the Aichi targets by naming 2011 to 2020 as the UN Decade on Biodiversity, an important period in which all sectors within the international community should work together to tackle biodiversity issues. The development of sustainable management for coastal areas is not just a domestic environmental policy issue, but also an international commitment. In this section, we propose future coastal environmental management policy measures based on major research results produced by S-13 projects.

8.3.1 Current Situation and Issues Regarding Coastal Environmental Management Policy

The various aspects of modern coastal regions can be organized into several elements.

Lifestyle and culture: Significant changes in the relationship between people and coastal areas in recent years.

Usage and development: In addition to the renewed recognition of the importance of the resources created by Satoumi in revitalizing regions that has accompanied the advancement of marine produce industries that use coastal areas, a new focus has been placed on the connection between satoyama community forests and the sea.

Management and technology: Because climate change continues to exert an effect on coastal areas, it has called attention to the importance of these areas for preventing and mitigating disasters and creating demand for their sustainable use.

Environment and conservation: It is important to practice integrated environmental conservation that includes forests, agricultural land, rivers, and seas.

In the 21st century, however, the elements that make up coastal areas are becoming more complex. To realize the sustainable management of coastal areas, rather than treating each coastal area as a venue for an individual marine product industry, we need to recognize them as integrated coastal management areas.

8.3.2 History and Issues Regarding Coastal Management

The McAteer-Petris Act, enacted in 1965 following strong pressure from people looking to save San Francisco Bay in California, played an important role in protecting the bay from previously unchecked land reclamation. The San Francisco Bay Conservation and Development Commission (BCDC) was established on September 17,

1965, as a provisional state legislative commission tasked with preparing a plan for the long-term use of San Francisco Bay, and in August 1969, an amendment established the commission as a permanent institution that contributes to policy in the San Francisco Bay Plan, which is part of California State legislature. San Francisco Bay has been protected by its local communities for more than 100 years, and the methods of coastal management seen in the BCDC's San Francisco Bay Plan have become models for coastal regions around the world.

The Coastal Zone Management Act (CZMA) is a piece of US federal legislation enacted in 1972 that clarifies coastal management as the responsibility of individual states and sets out procedures through which regulatory bodies established by a state to undertake the conservation and management of coastal areas can receive grants from the federal government. The goal of the CZMA is to encourage and support states to take responsibility for their own coastal regions effectively, as well as encourage coordination and cooperation between relevant federal, state, and local governments. CZMA policy includes preserving, protecting, developing, and, if possible, restoring and improving US coastal areas and coastal resources for current and future generations. It uses a voluntary system in which the federal government supports coastal states that decide to join the CZMA program. Through CZMA, US Congress delegates responsibility for the planning and management of coastal areas to the relevant coastal state. The decisions about the coastal management are also left to the state's discretion. In this way, as indicated by federal legislation, US government is acknowledging that the management of coastal areas is dependent on the regional characteristics of the areas themselves and buy-in from the local communities that have been conserving them for many years.

8.3.3 International Trends in Integrated Coastal Area Management

Integrated coastal area management includes measures that are being carried out by government agencies in Europe and North America, and academic and policy-related research that is being carried out by research institutions.

(1) Government Agencies

   1. United States

The Coastal Zone Management Act Performance Measurement System.

The Coastal Zone Management Act Performance Measurement System (CZMAPMS) measures the progress made on coastal area management plans and the National Estuarine Research Reserve System. To assess the success of Integrated Coastal Zone Management (ICZM), the National Oceanic and Atmospheric Administration (NOAA) established indicators with social, economic, and environmental perspectives (NOAA, 2010) through the NOAA Office of Ocean and Coastal Resource Management (OCRM) (National Oceanic and Atmospheric Administration (NOAA), 2010).

2. European Union

The Sustainable Tailored Integrated Care for Older People in Europe (SUSTAIN) Project provides partial grants through the European Regional Development Fund. The goal of this project is to "create a fully implementable policy tool to help coastal authorities and communities throughout Europe to deliver sustainability on Europe's coast." This tool applies to all 22 coastal states of the European Union and is based on a set of easily measurable sustainability indicators: governance (5), economy (4), environmental quality (8), and social well-being (5) (SUSTAIN, 2012). In 2008, the EU established the PEGASO Project to develop ICZM for the Mediterranean and the Black Sea. Its main goal is to "construct a shared ICZM Governance Platform with scientists, users, and decisionmakers linked with new models of governance." (PEGASO, 2008).

(2) Research Institutes

   1. MEDCOAST (Mediterranean Coastal Foundation)

Established in Turkey in 1993, MEDCOAST has continued as a research organization. Its goal is to conserve the sea and coasts of the Mediterranean and the Black Sea through ICZM (MEDCOAST, n.d.).

2. EMECS

The International EMECS Center is an organization that aims to solve issues related to the environmental conservation of enclosed coastal seas areas around the world, including the Seto Inland Sea, Chesapeake Bay (US), the Baltic Sea (Northern Europe), and the Mediterranean (Southern Europe). It was created with the goal of facilitating comprehensive, international exchange regarding a wide range of fields, including research, policy, citizen action, education, and industrial activity, covering not only coastal areas, but also their associated water catchment areas. It specifies ICZM as the sustainable use and development of coastal zone resources (EMECS, n.d.).

3. Science Council of Japan

On August 17, 2017, the Science Council of Japan's Committee on Food Science Subcommittee on Fisheries Science released a proposal titled "A vision for a sustainable marine produce industry in Japan: Marine resource management based on an ecological approach" (Science Council of Japan's Committee on Food Science Subcommittee on Fisheries Science, 2017). This proposal calls for the establishment of an ecological management approach with the aim of achieving a sustainable marine produce industry. Its content follows.

In proposal (1): An ecological approach to management, it says "making the continued monitoring of ecosystem structure and functions based on ecological management approach and the management of marine produce resources on a species and population level into the fundamentals of a future marine produce industry," "within an ecological approach to resources management, conduct analysis of the current status of marine produce industry and ecosystem evaluations and use the results of these as the basis for establishing future visions and targets," and "a method of backcasting should be used to establish sustainable marine resource management that develops in stages." It posits "the execution of marine resource management cannot be handled by science alone but requires governance from both a technocracy comprising scientists and policy makers possessing expert knowledge, and a democracy expressing the will of the people in a democratic manner."

In proposal (3): Regarding the conservation and restoration of coastal ecosystems, it says, "In essence, coastal areas are notable for the interaction between land and sea ecosystems, and they have the most productive marine ecosystems. In addition to the assessment of giant levees currently being conducted, there needs to be cooperation between relevant authorities (the Ministry of Agriculture, Forestry and Fisheries, the Ministry of Land, Infrastructure, Transport and Tourism, the Ministry of the Environment, and prefectural governments) to urgently restore coastal ecosystems suffering from attrition, including a review of excessive river development by humans, such as dams blocking the flow of rivers, rivers being paved with concrete on three sides, or the straightening of river bends. Furthermore, in addition to promoting scientific research into the conservation and restoration of coastal ecosystems, the results of this research should be used to make proposals to society related to legislation regarding the use of ecological services provided by coastal ecosystems, and visions for coastal areas that should be protected, from an interdisciplinary perspective incorporating both science and liberal arts subjects, including social science."

4. Japanese Association for Coastal Zone Studies (JACZS)

In December 2000, JACZS released "Appeal 2000: A proposal for sustainable use and environmental conservation of coastal areas." Appeal 2000 outlined the issues currently affecting coastal areas as follows (Japanese Association for Coastal Zone Studies (JACZS), 2000). "(1) While the importance of conserving coastal environments is being recognized, these environments are still deteriorating, (2) the frequency at which coastal areas are being used is increasing, and competition between the parties using these areas results in inefficient usage, and (3) there is no system for tackling these issues in an integrated manner, causing inefficient usage and delays to environmental conservation." The goal of the appeal is to "increase the level of management by proposing an integrated coastal area management method that realizes the comprehensive and systematic promotion of environmental management from a wider perspective including comprehensive coastal management in a way that incorporates the unique characteristics of individual coastal areas, and conservation and usage that incorporates adjustments to the way coastal areas are used with solutions to the issue of competition."

The appeal also contains groundbreaking content, including "our target is for no-net loss on 2000 environmental levels and net-gain on environmental recovery and creation. Also, while we do not completely reject the usage and development of coastal areas, this needs to be sustainable development that is essentially slow and maintains appropriate conditions, while fitting in as much as possible with environmental mechanisms," and "furthermore, we propose the enactment of a Comprehensive Coastal Zone Management Act, new legislation to ensure the authority and performance of management agencies and comprehensive management plans."

8.3.4 Main Suggestions for Environmental Policy Derived From the S-13 Project

This research was conducted to clarify how comprehensive coastal management should be, with the aim of achieving sustainable coastal seas through the Environment Research and Technology Development Fund S-13 provided by the Ministry of Environment (MOE) and the Environmental Restoration and Conservation Agency (ERCA) in its Development of Coastal Management Method to Realize the Sustainable Coastal Sea. It was based on the concept of Satoumi, from the integrated point of view of the natural sciences, social sciences, and humanities, and included establishing new conservation areas in the coastal seas, by considering resource use and water surface use in the coastal seas, which will be undertaken hereafter.

Based on the achievements of this research, we propose comprehensive coastal sea environmental management techniques in Japan, by making a comprehensive determination of the natural environments and human activities in the coastal seas of Japan and the land areas that constitute their hinterlands, including providing concrete suggestions about how the current situation should be changed to achieve optimal material circulation and ecotones.

The main suggestions for environmental policy derived from the research achievements follow.

Topic 1: Development of methods for managing nutrient concentrations in the Seto Inland Sea (enclosed coastal sea).

Outline of the research:

Efforts are underway to expand the current uniform method of water quality management in the Seto Inland Sea to bay and open sea management that takes into consideration social and geopolitical characteristics and seasonal fluctuations, and to preserve and restore nutrient management and biological habitat environments, in order to develop highly sustainable coastal management methods with the aim of achieving healthy substance circulation and high biological productivity that are not impaired by red tides or the like.

Achievements of the research:

1. The difference in regional natural environments associated with nutrient management, such as the Secchi depth specific to individual regions, was presented in terms of the characteristics of coastal sea areas required for future fine nutrient management based on the regional characteristics.

2. A method for estimating primary production from chlorophyll a concentrations and the Secchi depth was developed and used to estimate a quantitative determination of the primary production distribution in both temporal and spatial terms from historical monitoring data. Autumn primary production in the Seto Inland Sea has decreased by about 20% since the 1980s, but that decrease is small compared to the reduction of nutrient loads of nitrogen (40%) and phosphorus (61%) during the same period. It also was found that the sea areas in which primary production significantly decreased are those in which chlorophyll a concentration exceeded $10 \mu g L^{-1}$ in the 1980s.

3. The research determined that the transfer efficiency from primary production to secondary production decreases with increasing chlorophyll a concentration, and that the transfer efficiency is particularly low in sea areas in which chlorophyll a concentrations exceed $10 \mu g L^{-1}$. Therefore, it is estimated that a lowering of primary production in the sea area in which there is a high concentration of chlorophyll a does not lead to a decrease in the trophic levels that had been rising, such as of fish, but does contribute to an improvement of the bottom sediment that had been deteriorating.

4. In the Seto Inland Sea, there are still areas in which the concentration of chlorophyll a tends to be high, such as with the generation of red tides, and that receive loads of it from the land. Therefore, coastal areas with these environmental conservation issues were classified as highly vulnerable by using as an indicator the low salinity that relates to fresh water bringing loads of it from the land, and were identified as coastal areas for which intensive measures should be taken hereafter.

5. As measures for balancing environmental conservation and biological production, methods were proposed focusing on periods that should be managed intensively and differences in coastal areas. The periods-related method was to keep nutrient concentrations low to suppress excessive growth of phytoplankton during periods of elevated water temperature and to increase nutrient salt concentrations for biological production during periods of lower water temperature. With respect to the coastal areas, a scheme was proposed to use the eelgrass (*Zostera marina*) capability to absorb nutrients as a method to lower nutrient concentrations in coastal zones where these nutrients tend to be excessive. It was found that eelgrass increases the absorption of nutrients by almost doubling its internal nutrient concentrations when nutrient concentrations in the coastal areas increase because of rainfall. This shows that it is appropriate to use *Zostera marina* as a seaweed alga for nutrient management. In this approach, it will become possible to provide the nutrients required for biological production by lowering their concentrations through absorption by *Zostera marina* during phytoplankton blooms, without changing the total amount of nutrients flowing in from the land, and instead slowly supply nutrients in coastal areas through the breakdown of withered *Zostera marina* in and after autumn when the risk of red tides and other problems is low. Therefore, nutrient concentrations in coastal areas can be adjusted to a seasonally desirable level by creating, restoring, and conserving *Zostera* beds.

6. Planktivorous fish that have an important role in connecting lower and higher trophic levels in the Seto Inland Sea include sardines and sand eels, although their ecologies vary greatly. The stocks of Japanese anchovies that propagate with floating eggs have been increasing since 2000, but those of sand eels, whose estivation and

spawning grounds are sandy substrates and which depend heavily on bottom sediments, continue to decrease. The cause of their respective increase and decrease likely is because of a difference in their habitat environments. Sand eels in the Seto Inland Sea are composed of two phyletic groups in the western part of Bingo-nada and the Bisan-seto. Because this research identified no genetic differences between the groups, it is presumed it is possible to restore their stocks if the habitat is improved, even for the phyletic group of the western part of Bingo-nada, whose stocks are decreasing significantly. With respect to the phyletic group of the western part of Bingo-nada, we successfully identified the coastal sand eel spawning areas that should be conserved, by presuming estivation and spawning grounds through a particle tracking model and field study, and cross-checking those with gravel mining zones and the distribution of the sandy deposits preferred by sand eels.

Major contributions of the research achievements to environmental policy and recommendations in Topic 1 follows. Major contributions:

1) In accordance with the recommendations submitted to the Minister of the Environment by the Central Environment Council in December 2015, the target value of coastal transparency is supposed to be set as a regional environmental goal. This research determined the concrete Secchi depth specific to individual regions of the Seto Inland Sea by developing a method for estimating the background Secchi depth that assumes no phytoplankton in that coastal area. The relationship between the Secchi depth and phytoplankton concentrations was mathematized. Because nutrient management can control only the phytoplankton concentrations, the Secchi depth that is possibly achieved by minimizing the nutrient supply from land areas and so on, means transparency with zero concentration of phytoplankton, i.e., the Secchi depth specific to individual regions. Having this value as the basic parameter for a region by understanding the relationship between the Secchi depth and phytoplankton concentrations in that region, the regional goal for transparency can be highly achievable.

2) It is imperative to elucidate the structure of lower trophic level ecosystems starting with primary production to achieve goals related to the balance between conservation of water quality and biological production for "aiming to realize an abundant Seto Island Sea," and the bay and open sea management for "meticulous management tailored to regional characteristics," launched by the 2015's revision of the Act on Special Measures Concerning Conservation of the Environment of the Seto Inland Sea. We estimated primary production from data derived from water quality monitoring implemented to date by institutions, including the Ministry of Environment, and mapped spatial distributions and temporal changes of primary production in the Seto Inland Sea. On the basis of these data, the research examined the effectiveness of total emission reductions on water quality implemented thus far, and then identified that a considerable impact can be seen, particularly in coastal areas with high concentrations of phytoplankton (coastal areas that tend to generate red tides), and that suppressing red-tide generation improves bottom sediment and leads to the recovery of benthos. The nutrient management in coastal areas in which phytoplankton tends to bloom does not have a great impact on secondary production. Therefore, because fishery-targeted species that use secondary producers as bait are not adversely affected, the research indicated that nutrient management would be a possible measure to balance environmental conservation and biological production through suppressing red-tide generation and improving bottom sediment.

3) Because coastal areas in which phytoplankton tends to bloom are substantially affected by nutrient loads from the land area through rivers, the research identified that coastal areas can be classified by using salinity as an indicator showing the influence of fresh water, and that the salinity would be a new and effective control indicator for nutrient management rather than the "bay and open sea" based on topographical features.

4) As a measure for coastal areas in which phytoplankton tends to bloom, the research developed and indicated the effectiveness of a nutrient management method using the nutrient circulatory function of *Zostera* beds and tidal flats that absorb nutrients from spring to summer when the excessive phytoplankton blooms tend to occur, and release the absorbed nutrients over time.

Policy recommendations:

1) With respect to the setting of a regional goal for transparency, it is recommended that an appropriate goal be set, from the perspective of the necessity for conservation and regeneration of nursery grounds for marine aquatic plants comprising seaweed beds, and after careful consideration of the feasibility based on the difference between the achievable transparency in that coastal area, the Secchi depth specific to individual regions and the current transparency.

2) The research developed a method to identify coastal areas that are at high risk of red-tide generation with excessive phytoplankton blooms. It is recommended to proceed with this method for nutrient management with a focus on coastal areas that tend to have excessively high concentrations of phytoplankton.

3) It is recommended that a nutrient management measure be implemented by usng *Zostera marina* suitable for highly vulnerable coastal seas.

Topic 2: Development of coastal environmental management methods on the Sanriku Coast, which has a succession of open inner bays.

Outline of the research:

This research will monitor fluctuations in seaweed bed ecosystems on the Sanriku Coast to determine which human efforts are effective in restoring abundant coastal seas. Optimal aquaculture methods for oysters, scallops, wakame, etc., in the Sanriku open inner bays will be proposed.

Achievements of the research:

1. The research monitored changes in the coastal ecosystems on the Sanriku Coast consisting of inner bays open to the Pacific Ocean that were enormously damaged by the huge tsunami on March 11, 2011, and those in human activities in the sea such as aquacultures, and identified the status of ecotones, including salt marshes and seaweed beds that had disappeared because of reclamation before the tsunami but were restored by the tsunami, and use of the bay through satellite remote-sensing and field surveys. Field surveys revealed the circulation of materials in open inner bays, including land areas and the open sea.

2. As a result of a study about the temporal change of seaweed beds after the earthquake, it was found that the discontinuance of sea urchin harvesting around the bay had increased the abundance of sea urchins, and that it has influenced the spread of denuded beds of *Eisenia bicyclis* since 2014. This fact reveals that sea urchin harvesting is a Satoumi activity that conserves seaweed beds and marine biodiversity.

3. In Shizugawa Bay, we mapped distributions of macrophyte beds and different types of aquaculture facilities and measured the flesh weight of oysters. We observed marine environmental parameters such as water temperature, salinity, and current direction and speed throughout the year. By using these data derived from mapping and observing, it was verified that the Shizugawa Bay ecosystem model could explain growth of oysters and others in the bay. The scientific data of oyster cultures data were provided to the Tokura branch of the Miyagi Fisheries Cooperative to obtain an Aquaculture Stewardship Council (ASC) certificate that proves sustainable aquaculture that respects marine environments.

4. As well as quantifying the stocks of various primary producers (phytoplankton, benthic microalgae, periphytic microalgae, etc.), the primary production speeds of these organisms were measured, and the primary production capacity of the whole Shizugawa Bay was assessed. By analyzing nutrient circulation concerning oysters and sea squirts, the major aquaculture species in the bay, we recreated overviews of nutrient circulations for the four seasons.

5. The research identified the effects of land coverage and water use on dissolved organic matter (DOM) and dissolved iron in the river basins of Shizugawa Bay, and estimated the iron unit load from the respective land coverage. The role of land-derived iron and DOM for primary production in Shizugawa Bay was elucidated in each season. The unit load of particulate organic matter (POMs) from the respective land coverage was estimated from the relationship between the land coverage and use in river basins, and the dissolved iron and DOM in rivers.

6. The research examined oyster growth between two areas in the bay, where densities of buoy-and-rope type oyster culture facilities were different, and identified that oysters grew faster in the Tokura coastal area where the density of oyster aquaculture facilities is lower. Our mesocosm experiment showed that younger oysters excreted feces/pseudofeces with a relatively lower carbon content, which requires a lower oxygen consumption rate compared with older oysters. These findings demonstrate that oyster aquacultures with lower densities of facilities potentially enhance oyster growth and eventually lower the negative impacts of oysters' excretions on the surrounding environment.

7. The research involved a comprehensive examination of natural and human activities on the Sanriku Coast, which was destroyed by the huge tsunami caused by the Great East Japan Earthquake, and of the land areas that constitute the coastal hinterlands; proposals for the environmental management of coastal seas in Japan, including establishing conservation areas in the coastal seas and considering the resource use and water surface use; suggesting how these areas should be changed from their current state to ones with an optimal material circulation and ecotones; and, finally, what measures taken by local people are effective.

8. A local council was organized to discuss the future use of Shizugawa Bay with Miyagi Fisheries Cooperative members, local government officials of Miyagi-ken and Minamisanriku-cho, and Japanese staff and scientists from the WWF. The council discussed the appropriate number and placement of cultivation rafts for realizing the sound balance of aquaculture industry and marine environments based on scientific research results.

The major contributions of research achievements to environmental policy and recommendations in Topic 2 follow. Major contributions:

1) The research scientifically considered the methods of use and management of ecotones and the aquaculture management method associated with ecological transition after the huge tsunami in coastal areas of open inner bays, from the perspective of balancing sustainable fisheries and wholesome marine environments.

2) A council was established to consider the future marine environment in open inner bays with the local people concerned, where the specific placement of aquaculture facilities that balance sustainable fisheries and sound marine environments was proposed and discussed on the basis of scientific data.

3) The research provided scientific data in accordance with the purpose expressed for establishing the Sanriku Fukko (Reconstruction) National Park: "(1) To achieve a symbiotic society through the regeneration of nature and so on for restoring the ties between the forest, village, and the sea, while also taking advantage of locally rooted knowledge of symbiosis with nature. (2) To conduct a survey of the current status and a monitoring of secular changes to the natural environment affected by the tsunami."

4) The research provided the distribution data of seaweed beds required for their registration with the Ramsar Convention.

5) We comprehensively viewed natural and human activities in the coastal seas and land areas that constitute the coastal hinterland, targeting the ria inlet, which is an open inner bay about which there was little accumulated knowledge, and quantitatively determined the material circulation of nutrients, iron, and POC within the forest-river-sea areas.

6) We provided quantitative data in the research target area, Shizugawa Bay, showing that the iron concentration is maintained at sufficient levels for algal growth because of the dissolved iron from land at the head of the bay, while it is possibly lower than the level required for algal growth in the middle and outer part of the bay.

7) The research provided a scientific answer to the ongoing tree-planting project based on the hypothesis that is called "the forest is a sweetheart of the sea," and the grounds for assessing the impact of land areas on determining the marine environmental standards in open inner bays at the policy decision level, through quantitatively understanding the spatial variations of land-derived organic material transported to open inner bays from rivers, and the impact of land-derived organic material on inner bays.

8) Together with Topic 5, the research provided a diagnosis method to determine the soundness of the marine environment in open inner bays through developing the open inner bay ecosystem model including aquaculture.

Policy recommendations:

1) The methods of use and management of ecotones, and the aquaculture management methods associated with post-tsunami ecological transitions in coastal areas of open inner bays, must be considered scientifically from the perspective of balancing sustainable fisheries with a wholesome marine environment.

2) Because well-managed sea urchin fisheries promote the conservation of seaweed beds that cultivate biodiversity, it is recommended that sea urchins be removed manually in cases of rocky shore denudation they cause.

3) It is necessary to actively conserve the salt marshes and tidal flats, both of which were restored by the tsunami, considering their ecosystem services.

4) The role of forests in supplying nutrients, iron, and POC to the sea, which is assumed as the basis of tree planting among Satoumi activities, is small in open inner bays with narrow catchment areas, so it is necessary to consider forests' multifaceted functions such as control of sediment discharge and floods.

5) Because there is less organic material in feces/pseudofeces from young oysters than older oysters, and environmental loads can be reduced, it is recommended to lower the installation density of oyster aquaculture facilities, so that oyster growth can be accelerated and the shipment of younger oysters expected.

6) To realize a wholesome marine environment and sustainable fisheries in open inner bays, it is recommended that a council be established to enable discussions based on scientific data among local people engaged in fisheries, local government officials, NPO members, and scientists.

7) It is recommended that the open inner bay ecosystem model mainly for aquaculture be made universal by evolving the Shizugawa Bay ecosystem model, and that an optimal aquaculture diagnosis system be created, so that an appropriate marine environment and fishery yield can be estimated using different scenarios of numbers and placements of aquaculture rafts.

Topic 3: Development of ocean management of the Japan Sea, an international enclosed coastal sea that includes continental shelves and islands.

An outline of the research and achievements follows.

Outline of the research:

Global environmental changes and fluctuations in the East China Sea influence the environment of the Tsushima Warm Current medial zone and individual bays in the Japan Sea. This research project aims to understand the changes in the medial zone and common or unique aspects of the influence. In addition, on the basis of clarifying the role of the Marine Protected Areas (MPAs) designation in biodiversity conservation, the management method will be proposed for coordinating biotic conservation and fishing activities in MPAs. Integrated land-sea management of the Japan Sea and the Tsushima Warm Current in cooperation with China, South Korea, Russia and other nations also will be proposed.

Achievements of the research:

1. The research quantitatively showed that the environment and primary production in coastal zones of the Japan Sea, ranging widely from northern Kyushu to Tohoku, are strongly influenced by the East China Sea, through a lower trophic ecosystem model in the Japan Sea. The research analyzed the respective contribution ratio to primary production in the Japan Sea from the three nutrient origins; the East China Sea, the Japan Sea, and rivers in Japan, and the response of lower trophic ecosystems to the load variation. It was revealed that nutrients in the southern waters of the Japan Sea, including the coastal seas of Japan, are dependent upon nutrients transported from the East China Sea, and that a change in primary production in the southern waters of the Japan Sea occurs when the volume of nutrients from the East China Sea changes.

2. The research identified that the main origin of nutrients that flow into the Japan Sea through the Tsushima Strait is the lower layer of the Kuroshio Current and the Pacific Ocean. Although it was found that the influence of nutrients originating from rivers in China is smaller than expected, the research suggested there is a possibility that a qualitative aspect of the nutrients, such as the N/P ratio in the Japan Sea, will change.

3. Low-salinity water originating from Chinese rivers is transported to the offshore of Tohoku along the coast of Japan through the Tsushima Strait. Persistent organic pollutants are detected at relatively high concentrations in Chinese river waters and coastal zones of the East China Sea. The research showed that it is necessary to continue to monitor this influence from the perspective of pollution in the Japan Sea.

4. With respect to future fluctuations in the Japan Sea, changes were predicted in the physical environment, such as rising sea temperatures because of global warming and in the chemical environment such as with nutrients. The centurial increases in the temperature of the sea's surface were predicted to be about 1.0 °C and 2.4 °C based on the RCP 2.6 and 8.5 scenarios respectively, with this physical-biological coupled model of the Japan Sea. Moreover, the possibility of a substantial change was predicted in the nutrient environment in the northern waters of the Japan Sea associated with flow increases and changes in the flow path of the Tsushima Warm Current because of global warming.

5. It had been believed in the past that dissolved oxygen concentrations in the deep waters of the Japan Sea would decrease through global warming, but new findings show that the inflow of high-salinity water will sink to a deeper layer of the Japan Sea from a strengthened Tsushima Warm Current, so the oxygen supply to deep waters will be maintained.

6. The research found new information related to the conservation and sustainable use of low and high trophic ecosystems (such as Japanese common squid) that support abundant biodiversity in the Japan Sea. The research developed a transport survival model for the floating larvae of Japanese common squid and snow crab, which are two important fishery species in the Japan Sea, with consideration of spawning grounds, spawning stock biomass (SSB), egg mass, ocean currents, water temperatures, and bait. Based on this model, a candidate for a Marine Protected Area (MPA) was proposed, verifying the effectiveness of establishing an MPA in light of future changes in the marine environment through global warming, and specifying the marine areas and periods that contribute to more efficient and effective conservation of resources.

Major contributions to the research achievements of the environmental policy and recommendations in Topic 3 are as follows:

Major contributions:

1) The research revealed that the coastal zones of the Japan Sea are strongly influenced by the Yellow Sea and the East China Sea through the Tsushima Warm Current, and a desirable environment cannot be maintained only by the conventional land-sea integrated management. The research will be helpful in considering a new method of marine management in Japan, including the coverage of its coast and offshore, that cannot be handled by the conventional coastal zone management of individual coastal areas.

2) The research revealed that the Japan Sea is extremely vulnerable to global warming and environmental changes in the East China Sea. In order to detect their influence at an early stage and develop effective countermeasures, ocean monitoring candidates of coastal areas and items were illustrated as examples. This possibly will make a significant contribution to building a monitoring network in Japan in the future.

3) The Japan Sea is a special international enclosed sea area that is strongly influenced by substantial changes at the global and regional level, making international cooperation indispensable. The research provided information based on scientific knowledge, including specific examples of what kind of collaboration would be needed with neighboring countries, so that Japan could take the lead in promoting international cooperation.

4) Land and sea integrated management was examined for individual bays in coastal zones of the Japan Sea in light of common issues in coastal zones and individual regional features, based on the concept of the project promoted by the Ministry of Environment, "Connecting and Supporting Forests, Countryside, Rivers and Sea". The research presented the possibility of a new management of the forest-village-river-sea that includes the conservation and management of submarine ground water, whose importance has attracted attention in recent years.

5) With respect to the strategic target of conserving at least 10% of coastal and marine areas by 2020, which was one of the Aichi Biodiversity Targets, only 8.3% has been attained at this point. This project was conducted with a focus on the Japanese common squid and snow crab, which are two important fishery species in the Japan Sea. In particular, although the Japanese common squid is a key species that supports abundant ecosystems in the Japan Sea, a rapid decrease in stocks has been reported in recent years. Given that conservation of this species leads to conservation of biodiversity in the Japan Sea, a new Marine Protected Area (MPA) was suggested. In addition, because conventional conservation measures will not be able to adapt to future changes when considering the influence of global warming and other factors, the effectiveness of a dynamic MPA, which is a new idea, also was examined as compared to a conventional MPA. These new ideas could make a significant contribution to expanding MPAs and enhancing their management in Japan.

Policy recommendations:

1) In order to realize coastal zone management of the Japan Sea, it is proposed to introduce a three-layer management scheme consisting of three different geographical areas: large-scale, middle-scale, and small-scale.

2) It is proposed to establish a Tsushima-centered monitoring network for the Japan Sea, and to build a monitoring system to detect environmental changes at their early stages.

3) New management of the forest-village-river-sea, which includes management of groundwater, is recommended to create a sustainable coastal environment.

4) It is proposed to establish new MPAs, a dynamic MPA and a conventional MPA on the basis of future environmental changes such as global warming, and also to promote considering international joint management.

Topic 4: Proposed economic assessment and an integrated coastal management model for ecosystem services in coastal seas.

The outline and achievements of the research follows.

Outline of the research:

Economic and sustainability assessment methods for ecosystem services in coastal zones of Japan will be developed and quantified. An integrated management method needed for coastal zones to ensure their sustainable development will be proposed. The Satoumi stories needed to connect nonfishing residents with coastal areas and involve them in coastal area management will be discovered, created, and passed on. Methods for coordinating fishery activities in the Tsushima and Goto marine protected areas (MPAs) also will be proposed.

Achievements of the research:

1. Surveys of fisheries cooperatives and an economic assessment of ecosystem services were conducted in the Seto Inland Sea (Hiroshima Bay and Hinase Bay), the Sanriku Coast (Kesennuma Bay and Shizugawa Bay) and the Japan Sea (Toyama Bay and Nanao Bay). To the question awareness of successors that indicates the sustainability of the practices of fishermen and fisheries in the surveyed areas (Hinase Bay, Shizugawa Bay and Nanao Bay), the total percentage of respondents who already have a successor and have a person who will become a successor were 25.7% in Hinase Bay, 47.3% in Shizugawa Bay, and 29.1% in Nanao Bay, respectively. It was demonstrated that these results are higher than the national average of 16.5% (2013 Census of Fisheries).

2. The research developed an economic assessment approach for ecosystem services (AESCZ) and conducted the estimates of Hinase Bay (18.1 billion yen per annum), Shizugawa Bay (29.6 billion yen per annum), Nanao Bay (38.1 billion yen per annum), Hiroshima Bay (75.5 billion yen per annum), Kesennuma Bay (41.9 billion yen per annum) and Toyama Bay (122.7 billion yen per annum).

3. By targeting the head of Hiroshima Bay and by using the contingent valuation method (CVM), it was assumed that the seawater clarity (beauty) at bathing beaches, the standing stocks of cultivated oysters and the weight of shucked oysters (abundance) will be improved by controlling the discharge of nutrients from sewage treatment plants. In terms of abundance, the benefits received in the entire area in total as a beneficiary of abundant seas was estimated to be 30.1 million yen per month on the basis of the median value.

4. Long-term changes to the ecosystem services in the Seto Inland Sea were assessed. An online survey was conducted in 2015 targeting three types of environmental goods/services (three types of virtual plans) in the Seto Inland Sea, with the same methods used as in the CVM survey conducted in 1998 by Tsuge & Washida (2003), and the research compared the assessed values between 1998 and 2015.

The natural environment value of 594 trillion yen in 1998 increased to 2334 trillion yen (on the basis of the median value) in 2015.

1. The research developed a sustainability assessment method for coastal seas that integrated three approaches for Satoumi, ecosystem services, and inclusive wealth index (IWI), which calculates changes in current social values of capital (natural, artificial, and human), and applied the integrated method to the Seto Inland Sea on a trial basis. The change of IWI in the past 50 years was estimated as a 32% decrease from 4.7 trillion yen per annum to 3.2 trillion yen per annum.

2. A dynamic sustainability assessment is an assessment approach in which the situation of a target area is understood in a sustainability plan, capability is measured in order to achieve sustainability, and intention is reflected as a decision to implement a sustainability plan based on the capability. Dynamic sustainability assessments were applied to Hinase Bay, Shizugawa Bay, and Nanao Bay.

3. In Hinase Bay, marine environment education is run successfully in cooperation between the local fisheries cooperative and junior high schools, expanding the circle of activities to elementary schools and senior high schools. The research confirmed through interviews with junior high school students that their attachment to the sea and awareness of environmental conservation increases as they become upper graders who have more insightful marine environment education.

4. Integrated Coastal Zone Management (ICZM) was proposed as follows: (1) It is composed of three stages: the Satoumi that is formed around the scope of the prefectural coastal seas, the Satoumi network, and the provision of coastal zone infrastructure in relation to environment conservation, national land conservation, and use coordination. (2) Coastal seas beyond the three stages should be managed cooperatively among the prefectures. (3) Management principles cover all these stages. A theoretical framework for multistage management was constructed by identifying factors that inhibit the achievement of wide-area comprehensive management and points for improvement, and a general overview of these items in accordance with a governance hierarchy was prepared.

5. A database of the nationwide variety and popularity of Japan's fish-eating culture was constructed through a field survey of fish-eating habits in target areas, and a study about Japanese eating habits was conducted. The research revealed that the traditional Japanese fish-eating culture is in crisis and is being threatened by a loss of diversity in various parts of Japan. Recommendations were compiled of practical activities for revitalization of the Satoumi, in which a wide range of generations can participate with pleasure.

6. Collaborative oceanography was created to establish MPAs and coordinate fisheries based on scientific surveys, and a consensus-building system for establishing the Tsushima and Goto MPAs was developed. Collaborative oceanography started as environmental monitoring to observe ecological changes because of climate change and other factors. This is developing into the city's own marine environmental survey, taking advantage of the hometown tax payment program (which allows taxpayers to use part of their taxes for donating to their favorite local municipalities), such as citizens' participation in the creation of a fish picture book in Tsushima City, a survey of sustainable seaweed beds for fishermen, and a coastal culture survey for geopark registration in Goto City.

7. A regional-based survey also contributed to marine culture research. Japan's neighbor, South Korea, strengthened not only its natural sciences but also enriched its culture with the opening of the national marine museum in Busan. The need for a marine-specialized museum in Japan has been talked about in Kyushu, and momentum has been growing in recent years. This research, covering regional marine culture at its first stage, initiated interest in the marine environment field and strengthened exhibitions at national and quasi-national park facilities and folk museums and in Tsushima and Goto, and also provided excursions for residents.

8. Collaborative oceanography and local knowledge surveys also were helpful for responding to declining populations or depopulation in coastal zones. Because they are conducted on the basis of the participation of

multi-generations and various entities, a chain reaction from SDG Goal 14 to other goals has been seen. In Goto City, a community meeting is conducted and local knowledge that was about to be lost, such as management knowledge about seaweed beds, is leading to local revitalization. In Nagasaki Prefecture, Kaiyo High School has a project of making specialties from marubata (cultivated fields under natural coastal conditions) such as dried fish, and has started collaborations with the local economic community and high schools/ universities. In Tsushima, distribution improvement has started, including traceability implemented by adding environmental information to fish for which stock management is conducted, and cooperation with the Co-op and major supermarkets.

9. The recent rise in the water temperature of the Tsushima Warm Current, particularly in summer, has been a common key issue in coastal communities. There are concerns about the relationship of the deterioration of seaweed beds and the increase of heatwaves and intense rainfalls. Sustainability of coastal communities is attracting society's attention, since the island of Okinoshima and associated sites in Munakata City, Fukuoka Prefecture, and the Goto Islands, were registered as World Cultural Heritage Sites. Marine debris also has become a social problem. Individual management of the seashores and marine areas in each region has limitations, and ocean current system-based and cross-cutting cooperation has become necessary. Therefore, there is active human involvement in marine environmental issues in northern Kyushu, including Tsushima and Goto, and the possibility of an MPA network is indicated in the Tsushima Strait and surrounding area. Local knowledge about the seas in Tsushima and Goto areas was gathered, and its importance was assessed and systemized in terms of coastal environmental sciences. In particular, the research showed that a marine protected area (MPA) leads to comprehensive marine environmental policies through biodiversity conservation and sustainable use.

10. For internationalization of the MPA network in the East China Sea, discussions have started about academic and environmental activities such as convening an international symposium and consideration of joint surveys. Consideration started from an individual theme, and the need for broader cooperation has become evident. People are becoming aware that building a network that integrates existing individual considerations and connections is imperative in the basins of the Tsushima Warm Current and Kuroshio Current, and this move is gaining ground toward gradual consensus building. To illustrate by examples, these include a coastal organism, the horseshoe crab, marine debris, aquatic resource management for migratory fish, and geoparks. Internationalization is progressing; a direct flight exchange program between Goto City and Jeju Island in South Korea has begun.

Major contributions to research achievements in environmental policy and recommendations in Topic 4 follow: Major contributions:

1) By applying practical economic assessments of ecosystem services (supply, adjustment, culture, and habitats) in coastal seas, the position of fisheries and tourism in coastal seas can be estimated in the regional economy, and also their future growth.

2) By applying a sustainability assessment framework to coastal seas, indicators can be provided as useful information in order to realize a desirable Satoumi, and implement Integrated Coastal Zone Management (ICZM).

3) By applying a dynamic sustainability assessment method to coastal seas, important elements for sustainability of the assessed area can be identified, and challenges that should be considered for the area's future plans can be elucidated.

4) The multistage management method for coastal seas is a linking of four stages; the Satoumi, the Satoumi network, coastal zone infrastructure, and cooperation in marine areas, through an approach for the whole area, an approach by all the governments, and a support-type approach. This method proposes a way to develop multilayered organizations and systems for management at multiple stages in the nested structure, from an organization or system that manages the immediate environment in an area, to an organization or system that manages marine areas under prefectural jurisdiction or areas beyond that. A sustainability indicator will be a management indicator in this case. The research shows that this kind of multistage management can be applied not only to coastal zone management but also to the management of various environments.

5) To make the management of coastal seas a national movement, the research found that it is necessary to discover, build, and succeed with the Satoumi story that connects citizens and coastal seas, and that analysis of fish consumption in fishing villages throughout the country shows cultural aspects of ecosystem services.

6) An integrated perspective is necessary for marine environmental issues not only in the Japan Sea but also in the East China Sea. These sea areas are under the jurisdiction of the Coordinating Body on the Seas of East Asia (COBSEA) according to the United Nations Environment Programme (UNEP). Neighboring countries commit to both the

Northwest Pacific Action Plan (NOWPAP: Action Plan for the Protection, Management and Development of the Marine and Coastal Environment of the Northwest Pacific Region) and COBSEA, and have started to respond to continuous marine environmental issues in the Sea of Japan and the East China Sea. To handle the area of the East China Sea in which Japan's issues are concentrated, it is important to strengthen Japan's international marine environmental policy in the medium to long term. This covers various fields, including ocean water quality, biodiversity, waste, culture, marine and earth sciences, and meteorological phenomena, and also corresponds to a broad range of international conventions and domestic systems. What is characteristic about this is each country's viewpoint that coastal management progresses to ocean management, which in turn leads to an international network. Although this research also proposes multistage management, the viewpoint is being implemented throughout the world. Therefore, promotion of an integrated marine environmental policy also is becoming indispensable for Japan.

Policy recommendations:

1) As indicated by the results that sustainability of fishermen and fisheries in the target areas (Hinase Bay, Shizugawa Bay, and Nanao Bay) is related to the close relationship between the areas and the fisheries industry, and it is important for regional development to reassess ecosystem services in the coastal seas and to promote AFFrinnovation, 6th Industrialization of Agriculture, Forestry and Fisheries, on the basis of Satoumi, by taking advantage of their respective regional characteristics.

2) To establish a sustainable method of coastal sea management, it is important to identify issues by developing future scenarios based on economic assessments of ecosystem services through an understanding of the actual situation of Japan's coastal seas.

3) As indicated in Hinase Bay by the practice of the Satoumi conservation activities and the effectiveness of continuing marine environment education, it is necessary to have measures that strengthen human resource development in coastal zones.

4) To apply a multistage management method, it is necessary for relevant local governments to establish a cross-jurisdictional agreement (convention) with the idea of implementation of the basic ordinance for comprehensive management of coastal zones, to specify the multistage management framework and the roles of involved parties, and to oblige them to follow the PDCA cycle. In addition, together with promoting the enactment and implementation of the basic ordinance (convention) for comprehensive management of coastal zones, it is necessary to organize the structure of national measures to follow them, including the budget allocation needed to implement the basic ordinance (convention) through subsidies or grants. Furthermore, to push forward these measures, it is necessary for each prefecture to implement pilot programs and to try to disseminate them to other entities.

5) As for the fishery activity management based on scientific information that was verified in Shizugawa Bay, it is important to disseminate its significance to the world through the Forest Stewardship Council (FSC) or the Ramsar Convention.

6) It is necessary to spread the Satoumi story in light of the practice of revitalization of Satoumi through traditional food, as well as the discovery of historical and cultural local traditions.

7) Because the establishment of Marine Protected Areas (MPA) will enable the conservation and sustainable use of biodiversity in the Tsushima and Goto areas, it is necessary to promote the Tsushima Warm Current Basin MPA network. The MPA network corresponding to the ocean current system is particularly desirable to integrate into marine policy.

8) In the Tsushima area, the development of standards and planning for establishing an MPA helped the research and study of local knowledge and knowledge collection through resident participation. Thus, as the implementation of research achievements could be accelerated into actual practice, it is desirable for other areas to leverage the MPA system as well. To do so, an MPA policy panel consisting of the government, academic experts, local governments, NGO/CBO (civil organizations), and other entities hold meetings regularly and provide support, such as consultation about marine environmental policy, to local governments in their practice of consensus-building through panel meetings. With efforts to gather information, including dispatching experts to an international workshop where secretariats of the United Nations Environment Programme (UNEP) or the Convention on Biological Diversity (CBD) and so on attend, the release of practical examples from Japan that are acceptable by international standards also has significance.

9) It is necessary to strengthen and support the marine environmental policymaking of coastal local governments by creating an integrated framework for such policies and consider an integrated coastal management method that

can be established according to natural and social conditions in that area. Japan's enhanced contribution to international PA networking, and its development of international collaborative oceanography are effective.

Topic 5: Development of integrated numerical models for coastal sea management.
The outline and achievements of the research follows.
Outline of the research:

1. The overall goal of S-13 strategic project (topics 1–5) will is to "propose comprehensive coastal sea environmental management techniques in Japan," including coordination and promotion of collaboration of topics, the setting of numerical goals and indicators, the project's progress management, and informing the public and others about research details. In addition, research data will be gathered for each topic to develop integrated numerical models for coastal sea management techniques whose goal is to achieve sustainable coastal sea use in the 21st century.
2. Integrated models for Shizugawa Bay and the Seto Inland Sea will be developed.

Achievements of the research:

1. A sustainable coastal sea management method was developed by integrating a natural scientific management method for coastal seas with humanistic and social scientific management methods for coastal seas, including economic and cultural values.
2. Integrated numerical models for the distribution of water flows, water quality, nutrients, and the optimal aquaculture methods were developed for the Seto Inland Sea (enclosed coastal sea), the Sanriku Coast (open coastal sea), and the Japan Sea (enclosed coastal sea that requires international management).
3. A numerical model that integrated the Hiroshima Bay current and lower trophic level ecosystem model with an oyster cultivation model was constructed, and the oyster harvest yield and bottom layer DO concentrations beneath oyster rafts were recreated. An integrated current and lower trophic level ecosystem model for Osaka Bay was developed, and algorithms for estimating transparency using surface layer COD concentrations and chlorophyll *a* concentrations were established, and reproducibility was confirmed.
4. In the Shizugawa Bay integrated model, the optimal aquaculture calculations for oysters and seaweed and water quality prediction calculations were performed. The research made it possible to visualize these results by using geographical information and graphing.

A current and lower trophic level ecosystem model for Shizugawa Bay was developed, and by reducing oyster cultivation rafts by 30%, it was confirmed that the cultivation period was shortened from 18 months to 10 months, the accumulation of glycogen in the oyster's body was improved, and that the DO near the ocean floor would not decrease very much even in summer.

Major contributions of research achievements to environmental policy and recommendations in Topic 5 follow:
Major contributions:

1) Visualization of the policy effect for coastal sea conservation

By using the numerical model that clarified how a load of phosphorus and nitrogen, a major factor affecting changes in water quality and fish catches, is related to inflows from the land, the dissolution from bottom sediment, inflows from open seas, and biotic resources, a desirable management measure for the concentrations of phosphorus and nitrogen was considered. Visualizing the benefit by cost (B/C) of that measure's effect can deepen the public understanding of water quality conservation in coastal seas, and an effective nutrient management method can be instituted.

2) Promotion of the consistent enforcement of the Act on Special Measures Concerning Conservation of the Environment of the Seto Inland Sea

The administration can derive an effective measure from numerical calculations by using integrated numerical models that were developed. For example, the administration can propose and implement a new measure to develop an "abundant sea in which biological diversity and productivity are secured" as stipulated in the Basic Principles of the Revised Act on Special Measures Concerning Conservation of the Environment of the Seto Inland Sea issued in October 2015.

By using the numerical model that clarified how a load of phosphorus and nitrogen, a major factor affecting changes in water quality and fish catches, is related to inflows from land, dissolution from bottom sediment, inflows from open seas, and biotic resources, a desirable management measure for the concentrations of phosphorus and nitrogen is considered. Visualizing the benefit by cost of that measure's effect, can deepen the public understanding of water quality conservation in coastal seas, and an effective nutrient management method can be created.

Policy recommendations:

1) Integrated numerical models should be promoted for coastal sea management techniques to achieve sustainable coastal sea use in the 21st century.
2) We need to understand the actual situation of coastal seas and forecast the effects associated with the implemented measures, using the integrated numerical models developed for Hiroshima Bay, Shizugawa Bay, and Toyama Bay.

8.3.5 Prospects for Sustainable Coastal Sea Management

Developing a coastal sea management method to realize a sustainable coastal sea is a pressing issue when considering today's serious status of nature and ecosystems in coastal seas, as well as fisheries in coastal zones, with their decreasing numbers and the aging of fishery workers. Sustainable coastal sea management centered on ecosystem services in coastal seas will be a lever to recognize these coastal seas' true environmental value and to sustainably guarantee people's lives based on this value. In 1973 when the Seto Inland Sea was in a critical condition because of serious environmental pollution such as the red tides, the Act on Special Measures Concerning Conservation of the Environment of the Seto Inland Sea was enacted, and the pollutant load was reduced. We have experience and wisdom about how a sustainable community and environment should be in the 45 years that have passed since the enactment of the Act on Special Measures. To have a view of future sustainable coastal sea management through dialogue between scientists and citizens, the following challenges are important:

(a) Formulation of environmental policy that integrates specialized fields and can obtain the public's understanding
     This project aims to achieve a fusion of policies based on an area's characteristics, not only in the natural sciences, but also in the fields of social science and humanities. Therefore, unlike past administration-driven measures that focused mainly on water quality control, we propose comprehensive measures that also consider suggestions based on economic and cultural aspects, so the best combination of measures suitable for promoting a new administrative policy can be explored, and the research can contribute to formulation of environmental policy for which the public's understanding can be gained.
(b) Promotion of an environmental administration that balances water quality conservation and marine produce industries
     The research can contribute to promotion of a comprehensive environmental administration that balances water quality conservation and marine produce industries by scientifically demonstrating that a "beautiful sea and abundant sea can co-exist" through integrating research achievements for each topic, and sharing knowledge with the administration and those employed in marine produce industries. When we explained the results of an oyster aquaculture model for Shizugawa Bay to those employed in local marine produce industries in Minamisanriku-cho in May 2016, the results were consistent with their impressions of the local area, and this enables confidence in the research to become greater. Research achievements will be shared among local concerned parties, such as those employed in marine produce industries, and a new policy system will be developed. At the same time, collaborative efforts among the concerned parties can be expected.
(c) Contribution to SDGs and image building of coastal zones management in 2050

Since 2016, various sustainable development goals (SDGs) related activities have been conducted toward 2030. As also indicated in the "Conserve and sustainably use the oceans, seas and marine resources for sustainable development" (SDGs: Goal 14: Conservation of Marine Resources), the significance of the ocean is recognized. We anticipate through this recognition that the development of coastal sea management centered on ecosystem services will be the key to a new regional revitalization. For sustainable coastal sea management in 2050, it is hoped that we will create a model of sustainable regional revitalization by presenting a coastal sea management law based upon population decline and an aging society in coastal zones, as well as in response to global challenges, such as global warming.

References

EMECS, n.d. http://www.emecs.or.jp/s-13/en/.
Japanese Association for Coastal Zone Studies (JACZS), 2000. Appeal 2000: A Proposal for Sustainable Use and Environmental Conservation of Coastal Areas.
MEDCOAST, n.d. https://www.medcoast.net/

National Oceanic and Atmospheric Administration (NOAA). 2010. Coastal Zone Management Act Performance Measurement System: Contextual Indicators Manual. U.S. Department of Commerce & National Oceanic and Atmospheric Administration & National Ocean Service & Office of Ocean and Coastal Resource Management. http://www.noaa.gov/.

PEGASO (2008), The Pegaso CASEs Building Capacity and Sharing Experiences for Integrated Coastal Zone Management (ICZM). http://www.vliz.be/projects/pegaso/

Science Council of Japan's Committee on Food Science Subcommittee on Fisheries Science. 2017. A Vision for a Sustainable Marine Produce Industry in Japan—Marine Resource Management Based on an Ecological Approach.

SUSTAIN, 2012. The SUSTAIN Indicator Set, A Set of Easily Measurable Sustainability Indicators. INTERREG IVC & European Union.www.sustain-eu.net/what_are_we.../sustain_indicator_set.pdf.

Tsuge, T., Washida, T., 2003. "Economic valuation of the Seto Inland Sea by using an Internet CV survey" Marine Pollution Bulletin 47 (1–6): 230–236.

Index

Note: Page numbers followed by *f* indicate figures and *t* indicate tables.

Printed in the United States
By Bookmasters